W0043406

F. Kajiya G.A. Klassen J.A.E. Spaan
J.I.E. Hoffman (Eds.)

Coronary Circulation

Basic Mechanism and Clinical Relevance

With 141 Figures, Including 4 in Color

Springer-Verlag Tokyo Berlin Heidelberg New York
London Paris Hong Kong

Dr. Fumihiko Kajiya
Department of Medical Engineering and Systems Cardiology
Kawasaki Medical School
Kurashiki, 701-01 Japan

Dr. Gerald A. Klassen
Departments of Medicine, Physiology and Biophysics
Dalhousie University
and Maritime Heart Centre
Victoria General Hospital
Halifax, N.S. B3H 2Y9, Canada

Prof. Jos A. E. Spaan
Department of Medical Physics
University of Amsterdam
1105 AZ Amsterdam,
The Netherlands

Dr. Julien I. E. Hoffman
Department of Pediatrics and Cardiovascular Research Institute
University of California
San Francisco, CA 94143, USA

Associate Editors:
K. Onodera, Y. Maruyama, H. Tomoike, K. Tsujioka

ISBN-13: 978-4-431-68089-5 e-ISBN-13:978-4-431-68087-1
DOI:10.1007/978-4-431-68087-1

Library of Congress Cataloging-in-Publication Data
Coronary circulation: basic mechanism and clinical relevance/F. Kajiya... [et al.] (eds.).
p. cm. Includes index. ISBN-13: 978-4-431-68089-5 (U.S.)
1. Coronary circulation. 2. Coronary heart disease. I. Kajiya, Fumihiko. [DNLM: 1. Coronary Circulation. WG 300 C82154] QP108.C66 1990 612. 1'7--dc20 DNLM/DLC

Softcover reprint of the hardcover 1st edition 1990

Typesetting: Asco Trade Typesetting Ltd., Hong Kong

Preface

Coronary heart disease is one of the major health problems in industrialized nations because of its high incidence and severity. Recent innovations in medical and surgical treatment of coronary heart disease have increased the importance of accurate diagnostic methods for determining the severity of coronary disease, identifying potential treatment alternatives, and evaluating the results of treatment. Great advances have also been made in basic research on coronary circulation and its interaction with myocardial contraction and relaxation and neural and humoral control.

With respect to these developments, the papers included in the present monograph deal with important topics concerned with the basic mechanism of coronary circulation as well as ones of clinical relevance. It is directed toward physicians (cardiologists, cardiac surgeons, cardiac radiologists, anestheologists, and others) and basic scientists (e.g., physiologists, bioengineers). We would like to emphasize the importance of a multidisciplinary approach in which basic scientists and clinicians work closely together.

This volume consists of nine chapters. Chapter 1 contains macroscopic and microscopic descriptions of coronary vascular anatomy, which is closely related to the functions of coronary circulation. In Chap. 2 two methods in current use to evaluate phasic coronary blood velocity waveforms are presented, i.e., the laser Doppler and ultrasound Doppler methods. Chapter 3 describes the mechanical properties of coronary circulation, a knowledge of which is indispensable for an understanding of coronary arterial and venous blood flow velocity waveforms in relation to cardiac contraction and relaxation. The diastolic pressure-flow relationship is also discussed in this chapter. Chapter 4 is concerned with the transmural distribution of coronary flow with special reference to the higher vulnerability of the endomyocardium to ischemia, while Chap. 5 focuses on the pressure and flow in coronary microcirculation concerning one recent topic of interest, "vascular segmentation." Chapter 6, which is on the control of coronary circulation, presents new information regarding the sympathetic control of coronary circulation and discusses some current topics of humoral control and models based on the mass balance between O_2 supply and demand. Animal models used for investigation of coronary circulation in cardiac hypertrophy and for provocation of coronary spasm are dealt with in Chap. 7. Chapter 8 focuses on coronary collateral circulation, including anatomical considerations, the mechanism for collateral growth, the rate of development of collateral perfusion, and the effects of interventions on collateral hemodynamics. The five papers constituting Chap. 9 present examples of clinical evaluation of coronary circulation, including coronary reserve, the hemodynamics of coronary stenosis, aortic valve diseases, syndrome X, and velocity profiles in coronary bypass grafts.

Many of the contributors to this book were participants in the 8th International Conference of the Cardiovascular System Dynamics Society (CSDS) held in Osaka, Japan in 1987 (chairman, F. Kajiya; secretary general, M. Hori). On that occasion, Drs. F. Kajiya, G. A. Klassen, and J. A. E. Spaan discussed the publication of a monograph on coronary circulation and outlined the contents. Dr. J. I. E. Hoffman, who could not attend the Osaka Conference, was invited to be the editor of this book, and Drs. K. Onodera (chairman of the satellite symposium), H. Tomoike, Y. Maruyama, and K. Tsujioka were asked to serve as associate editors. Whereupon, the contents of this volume were finalized. The editors are greatly indebted to all the authors who have contributed to this book in spite of their busy schedules. Special thanks are also due to Mayumi Yokomizo, Mari Tanaka, and Yukiko Sato for their excellent secretarial work.

The editors are also grateful to the members of the CSDS (president: Professor T. Kenner, University of Graz) who gave us the opportunity to publish this book. We especially wish to acknowledge our gratitude to the late Dr. Kiichi Sagawa, Department of Biomedical Engineering, Johns Hopkins University for his continuous guidance, and we would like to express our sincerest condolences to his family.

Finally, it has been a pleasure to work with Springer-Verlag Tokyo. They provided us with excellent advice regarding many aspects of this project and provided us with an outstanding language editor.

Okayama, April 1990 FUMIHIKO KAJIYA

In Memorial

We were saddened to learn of the death of Dr. Kiichi Sagawa, a valued member of the CSDS and a bioengineer of extraordinary ability. His studies on ventricular mechanics culminated in the development of the concepts of ventricular elastance, one of the most fruitful concepts ever developed for understanding the nature of myocardial contractility. His students can be found all over the world and can testify to his ability and unfailing kindness to everyone. We are all the poorer for his loss and would like to offer our sympathy to his family.

As the time of publication of this volume approached, we also learned with great regret of the death of Dr. Melvin L. Marcus. The author of the outstanding book *The Coronary Circulation in Health and Disease*, Dr. Marcus was one of the most eminent investigators of the coronary circulation. He was noted for his critical ability, for knowing what studies would prove to be important, and for developing the techniques needed to solve major problems in the coronary circulation. He led one of the most successful groups of cardiovascular investigators in the world and was responsible for the training of many of the current leaders in the field. He will be missed by all his friends and collaborators, and the editors offer their sympathy to his family on their great loss.

Editors

List of Contributors

Table of Contents

1. Anatomy of Coronary Circulation

Anatomical Principles of the Coronary Circulation

Robert J. Tomanek[1]

Summary. This review considers structure-function relationships of the coronary circulation. The coronary vasculature begins developing during late embryonic life via vascular sprouts and ultimately is expressed as a hierarchy of vessels. Since the microvasculature is the key to our understanding of coronary flow and the exchange of nutrients, considerable exploration of this component of the coronary circulation is needed. Perhaps the most evident gap in our knowledge in this area is the angiogenic response during development and during myocardial hypertrophy. While growth factors have been identified, the regulation and stimulation of these molecules need to be elucidated. A second deficiency concerns transcapillary phenomena, especially those regarding the role of vesicles in the transportation of large molecules. Finally, the concept of capillary recruitment needs to be verified and defined.

Key words: Microcirculation—Angiogenesis—Vascular hierarchy—Histology—Ultrastructure

Introduction

In several respects, the coronary vascular circuit has unique properties which underlie the performance and adaptability of the coronary circulation. Transmural perfusion occurs mostly during diastole, since vascular compression limits flow during systole. Because of this unique feature, epicardial arteries, as well as the venous system, function in coronary capacitance. This role of epicardial arteries was demonstrated by Vatner and colleagues [1] who noted that the diameter of the canine circumflex artery increases approximately 10% during early systole. Because of high metabolic demands placed on the myocardium and because of the tissue's dependence on oxidative metabolism, the heart is dependent upon not only a relatively high blood flow, but also upon relatively short O_2 diffusion distances. The latter attests to the special importance of the rich coronary capillary network charateristic of the myocardium.

Another important challenge which faces the coronary circulation is the necessity to alter perfusion rapidly to meet changes in cardiac work. Since perfusion requirements may vary regionally, an effective system of regional control is

[1]Department of Anatomy, College of Medicine, The University of Iowa, Iowa City, IA 52242, USA

necessary. Finally, the coronary vasculature needs to adapt chronically to alterations in preload, afterload, neural and hormonal inputs, and to normal growth and aging processes. These considerations may involve angiogenic factors which, at this time, are not well understood. Angiogenesis, however, is an important consideration in a variety of diseases which affect the coronary circulation, e.g., coronary artery occlusion and cardiac hypertrophy.

This chapter is aimed at providing an anatomical basis for understanding the coronary circulation. The specific goals of this communication are: (1) to review macroscopic and microscopic principles, (2) to provide some functional correlates to anatomical features, and (3) to emphasize the dynamic aspects of the coronary vasculature.

Development

Nourishment of the cells of the primitive cardiac tube occurs directly from the cavity. Epicardial vascular buds coalesce with the developing coronary arteries, which originate, bilaterally, from endothelial buds of the truncus arteriosus [2]. As the chamber walls thicken, intertrabecular spaces form intramural channels. Vascular sprouts are accompanied by smooth muscle cells which are believed to have differentiated from epicardial mesenchymal cells [3]. Capillaries lacking a basal lamina appear as outgrowths of both coronary vessels and the intertrabecular spaces [4]. It has also been suggested that cardiac chamber endothelial cells contribute to the transmural capillary system, as evidenced by observations that the first blood vessels appear in embryonic rat hearts as invaginations of the endocardium in the sinuses [5]. This study also concluded that growth of capillaries precedes their attachment to coronary arterioles.

Vessels begin forming at the end of the embryonic period in many species [4] including man [2]. However, species vary in the timing of developmental events of the coronary vasculature. Rakusan [6] has reviewed the development of the rat's coronary bed while Hudlicka and Tyler [4] have reviewed the development of coronary vessels in various species. A coronary circulation is established at an earlier gestation stage in humans than in rats, although capillary growth is marked in both species, especially during early postnatal life [4].

Although relatively little is known about the details of the events of coronary vessel development, even less is known about the regulation of this process. During the last decade major advances in the field of angiogenesis have included the discovery and characterization of growth factors. A number of such substances have been isolated and their mitogenic activity demonstrated [7]. It is expected that the utilization of this more recently acquired knowledge and the application of tools of molecular and cellular biology should advance the state of knowledge of coronary vessel development.

Distribution of Coronary Arteries and Veins

While considerable information on the distribution of the coronary arteries has been published, this chapter includes but a brief synopsis of this topic. The right and left coronary arteries arise just above the root of the aorta. A uniform inflow into these vessels is aided by the turbulent flow characteristics of the sinuses of Valsalva which lie just below the coronary ostia. Coursing through the atrioventricular groove (coronary sulcus), the right coronary artery passes inferiorly and rightward on the right ventricular surface as it gives rise to branches. While these branches serve mainly the right ventricle and the atria, there clearly is a contribution of the right coronary artery to portions of the left ventricle. This artery gives off conus branches anteriorly (the marginal branch being the largest) and the sinoatrial artery posteriorly. Several branches supply the atria. Coursing posteriorly in the atrioventricular groove the right coronary artery produces many branches, the posterior descending being the largest.

The left coronary artery usually bifurcates into the left anterior descending (LAD) and circumflex arteries. The former courses along the anterior interventricular groove and supplies primarily the anterior left ventricle, but also has branches which extend to the lateral left ventricular wall, and to the anterior right ventricular wall. Septal branches, which vary in number, penetrate into the interventricular septum and, therefore, constitute intramyocardial arteries in contrast with the other major coronary arteries which are epicardial. In some cases the septal artery originates directly from the left main coronary artery. Running laterally in the atrioventricular groove, the circumflex artery provides numerous branches along its course. This artery provides the main blood supply to the posterior and lateral left ventricular wall, and also supplies the left atrium.

Anastomoses between coronary arteries have long been recognized. Such anastomoses in man are evident between the left anterior descending, circumflex, and right coronary arteries. Most commonly epicardial coronary collateral vessels have been described, but intramyocardial anastomoses in both normal and diseased hearts have been well documented [8]. Collaterals may connect different branches of the same coronary artery ("homocoronary anastomoses") or branches of different coronary arteries ("intercoronary anastomoses") as demonstrated in coronary cast replicas [9]. These vascular connections appear to range in size from about 20 μm to 1mm. Native (pre-existing) collateral vessels are usually small and enlarge dramatically in response to a gradual occlusion of a parent vessel [10]. These remodelled collaterals then become medium or even large coronary arteries, with the potential for a markedly increased flow. That anastomoses between smaller precapillary vessels such as arterioles may also exist has been suggested [11]. However, there are important species variations regarding collateral vessels. For example, primate and canine hearts have a discernable collateral system, while porcine hearts lack major connections between large arteries.

Smaller divisions of the epicardial arteries may supply any layer of the ventricle. Some intramural channels are relatively short and course to the middle layer of the ventricular wall while others extend, apparently without branching, to the endomyocardium. Thus, the transmural channels follow a centripetal arrangement. Further details of these vessels and their contributions to subendocardial plexus have been described by Fulton [8]. As can be appreciated in arteriograms, transmural arteries and arterioles form dense vascular networks.

Three large veins are usually evident on the heart's epicardial surface: the great cardiac vein which follows the LAD and circumflex arteries, and the middle and posterior cardiac veins which lie parallel to each other on the posterior surface of the heart. A small cardiac vein follows the right coronary sulcus. These vessels drain into the coronary sinus, as does the oblique vein of the left atrium. Anterior cardiac veins from the right ventricle empty directly into the right atrium. Some smaller veins or venules in the ventricles and the atrial septum are components of a system of thebesian veins which open directly into the cardiac chambers. Because these vessels comprise a small percentage of the venous drainage and most of these veins empty into the right ventricle, the O_2 saturation of the left ventricular chamber is not significantly affected. Anastomotic connections between large coronary veins have been described in dog and man. In the dog, selective occlusion of a major cardiac vein does not markedly alter venous drainage, a finding which confirms the presence of alternate drainage pathways [12].

Vascular Histology and Ultrastructure

Coronary vascular morphology is, in general, similar to that of other vascular beds. A good, and detailed, description can be found in reference [13]. Figures 1 and 2 include canine coronary arteries and arterioles. The distinction between these two types of precapillary vessels is usually arbitrary, although arterioles are usually considered to be less than 200 or 250 μm in diameter. In precapillary vessels, with the exception of the smallest arterioles, the muscular media and the thin intima are separated by an internal elastic membrane (lamina). In large epicardial arteries, smooth muscle cells are usually separated by collagen, whereas in smaller arteries and arterioles, little collagen is present in the media (Figs. 1 and 2). Thus, in these smaller vessels, smooth muscle cells are separated only by a basal lamina. Endothelial cells of the intima are linked by occluding and tight junctions. Vasa vasora occur in the adventitia of both coronary arteries and veins. With decreasing vessel size the medial thickness/lumen diameter ratio increases. For example, this ratio in large epicardial arteries in dogs is about 0.04, in small arteries about 0.06, and in arterioles approximately 0.09–0.25 [14]. These data are based on vessels fixed by vascular perfusion after the administration of procaine. Therefore, they reflect dimensions of dilated vessels from diastolic hearts.

That both noradrenergic and cholinergic nerves innervate large coronary arteries is generally accepted, although their functional roles in flow regulation are not easily understood. Although earlier studies reported that nerve endings penetrate the media [15], it appears that these structures are limited to the adventitia [16, 17]. Both adrenergic and cholinergic fibers terminate at the adventitial-medial border in arteries and arterioles. The former are pathways for eliciting vasoconstriction while the latter have been shown to affect vasodilation [18]. Cholinergic fibers have their greatest densities in arterioles and small arteries as demonstrated histochemically by the acetylcholinesterase technique [19]. Coronary afferents course from the heart through both sympathetic nerves and the vagus. The former includes pain fibers which have been shown to be trig-

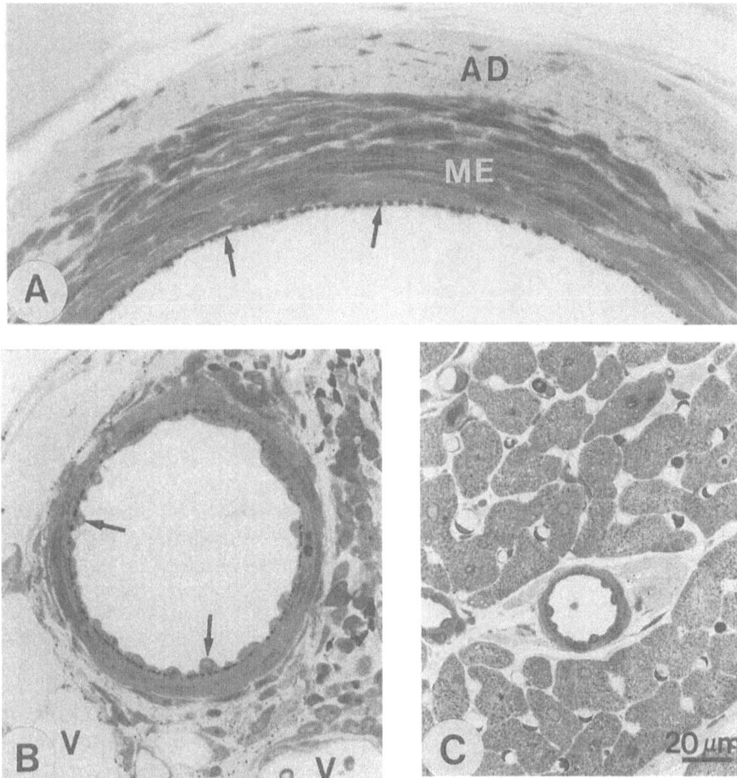

Fig. 1 A–C. Light micrographs of coronary precapillary vessels. **A** Portion of an artery (lumen diameter = 380 µm). The darker staining media (*ME*) contrasts with the adventitia (**AD**); the internal elastic lamina appears as a narrow dark band (*arrows*) with numerous interruptions (fenestrae). **B** A medium size arteriole (lumen diameter = 83 µm); endothelial cell nuclei bulge into the lumen (*arrows*) and several venules (*V*) are seen in the field. **C** A small arteriole (lumen diameter = 24µm) surrounded by cardiocytes and capillaries

gered by ischemia, i.e., occlusion of a coronary artery in cats and monkeys [20]. Neurons which are responsive to coronary occlusion are components of spinothalamic and spinoreticular tracts. Although less is known concerning vagal afferents, they appear to include pain fibers and mechanoreceptors [2]. Presumably, afferent vagal fibers contribute to baroreceptor and chemoreceptor reflexes.

Pericytic venules, the first postcapillary vessels, are composed of an endothelial layer, basal lamina, and a variable (usually small) amount of collagen. These vessels are termined "pericytic" because they contain slender phagocytic cells called pericytes as an adventitial component. As venular size increases, an incomplete layer of smooth muscle cells may be present. Only the larger coronary veins include a complete muscle layer.

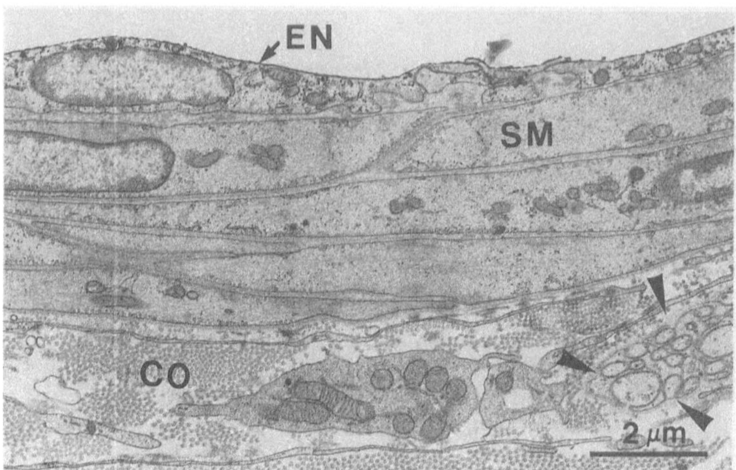

Fig. 2. Electron micrograph of the wall of a small artery; closely packed smooth muscle cells (*SM*) separated only by a basal lamina lie beneath the endothelium (*EN*). Collagen fibrils (*CO*) comprise most of the adventitia; nerve axons lie in the adventitia (*between arrowheads*)

Myocardial capillaries indent into the cardiocytes they supply and consist of a very thin layer of endothelium surrounded by a basal lamina (Fig. 3). Pericytes are seen, on occasion, outside the basal lamina. Henquell et al. [21] using in vivo techniques showed that capillary diameter is approximately 5 μm in the diastolic heart, but decreases to approximately 4 μm during systole, due to the compressive force of contraction. Capillary diameter in arrested hearts is about 10% greater than that of diastole. Accordingly, the contractile state of the myocardium clearly affects capillary lumen size. The ultrastructural features related to transcytosis and the quantitative aspects of the capillary bed are discussed in the next section.

Microvessels

Small precapillary and postcapillary vessels together with the capillaries constitute the microvasculature (Fig. 4 illustrates capillaries connecting to venules). Myocardial arterioles branch variably and, therefore, it is difficult to classify "orders" according to lumen diameter. The variable branching pattern of precapillary vessels, based on coronary casts from rabbits, is summarized in Fig. 5 [14]. This work suggests that: (1) arteries give rise to branches which may differ widely in their diameters and (2) the number of precapillary vessel generations is variable. Thus, a terminal arteriole (an arteriole from which capillaries arise), usually less than 30 μm in diameter, may be preceded by several generations of arterioles, or may arise directly from a relatively large vessel, e.g., 200 μm or larger. The extremely low numerical density of arterioles, 1–5 per mm^2, is related, in part, to the fact that the more numerous terminal arterioles, which

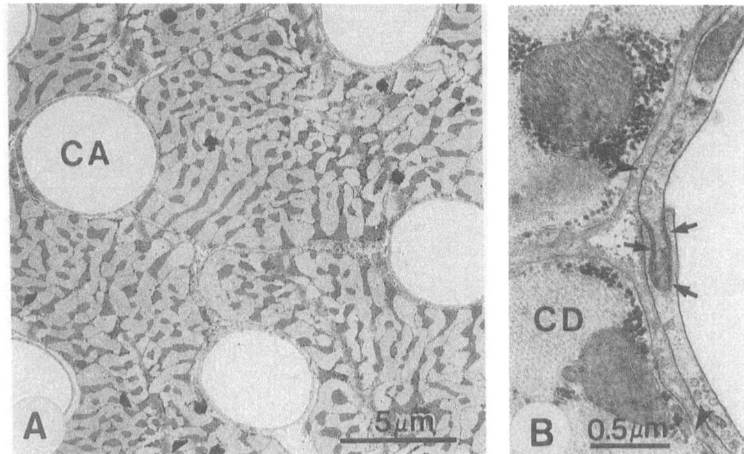

Fig. 3. **A** Electron micrograph of capillaries (*CA*) surrounded by cardiocytes; note the narrow intervening extracellular region. **B** Higher-power electron micrograph illustrating capillary endothelial cleft (*arrows*); note the basal lamina (*arrowheads*) between endothelium and cardiocytes (*CD*). Micrograph **B** courtesy of Kimberly Schalk. Reprinted by permission of Kluwer Academic Publishers

account for more than 50% of the arteriolar population (vessels less than 200 μm), are relatively short, i.e., about 100 μm [14].

Arterioles are the prime regulators of myocardial perfusion, as can be appreciated by the fact that there is a negligible (less than 10%) drop in pressure from the aorta to 300 μm diameter coronary arteries, but a steep pressure drop as arteriolar diameter decreases from 200 μm to 50 μm [22]. Even a relatively modest decrease in arteriolar diameter has a marked effect on resistance, since flow is related to the fourth power of the vessel's lumen diameter. Flow through a capillary can be altered by arterioles in two ways: (1) non–occlusive vasoconstriction can obviously reduce flow, while (2) occlusive constriction can stop flow. The latter, whether the consequence of closure of precapillary sphincters or substantial constriction of arterioles, would allow for intermittent flow through capillaries. Thus, flow during rest is probably cyclic. Evidence that capillary recruitment occurs in the coronary circulation is based on direct visualization of the capillary vessels on the surface of the heart [23] and from intravital fluorescent dyes [24]. These studies suggest that under resting conditions about 50% of the capillaries are open to plasma flow at any time in the heart. In contrast, Tillmanns and colleagues [25], using an in vivo microscopic technique, suggest that plasma flow through the capillaries is not interrupted, but that capillary recruitment does affect erythrocyte passage. This position tends to agree with that of Vetterlein and colleagues [26] who found that all capillaries show fluorescent dye within 10s of injection. Thus, the nature and role of capillary recruitment as an expression of coronary reserve is yet to be settled. Moreover, the preferential method of capillary closure, i.e., via precapillary sphincter or reduction in arteriolar diameter, is uncertain.

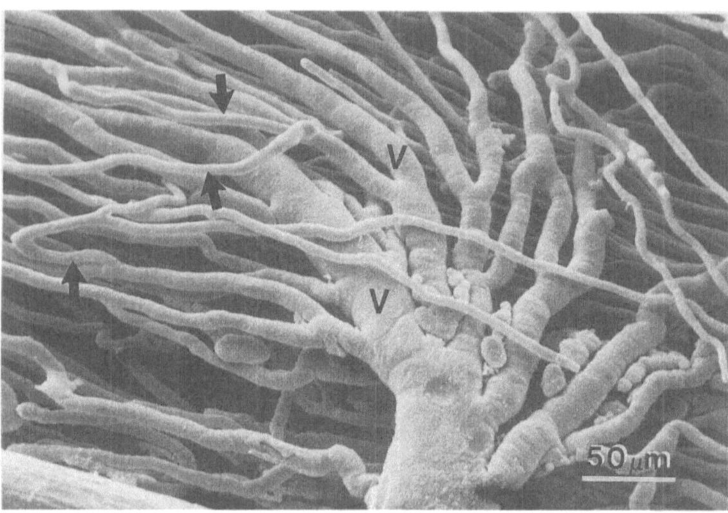

Fig. 4. Coronary cast replica of capillaries (*arrows*) draining into venules (*V*) which form a dense plexus

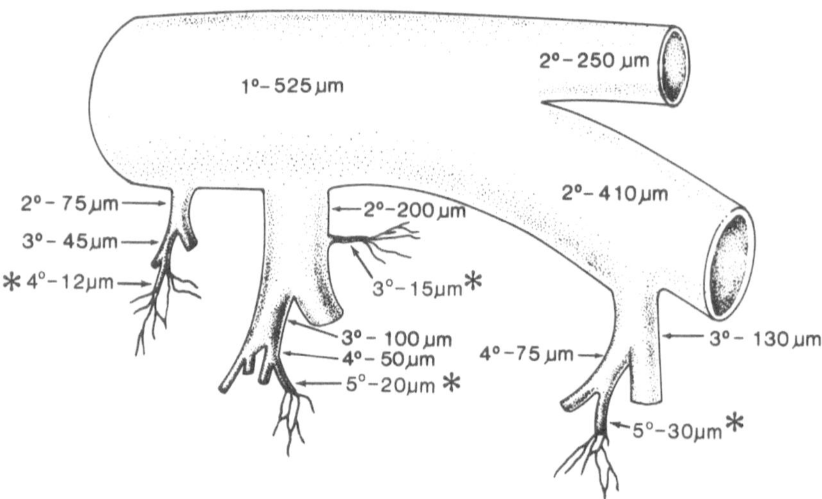

Fig. 5. Diagram illustrating variety of branching patterns in precapillary vessels (rabbit). *1°* indicates a primary branch of the left anterior descending artery; *2°* is a secondary branch, and so forth. Terminal arterioles (*) are 30 μm or less in diameter and vary as to the number of vessels preceding them. Reproduced from [14], with permission

As can be appreciated in histological sections, coronary venules are more numerous than arterioles. Bassingthwaighte et al. [27] estimated that venules were approximately 2–4 times as numerous as arterioles. The complex system of venules which drain capillaries can be appreciated in the coronary cast replica micrograph illustrated in Fig. 4. These vessels also increase rapidly in size as they course from the capillary bed, an architectural feature consistent with their low flow and capacitance function.

While the arterioles regulate the volume of flow, the anatomical and geometric characteristics of the capillary network are important determinants of tissue oxygenation. Anisotropy is a characteristic of myocardial capillaries since their orientation is primarily parallel to the long axes of the cardiocytes, but branching and anastomoses are common, as can be appreciated by coronary cast replicas [27, 28] or after silicone rubber infusion [11].

Capillary anatomical density is extremely high in the mammalian myocardium. When measurements are performed on semi-thin or thin sections of hearts fixed by vascular perfusion (after maximal dilation), capillary density approximates 3 500–4 200/mm^2 in both dogs [29] and rats [28, 30]. This range of densities predicts that average diffusion distances (one half the distance between the outer edge of two adjacent capillaries) are about 5.5–6.3 μm. These distances are calculated means, based on the formula:

$$\text{Diffusion distance} = \left(\sqrt{\frac{2A}{N\sqrt{3}}} - \bar{X}d\right) ½$$

where A = the area of the tissue field, N = number of capillaries and $\bar{X}d$ = mean capillary diameter (5.5 μm). In reality the distances between capillaries are heterogeneous; thus, their perfusion domains, and consequently tissue PO_2 values, are variable. Capillary spacing distributions can be determined from x and y axes plots (by digitization) and an estimate of the heterogeneity can be obtained from the log SD of the domains [31]. It can be shown that a wide range of PO_2 values exists in myocardial tissue, due to the heterogeneous capillary distribution.

Capillary function is closely linked with its permeability characteristics. Myocardial capillaries are of the continuous variety, i.e., they are characterized by endothelial cells which are connected by tight junctions (Fig. 3). Intercellular clefts, where the two adjacent endothelial cell membranes are separated by a space about 10–20 nm wide, form winding channels between tight junctions. A search for the anatomical equivalent of the physiologist's "small" and "large" pore systems has provided some candidates as demonstrated by electron dense tracer studies. Plasmalemmal vesicles have been for some time regarded as the transcytotic route of molecules less than 10 nm in diameter [13]. Although vesicular transport has generally been regarded as the mechanism for transcytosis of these large molecules, more recent evidence contradicts the idea that plasmalemmal vesicles travel from one surface of the endothelial cell to the other. Serial section reconstruction of rat myocardial capillaries provides evidence that all plasmalemmal vesicular profiles are part of a vesicular chain which forms a channel between the luminal and tissue surfaces of the cell [32]. This finding, then, supports the existence of "endothelial plasmalemmal invaginations" (actually a chain of interconnected vesicles) in myocardial capillaries as opposed to compartmentalized vesicles. The way in which the passage of molecules through these channels is regulated is not yet clear. An anatomical small pore system is not unequivocally established at this time. Since tracer molecules less than 10 nm have been shown to traverse the intercellular clefts, this route appears to be a likely candidate. Yet there is also evidence that small hemopeptides with diameters approximately 2 nm preferentially course through the vesicular chains or endothelial plasmalemmal invaginations [13].

An important feature, as it relates to permeability, of the endothelium is the cell coat which contains glycolipids, glycoproteins, and proteoglycans. These surface membrane components may serve as binding sites for plasma molecules. Vesicles have been shown to contain generous amounts of galactose and *N*-acetylglucosamine and favor the passage of anionic proteins [13].

Angiogenesis

In addition to its occurrence during development, coronary angiogenesis is of importance in the adult in certain situations, e.g., cardiac hypertrophy and myocardial ischemia. Yet an inadequate growth of vessels often occurs, a situation which limits maximal myocardial perfusion and O_2 delivery. Similarly, collateral vessel development appears to be variable and even inconsistent. It is evident that factors which precipitate angiogenesis are operable in some circumstances but not in others. For example, left ventricular hypertrophy secondary to exercise training or thyrotoxicosis is usually accompanied by adequate growth of the coronary capillary bed, while pressure-overload hypertrophy is usually characterized by a decrement in capillary density, i.e., number/mm^2. Thus, the stimulus for hypertrophy is one determinant of angiogenesis. A second consideration is the duration of the hypertrophy. Data from my laboratory demonstrate that vascular growth over time may be significant to normalize minimal coronary vascular resistance (based on pharmacologically induced maximal coronary perfusion) in spontaneously hypertensive rats and dogs with one-kidney, one-clip renal hypertension. Anatomical data on the hearts of the dogs with seven months' hypertension (one-kidney, one-clip) demonstrate a normal arteriolar density in spite of a 47% increase in left ventricular mass.

Such findings call the question: what is the stimulus (or stimuli) which triggers an angiogenic response? Hudlicka and Tyler [4] suggest that endothelial cell proliferation is stimulated by increases in flow or turbulence. Support for this hypothesis comes from experiments which have elevated coronary blood flow by chronic administration of dipyridamole [33], adenosine, or a xanthine derivative, HWA 285 [4]. It is postulated that increased wall tension and sheer stress (especially tangential wall stress), and intravascular pressure leading to stretch, constitute consequences of enhanced flow which may serve as mechanical stimuli for endothelial proliferation [4]. Precisely how these phenomena "turn-on" the intracellular signals resulting in the endothelial cell's growth response is not clear. Moreover, the basis for the recruitment of angiogenic factors, e.g., fibroblast growth factors, endothelial cell growth factor, and angiogenin is not understood, although it is now known that heparin from mast cells regulates the process of angiogenesis [7].

Angiogenic factors may also be involved in the development of collateral vessels. Whether the growth of collateral vessels is *de novo* or by expansion of native vessels, angiogenesis is obviously involved and in the latter case requires extensive remodelling. The classic work of Schaper [10] showed that the internal diameter of a 40 μm vessel increases to nearly 600 μm after 8 weeks of progressive narrowing of a large coronary artery. Medial thickening occurs at a slower rate, but after 26 weeks the number of medial smooth muscle layers increases approximately 5-fold. Damage to a native collateral vessel precedes its growth and could provide the signal for the release of growth factors, many of which are stored in the basal lamina. Whether exercise or pharmacological interventions stimulate coronary vessel growth remains controversial (see [4] for review).

Atherosclerosis is another process which involves vascular proliferation and remodelling. This lesion encompasses a cascade of events reviewed by Ross [34] including the proliferation of endothelial and smooth muscle cells. A number of growth factors may be involved in the process, including platelet-derived, transforming-β, fibroblastic, and epidermal growth factors. Neovascularization also may occur in atherosclerosis when adventitial blood vessels penetrate into the intima.

In conclusion, it should be emphasized that coronary vessels have the potential to undergo structural and architectural adaptations to chronic changes. The common occurrence of clinical situations which require such adaptations (e.g., myocardial ischemia, hypertension, cardiac hypertrophy, atherosclerosis) should underline the importance of research aimed at pinpointing the stimuli and mechanisms of coronary vascular remodelling and growth. This goal requires a considerable effort involving a variety of techniques using both in vivo and in vitro approaches.

References

1. Vatner SF, Pagani M, Manders WT, Pasipoularides AD (1980) Alpha adrenergic vasoconstriction and nitroglycerin vasodilation of large coronary arteries in the conscious dog. J Clin Invest 65: 5–14
2. Boucek RJ, Morales AR, Romanelli R, Judkins MP (1984) Coronary artery disease. Pathological and clinical assessment. Williams and Wilkins, Baltimore, pp 86–107
3. Manasek FJ (1969) Embryonic development of the heart. II. Formation of the epicardium. J Embryol Exp Morphol 22: 233–248
4. Hudicka O, Tyler KR (1986) Angiogenesis. Academic Press, London
5. Virgh S, Challice CE (1981) The origin of the epicardium and the embryonic myocardial circulation in the mouse. Anat Rec 201: 157–168
6. Rakusan K (1984) Cardiac growth, maturaion and aging. In: Zak R (ed) Growth of the heart in health and disease. Raven Press, New York, pp 131–164
7. Folkman J (1986) Angiogenesis: What makes blood vessels grow? News Physiol Sci 1: 199–202
8. Fulton WFM (1982) Morphology of the myocardial microciulation. In: Tillmanns H, Kubler W, Zebs H (eds) Microcirculation of the heart. Springer–Verlag, Berlin, pp 15–25
9. Baroldi G (1981) The coronary circulation in man. In: Schwartz CJ, Werthessen NT, Wolf S (eds) Structure and Function of the Circulation. Plenum Press, New York, pp 77–190
10. Schaper W (1971) The collateral circulation of the heart. Elsevier, New York
11. Grayson J (1982) Functional morphology of the coronary circulation in the coronary artery. In: Kalsner S (ed) The coronary artery. Croom Helm, London, pp 343–364
12. Gregg DE, Shipley RE (1947) Studies of the venous drainage of the heart. Am J Physiol 151: 13–25

13. Simionescu N, Simionescu M (1988) The cardiovascular system. In: Weiss L (ed) Cell and tissue biology. Urban and Schwarzenberg, Baltimore, pp 355–400
14. Tomanek RJ (1987) Microanatomy of the coronary circulation. In: Spaan JAE, Bruschke AVG, Gittenberger AC, De Groot (eds) Coronary circulation: From basic mechanisms to clinical implications. Martinus Nijhoff, Dordrecht, pp 3–12
15. Abraham A (1969) Microscopic innervation of the heart and blood vessel in vertebrates including man. Pergamon Press, New York, p 342
16. Randall WC (1984) Anatomy of blood circulation. In: Abrahamson DI, Dolrin PB (eds) Blood vessels and lymphatics in organ systems. Academic Press, Orlando, pp 319–326
17. Kristek F, Gerova M (1987) Autonomic nerve terminals in relation to contractile and non-contractile structures in the conduit coronary artery of the dog. Acta Anat (Basel) 129: 149–154
18. Feigel EO (1969) Parasympathetic control of coronary blood flow in dogs. Circ Res 25: 509–519
19. Gerova M (1982) Autonomic innervation of the coronary vasculature. In: Kalsner S (ed) The coronary artery. Croom Helm, London, pp 189–215
20. Blair RW, Ammons WS, Foreman RD (1984) Response of thoracic spinothalamic and spinoreticular cells to coronary artery occlusion. J Neurophysiol 51: 636–648
21. Henquell L, Lacelle PL, Honig CA (1976) Capillary diameter in rat heart in situ: Relation to erythrocyte deformability, O_2 transport and transmural O_2 gradients. Microvasc Res 12: 259–274
22. Chilian WM, Eastham CL, Marcus ML (1986) Microvascular distribution of coronary vascular resistance in beating left ventricle. Am J Physiol 251: H779
23. Feldstein ML, Henquell L, Hoing CR (1978) Frequency analysis of coronary intercapillary distances. Site of capillary control. Am J Physiol 235: H321–H325
24. Weiss HR, Conway RS (1985) Morphometric study of the total and perfused arteriolar and capillary network of the rabbit left ventricle. Cardiovasc Res 19: 343–354
25. Tillmanns H, Leinberger H, Neumann FJ, Steinhausen M, Parekh N, Zimmermann R, Dussel R, Kuebler W (1987) Myocardial microcirculation in the beating heart—in vivo microscopic studies. In: Spaan JAE, Bruschke AVG, Gittenberger AC, De Groot (eds) Coronary circulation. Martinus Nijhoff, Dordrecht, pp 88–94
26. Vetterlein F, Ri HD, Schmidt G (1982) Capillary density in rat myocardium during plasma staining. Am J Physiol 242: H133–H141
27. Bassingthwaighte JB, Yipintsoi T, Harvey RB (1974) Microvasculature of the dog left ventricular myocardium. Microvas Res 7: 229–249
28. Tomanek RJ, Searls JC, Lachenbruch PA (1982) Quantitative changes in the capillary bed during developing, peak and stabilized cardiac hypertrophy in the spontaneously hypertensive rat. Circ Res 51: 295–304
29. Tomanek RJ, Palmer PJ, Peiffer GW, Schreiber K, Eastham CL, Marcus ML (1986) Morphometry of canine coronary arteries, arterioles, and capillaries during hypertension and left ventricular hypertrophy. Circ Res 58: 38–46
30. Anversa P, Ricci R, Olivette G (1986) Coronary capillaries during normal and pathological growth. Can J Cardiol 2: 104–113
31. Turek Z, Hoofd L, Rakusan K (1986) Myocardial capillaries and tissue oxygenation. Can J Cardiol 2: 98–103
32. Bundgaard M, Hagman P, Crone C (1983) The three-dimensional organization of plasmalemmal vascular profiles in the endothelium of rat heart capillaries. Microvasc Res 25: 358–368
33. Mall G, Schikora I, Mattfeldt T, Bodle R (1987) Dipyridamole-induced neoformation of capillaries in the rat heart. Quantitative stereological study on papillary muscles. Lab Invest 57: 86–93
34. Ross R (1987) Growth factors in the pathogenesis of atherosclerosis. Acta Med Scand [Suppl] 715: 33–38

Flow Strategy and Functional Design of the Coronary Network

Mair Zamir[1]

Summary. Blood supply to the heart for its own metabolic needs is discussed in terms of the vascular system which provides this supply within the framework of the systemic circulation. A brief description of these (coronary) vessels is given in the classical/anatomical tradition, but more emphasis is placed on their functional arrangement. A possible functional design of the coronary system underlying the wide variability in its anatomical details is discussed. Other features also discussed are left/right dominance, collateral circulation, and the special flow conditions under which the coronary vessels function.

Key words: Coronary arteries—Coronary heart disease—Heart—Hemodynamics—Cardiovascular system

Introduction

In the human cardiovascular system, as blood leaves the left ventricle for distribution to the entire organism, the first destination is the heart itself. The first branches of the ascending aorta, which arise just downstream from the aortic valve, serve this purpose. In rare cases there is only one such branch, in most other cases there are two major branches and up to three smaller ones. The major two provide the bulk of blood supply to the myocardium and they are commonly known as the left and right coronary arteries. The other one to three minor branches may be referred to as "additional" coronary arteries in the sense that they are not always present. They are small compared with the left and right coronary arteries, and their ostia are usually close to the ostium of the right coronary artery (Fig. 1). From the point of view of variability, additional coronary arteries appear to be branches of the right coronary artery whose points of origin have migrated from positions on the proximal segment of the right coronary artery to nearby positions on the ascending aorta. This is an oversimplification, however, since in some cases the two major branches of the left coronary artery may arise separately from the aorta, thus producing a third coronary

[1]Departments of Applied Mathematics, Medical Biophysics, and Pathology, University of Western Ontario, London, Ontario, N6A 5B9, Canada, and Heart and Circulation Group, John P. Robarts Research Institute, London, Canada

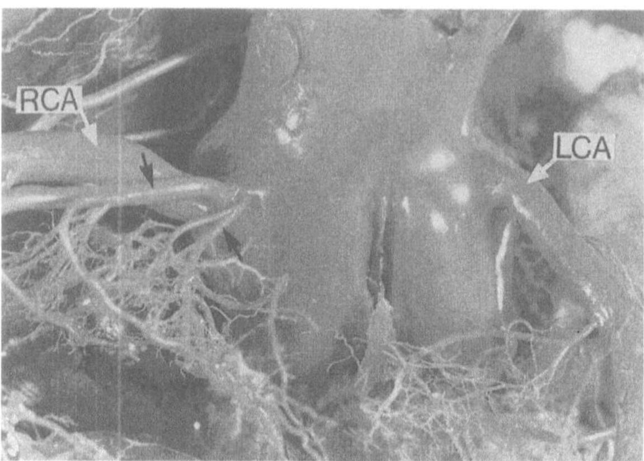

Fig. 1. Two "additional" coronary arteries (*black arrows*) arising, as they typically do, near the origin of the right coronary artery (*RCA*). From a resin cast of a human coronary network. *LCA*, left coronary artery

artery which in this case is neither small nor related to the right coronary artery. The number of possible variations is in fact very large and defies simple classification. Nevertheless, the model of two major coronary arteries and up to three minor ones is representative of the overwhelming majority of cases.

While the term "coronary" was presumably intended to describe the way in which the two major coronary arteries "encircle" the heart, today it is used in a more general sense to refer to all levels of the heart's vasculature, to large and small arteries as well as veins, or simply to any blood vessel whose sole purpose is to serve the heart tissue. Traditionally some coronary vessels have been referred to as "arteries" and some as "branches," but this is rather arbitrary and may lead to ambiguity, since in the hierarchy of an arterial tree every vessel is both an artery and a branch. Whether a vessel is designated as one or the other must clearly depend on the context, and this is the convention we shall use here. The term coronary "network" shall be used to refer collectively to the entire vasculature of the heart. This term may have the unintended connotation that the coronary vasculature is in the form of an inter-connected mesh as opposed to a tree-like structure. We shall use it without this connotation.

The coronary circulation, like that of other organs, extends between the arterial and venous sides of the systemic circulation, but it differs in that it empties directly into the right atrium without going through the vena cava. Thus blood flow enters the coronary ostia at the base of the ascending aorta and proceeds through many generations of branches and sub-branches which terminate in the capillary beds of the myocardium. From there it is collected by a reverse hierarchy of venules, small and large veins. Most of the latter converge onto a main venous channel known as the coronary sinus, which empties directly into the right atrium. Others empty directly into the right atrium without going through the coronary sinus.

The coronary branches of the ascending aorta are normally the only feeding vessels as well as the only extra-cardiac connections of the coronary network. If the ascending aorta is cut above the orifices of the coronary arteries and the heart is removed from the body, the coronary network would normally be intact and can be perfused under pressure. Other extra-cardiac connections have been reported, however, most notably through branches of the bronchial and internal mammary arteries that come close to the heart [1]. Information in this area is incomplete, particularly as it relates to the frequency and functional significance of these anastomoses. The subject is usually discussed in the context of collateral circulation of the heart, although the latter refers mostly to intracardiac anastomoses between one coronary artery and another.

The primary purpose of the vascular network of an organ is to serve the metabolic needs of the organ. A secondary purpose, in some cases, is to bring blood to and from the organ for some processing function of the blood itself, as in the lungs and kidneys for example. The heart too has a blood-processing function, namely that of pumping blood continuously to maintain pressure on the arterial side of the circulation, and blood must be brought to and from the heart constantly for this purpose. The design by which a vascular network brings blood to and from an organ and circulates it within it depends on whether the metabolic and the processing activities of the organ are to be served together or separately. In the lungs and kidneys the two needs are coupled and are served by the same vascular network. In the heart they are uncoupled. The coronary network serves only the metabolic needs of the heart and it is entirely separate from the network of vessels that bring blood to and from the heart for pumping purposes. In this sense we may say that the coronary network is a purely "metabolic" network.

Per gram of tissue the metabolic needs of the heart are more intense and more variable than the needs of other organs. The vasculature of the myocardium is perhaps more dense than that in any other purely metabolic network in the body. Much of this vasculature is embedded within the myocardium; thus flow through the coronary network is greatly affected by the pumping action of the heart as the cardiac muscle continuously contracts and relaxes. A study of the coronary network must therefore deal not merely with its static anatomical details but with the manner by which the coronary network is designed to meet these special hemodynamic challenges. It is a study of how the network works rather than how it looks, and of why the network fails in its mission more often than do other networks in the cardiovascular system. The pathogenesis of atherosclerosis may explain a common mechanism by which the failure is mediated, but it does not explain why the mechanism is so highly effective in the coronary network.

Orientation

The terms "left" and "right" are widely used in the descriptions of the heart and the coronary network, often with some ambiguity. While the left and right sides of the human body are well-defined in terms of a geometrical frame of reference, the left and right sides of the heart and of the coronary network are purely functional entities. In fact, in the heart and coronary network these terms would be more accurately regarded as mere names than as indicators of geometry or orientation.

When the human body is in the anatomical position the median plane divides it accurately into its left and right sides, but it does not divide the heart into what we call its left and right side [2–4]. In fact the heart does not have a median plane since it lacks the required symmetry. A plane through the interventricular septum separates the left and right ventricles and only approximately the left and right atria. Furthermore, when the heart is *in situ* this plane is oblique to the median plane of the body (Fig. 2). Similarly, a plane through the atrioventricular septum separates the atria from the ventricles, but this plane is oblique to the transverse plane of the body when the heart is *in situ*.

From the point of view of the coronary network a useful position of the heart is that in which the heart is rotated until the atrioventricular septum coincides with the transverse plane of the body and the interventricular septum coincides with the median plane (Fig. 2). This may be referred to as the "upright" position, in contrast with the "anatomical" position when the heart is *in situ*. Clearly, since the atrioventricular and interventricular septa are not exactly perpendicular to each other, and since neither is exactly a plane, their coincidence with the planes of the body when the heart is in the upright position is only approximate. Nevertheless, in that position the left side of the heart is approximately on the left side of the body's median plane, the right side is approximately on the right, and the terms "left" and "right" come close to being justified. When the heart is in the upright position its apex is pointing down and the atrial cap is vertically above it, which is a convenient orientation for describing the coronary network. In the anatomical (*in situ*) position the heart is lying on its right side with its apex pointing obliquely to the left and the atrial cap is lying on a slant (Fig. 2).

The two major coronary arteries circle the heart in the plane of the atrioventricular septum, which may therefore be referred to as the coronary plane. When the heart is in the upright position the coronary plane may be considered as being horizontal, in the sense that it is parallel to the transverse plane of the body. This is a useful orientation for describing the course and branches of the two major coronary arteries. Branches destined for the atrial walls go "up" and those for ventricular walls go "down." What we call the right coronary artery arises from an anterior position on the coronary plane and courses to the right, circling the right side of the heart in a clockwise direction (Fig. 3). What we call the left coronary artery arises from a posterolateral position on the cononary plane [5, 6] and divides almost immediately into two major branches: the circumflex branch, which circles the heart in an anticlockwise direction, and the anterior descending branch, which descends diagonally down along the anterior interventricular sulcus. When the heart is in the anatomical position, because the coronary plane is then not horizontal, the initial course of the right coronary artery is not right but diagonally down, and that of the circumflex is not left but diagonally up.

Dominance

A more important left/right question arises in relation to the manner in which the two major coronary arteries share the task of blood supply to the left and right sides of the heart. In this context, discussion centers mostly on blood sup-

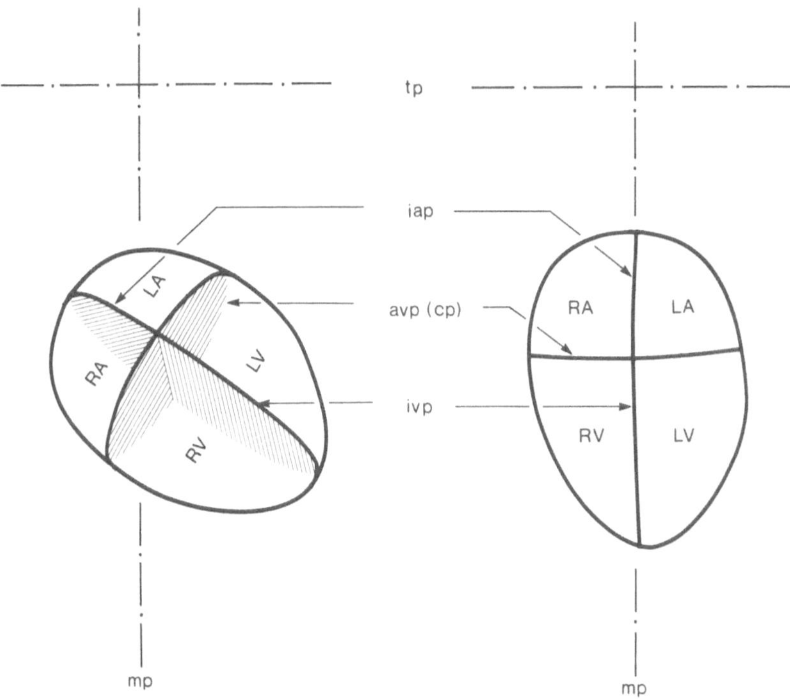

Fig. 2. A schematic diagram showing approximately how the reference planes of the heart relate to the reference planes of the body when the heart is *in situ* (*left*) and in the upright position (*right*). *mp*, median plane; *tp*, transverse plane; *iap*, interatrial plane; *avp* (*cp*), atrioventricular plane (coronary plane); *ivp*, interventricular plane; *RA*, right atrium; *LA*, left atrium; *RV*, right ventricle; *LV*, left ventricle

ply to the ventricles since supply to the atria and to the atrioventricular septum is very small compared with that to the ventricles. Supply to the ventricles is considered in terms of the 5 walls of the myocardium: the interventricular septum, the left lateral and left posterior walls, which comprise the left ventricle; and the right posterior and right anterior walls, which comprise the right ventricle. The posterior interventricular sulcus is the line separating left and right posterior walls, and the point at which this line meets the posterior atrioventricular sulcus is the "crux," which is an important landmark in this context (Fig. 4).

As the right coronary artery circles the right side of the heart along the atrioventricular sulcus, it serves the anterior and posterior walls of the right ventricle. The left coronary artery through its anterior descending branch serves the anterior half of the interventricular septum and the lateral wall of the left ventricle, and through its circumflex branch serves the posterior wall of the left ventricle (Figs. 3, 4). To this point the left and right coronary arteries serve the left and right ventricles respectively. As the right coronary artery and the left circumflex artery proceed towards the crux, however, only one of them usually reaches it and gives rise to the posterior descending branch. The latter is an important coronary artery which serves the posterior half of the interventricular septum.

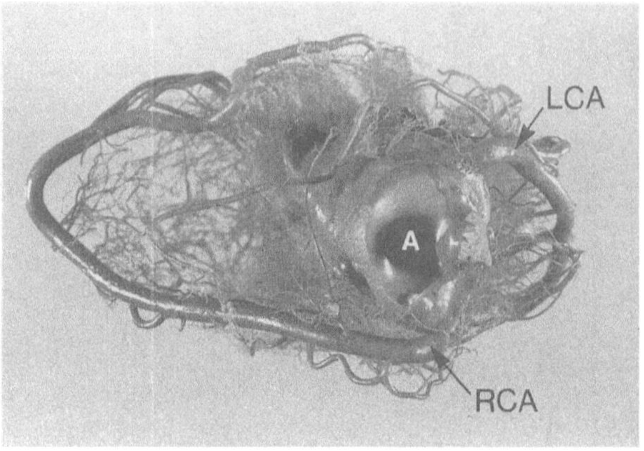

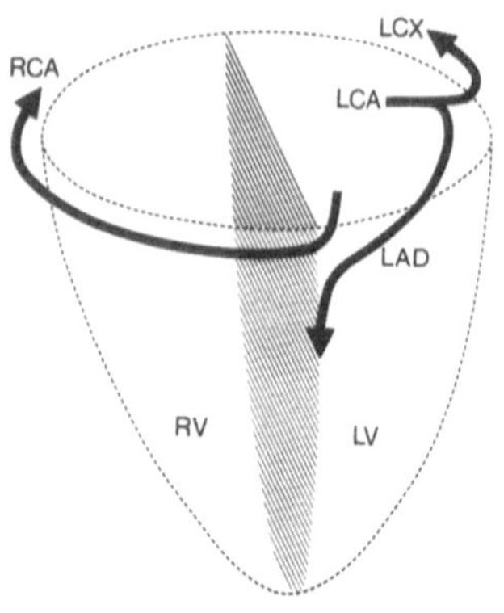

b

Fig. 3a, b. The origin and initial course of the major coronary arteries: **a** resin cast of a human coronary network in an upright/superior view, **b** schematic diagram showing oblique view. The right coronary artery (*RCA*) arises from an anterior position and circles the heart in a clockwise direction. The left coronary artery (*LCA*) arises from a posterolateral position and bifurcates into the left circumflex artery (*LCX*) which circles the heart in an anticlockwise direction, and the left anterior descending artery (*LAD*) which descends along the anterior interventricular sulcus. *A*, aorta; *RV*, right ventricle; *LV*, left ventricle

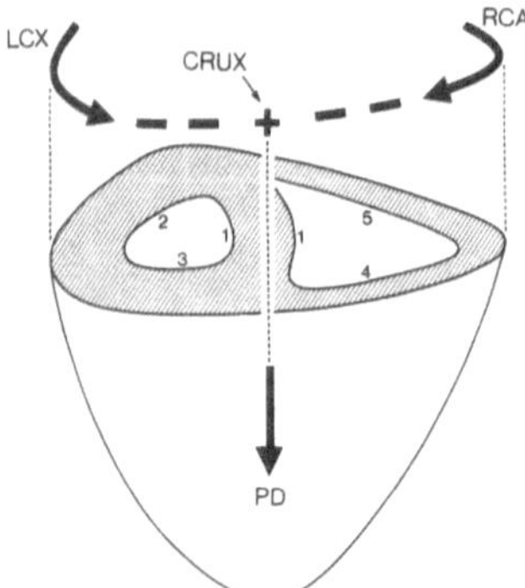

Fig. 4. Subsequent course of the major coronary arteries in an upright/posterior view and in relation to the 5 walls of the myocardium. Either the right coronary artery (*RCA*) or the left circumflex artery (*LCX*) reaches the crux and gives rise to a posterior descending branch (*PD*)—most frequently the former, much less frequently the latter, occasionally both. *1*, interventricular septal wall; *2*, lateral left ventricular wall; *3*, posterior left ventricular wall; *4*, posterior right ventricular wall; *5*, anterior right ventricular wall

There is a considerable amount of variability in the final courses of the left circumflex and right coronary arteries, with regard to which of the two vessels reaches the crux and whether that vessel continues beyond the crux. The concept of left/right "dominance" was developed presumably in order to deal with this variability. The concept is based on 3 categories of coronary architecture, defined as follows:

1. The left circumflex artery reaches the crux, gives rise to the posterior descending branch, and terminates at that point. The left coronary artery is then said to be dominant, and the heart and the coronary network are said to be "left dominant."

2. The right coronary artery reaches the crux, gives rise to the posterior descending branch, and terminates at that point. The left and right coronary arteries, as well as the heart and the coronary network, are said to be "balanced."
3. The right coronary artery reaches the crux, gives rise to the posterior descending branch, and continues beyond the crux to serve part of the posterior wall of the left ventricle. The right coronary artery is said to be dominant, and the heart and the coronary network are said to be "right dominant."

Statistics indicate that the incidence of left dominance in the human heart is approximately 10%, right dominance 70%, and balanced 20% [5–7]. In a left-dominant case, the left ventricle (including the interventricular septum) depends entirely on the left coronary artery for its blood supply. In a balanced case, the posterior half of the interventricular septum receives its blood supply from the right coronary artery while the rest of the left ventricle is supplied by the left coronary artery. In a right-dominant case, the right coronary artery supplies the posterior half of the interventricular septum and part of the posterior left ventricular wall, the remainder of the left ventricle being supplied by the left coronary artery.

Thus in approximately 90% of human hearts, what we call the left and right coronary arteries do not in fact serve exclusively the left and right sides of the heart as the names may imply. Also, in almost all cases the amount of blood supply provided by the left coronary artery is usually higher than that provided by the right [5, 6]. Therefore, while the distinction between the above three categories is important, the use of the term "dominance" is somewhat misleading. "Dependence" would be a more appropriate term: in a left-dominant case the left ventricle depends entirely on the left coronary artery, in a balanced case it depends to a small degree on the right coronary artery, while in a right-dominant case it may depend to equal degrees on the left and right coronary arteries. Thus in terms of *dependence* the three categories may be referred to as "totally left-dependent," "mostly left-dependent," and "balanced."

While the concept of dominance does not deal rigorously with the wide variability in coronary architecture, and while its usefulness has often been clouded by confusion and misunderstading, it is a valuable reminder of the need to deal wih this variability. The range of variability is in fact quite large. This is illustrated in Fig. 5 in terms of the extent to which the right coronary artery may contribute to blood supply to the left ventricle. In the face of such wide variability, to assess the role of a coronary artery in the scheme of blood supply to an ischemic heart is as important as it is to assess the degree of stenosis of that vessel.

Descriptive anatomy

We present here a brief description of the coronary network in the classical tradition, whereby vessels are identified by anatomical names relating usually to their dispositions or destinations. There is a considerable amount of variability

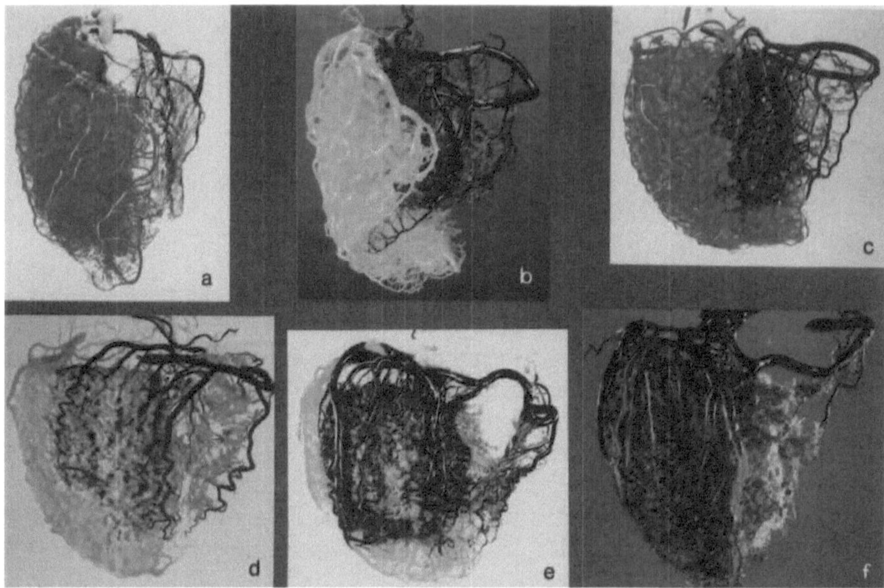

Fig. 5a–f. A sequence of resin casts of human coronary networks in an upright/posterior view to show the highly variable extent to which the right coronary artery may contribute to blood supply to the left ventricle. In (**a**), (**c**), and (**f**) the distribution of the left coronary artery and its branches is red and that of the right is blue. In (**b**), (**d**), and (**e**) the corresponding colours are yellow and green. At one extreme (**a**) the left ventricle is supplied entirely by the left side of the coronary network, while at the other (**f**) the right coronary artery supplies the posterior wall of the left ventricle and the posterior part of the interventricular septum

in the branching architecture of the coronary network and it is neither possible nor meaningful to provide an "accurate" description of this architecture. The number and size of diagonal branches of the left anterior descending artery, for example, may vary considerably from heart to heart. The left circumflex artery in one heart may be insignificantly small while in another it may have a much longer course and many more branches. The description below is not intended to deal with these variations in a statistical or geometrical manner. There are many excellent descriptions which do so and to which the reader may refer for more details [5, 7–16]. The emphasis here is on the place and functional role of individual arteries within the coronary network, rather than on the exact topology of their path.

From the point of view of branching hierarchy, all vessels on the arterial side of the coronary network are branches or subbranches of the two main feeding arteries. From a functional point of view, however, some of these vessels are more important than others and there has been a tendency to refer to them as "arteries" rather than "branches." In coronary heart disease, for example, the right main coronary, the left anterior descending, and the left circumflex are all regarded as arteries by virtue of their functional importance, although the last two are branches of the left main coronary artery. At the same time a very small additional coronary artery of only small functional significance is usually also

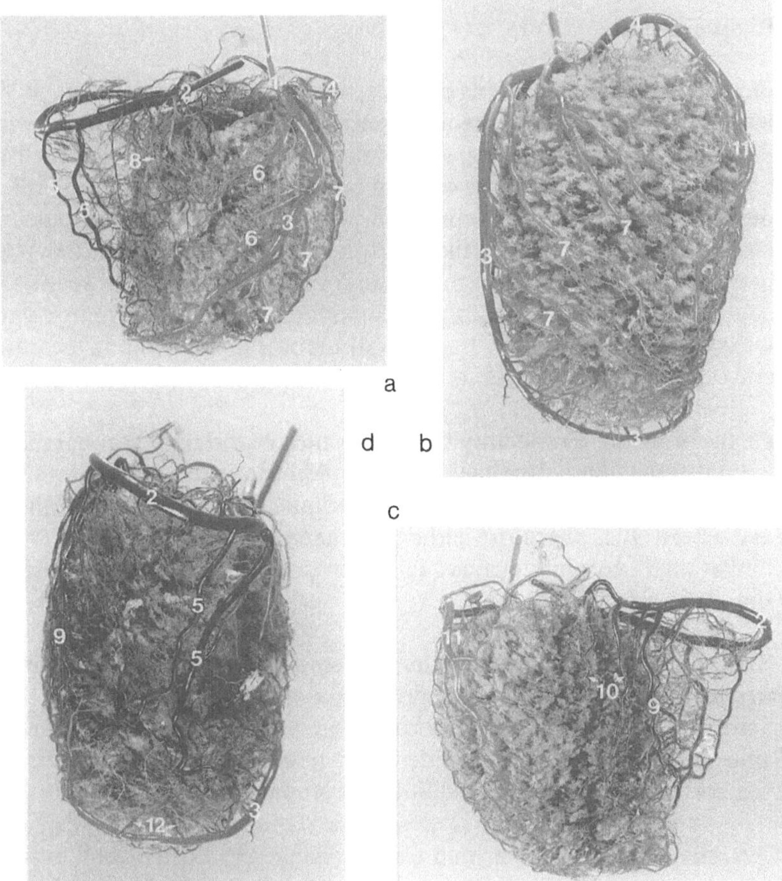

Fig. 6a–d. The human coronary network in an upright position and **a** anterior, **b** left lateral, **c** posterior, and **d** right lateral view. A typical selection of vessels is numbered: *1*, left (main); *2*, right (main); *3*, anterior descending; *4*, circumflex; *5*, acute marginal; *6*, septal; *7*, diagonal; *8*, anterior right ventricular; *9*, posterior descending; *10*, posterior left ventricular; *11*, obtuse marginal; *12*, apical

referred to as an artery, in this case because it is primary on the scale of branching hierarchy. There is thus some confusion about what should be referred to as an artery and what should be referred to as a branch in the coronary network and no standard convention exists in the literature. The convention we shall use here is to let the branching context dictate, as it should, which of the two terms to use, without implying any degree of functional importance. Thus the left circumflex for example may be referred to as a "branch" in relation to the left coronary artery and an "artery" in relation to it own branches.

The vessels listed below are those commonly named in the coronary nctwork. Branches are listed in the order in which they would normally be encountered along the streamwise course of the parent vessel. Only the most common variations are included. The heart is assumed to be in the upright position when describing the course or orientation of a vessel. Typical examples of some of the vessels are shown in Fig. 6.

Right Coronary Artery

One of the two main feeding arteries, arising from the right sinus of Valsalva at a level above the plane of the atrioventricular septum. Despite their names, both the sinus and the initial segment of the artery are directed as much towards the anterior aspect of the heart as they are towards the right [5, 6, 16]. The vessel descends immediately towards the anterior atrioventricular sulcus and turns sharply right on it to circle the heart in a clockwise direction. How far it reaches may be anywhere between the left and right margins of the heart, and along its course it gives off branches to destinations on both the ventricular and atrial walls of the myocardium. Its share of blood supply to the myocardium and the corresponding territory of its distribution are thus highly variable.

Conus branch. Frequently the first branch of the right coronary artery, arising from its immediate proximal segment. Almost as frequently, however, it arises directly from the ascending aorta at a point close to the origin of the right coronary artery. It is therefore either a branch of the right coronary artery or an "additional" coronary artery. In either case it is small and is destined to serve the conus arteriosus.

Sinus-node branch. Usually the next small branch of the right coronary artery, destined to serve the sinus node. Frequently it may arise from the circumflex artery on the left side of the coronary network. In rare cases there may be one sinus node artery from each side of the network, or one may arise directly from the aorta in the form of an additional coronary artery.

Anterior right ventricular and atrial branches. Mostly small branches, arising at different points along the anterior segment of the right coronary artery. Their number and size are highly variable and, again, one of these may on occasion arise directly from the ascending aorta in the form of an additional coronary artery or a second conus branch. They serve the anterior right ventricular and atrial walls.

Acute marginal branch. A major branch of the right coronary artery and the main supplying artery to the right ventricular wall. It usually arises as the right coronary artery approaches the right side of the heart. Its course runs along the acute margin of the heart, hence its name.

Posterior right ventricular and atrial branches. Mostly small branches, arising at different points along the posterior segment of the right coronary artery. They are highly variable in size and number, and are destined to serve the posterior right ventricular and atrial walls.

AV-node branch. A small but important and fairly constant branch, destined to serve the atrioventricular node region. It arises from the posterior segment of the right coronary artery, usually as the latter nears the crux.

Posterior descending branch. The largest and most important branch of the right coronary artery, destined to serve the posterior part of the interventricular septum. It arises at a point near the crux and descends along the posterior interventricular sulcus towards the apex of the heart, hence its name. In a small percentage of hearts the right coronary artery may terminate before it reaches the crux and hence does not give rise to the posterior descending artery. In such cases a posterior descending branch arises from the left circumflex artery. In some cases there may be a posterior descending branch from each side of the coronary network. Most commonly, however, the right coronary artery reaches the crux and gives rise to this branch and then either terminates at that point or continues beyond the crux along the left posterior atrioventricular sulcus.

Posterior left ventricular and atrial branches. Variable in size and number but ventricular branches may be fairly large. Destined to serve the posterior left ventricular and atrial walls. They arise at different points along the segment of the right coronary artery which runs along the posterior left atrioventricular sulcus when the right coronary artery reaches that far. The right coronary artery then terminates in the form of the last of these ventricular branches.

Left (Main) Coronary Artery

The second main feeding artery in the coronary network, arising from the left sinus of Valsalva at a level above the plane of the atrioventricular septum. The sinus and the artery are in fact directed towards the postero(left)lateral aspect of the heart [5, 6, 16]. Following a very short course towards the posterolateral corner of the coronary plane, the left coronary artery bifurcates into two major branches.

Left anterior descending artery. One of the two branches of the left coronary artery at its major bifurcation. It turns sharply towards the anterior interventricular sulcus and runs along it towards the apex of the heart, and then circles the apex to reach the posterior side of the heart. It is one of the most important vessels in the coronary network, destined to serve the anterior part of the interventricular septum and the lateral wall of the left ventricle. It has a number of fairly constant branches.

Septal branches: Variable in size and number, but at least one or two are fairly large. Destined to serve the anterior part of the interventricular septum, they arise at different points along the descending segment of the left anterior descending artery and are directed straight into the septal wall.

Diagonal branches: Variable in size and number, but are generally large. Destined to serve the lateral wall of the left ventricle, they arise at different points along the descending segment of the left anterior descending artery and are directed diagonally along the surface of that wall. The first diagonal branch may arise very close to or actually at the major bifurcation of the left coronary artery, thus turning it into a trifurcation.

Apical branches: Small in size, they arise at fairly regular intervals from the apical segment of the left anterior descending artery as it circles the apex of the heart. They are directed into the thickness of the apex to serve that part of the myocardium. The left anterior descending artery terminates as it gives off the last of these branches, on the posterior side of the heart.

Circumflex artery. The second branch of the left coronary artery at its major bifurcation. It is directed towards the left lateral segment of the atrioventricular sulcus and turns sharply on it to circle the heart in an anticlockwise direction. While it is a major coronary artery, the length and size of the circumflex artery are highly variable and therefore so is the number of its branches. In a small percentage of hearts it is fairly large, reaching the crux on the posterior side of the heart and giving rise to the posterior descending artery. Most frequently, however, it does not reach the crux and in some cases it may not go far beyond the left (obtuse) margin of the heart. The sinus node artery may frequently be the first branch of the circumflex artery, although more frequently it is a branch of the right coronary artery.

Obtuse marginal branch: The only constant branch of the circumflex artery, destined to serve the lateral and posterior walls of the left ventricle. It runs down along the obtuse margin of the heart, hence its name. The circumflex artery may actually terminate in the form of this branch or, more frequently, it may continue in a fairly diminished form after giving rise to it.

Lateral left ventricular and atrial branches: Highly variable in size and number, destined to serve the lateral left ventricular and atrial walls. The lateral left ventricular wall is served mostly by diagonal branches of the left anterior descending artery.

Posterior left ventricular and atrial branches: Variable in size and number but ventricular branches may be fairly large. Destined to serve the posterior left ventricular and atrial walls. In some cases this territory may be supplied entirely by the right coronary artery. However, most frequently it is shared between the right coronary and circumflex arteries as they both reach the left posterior wall of the heart, coming from different directions along the atrioventricular sulcus. They both give rise to some posterior left ventricular and atrial branches and then terminate in the form of their last ventricular branch.
Rarely, when the circumflex artery goes as far as the crux and gives rise to the posterior descending branch, it may also give rise to the AV-node branch.

Coronary Veins

The venous part of the coronary network parallels much of the arterial part. Schematically, the venous system consists of anterior and posterior descending veins which accompany the anterior and posterior descending arteries respectively and which empty into a large vein known as the "great cardiac vein" (Fig. 7). The latter circles the left side of the heart along the atrioventricular sulcus from an anterior point to a posterior one. At the anterior end the great cardiac

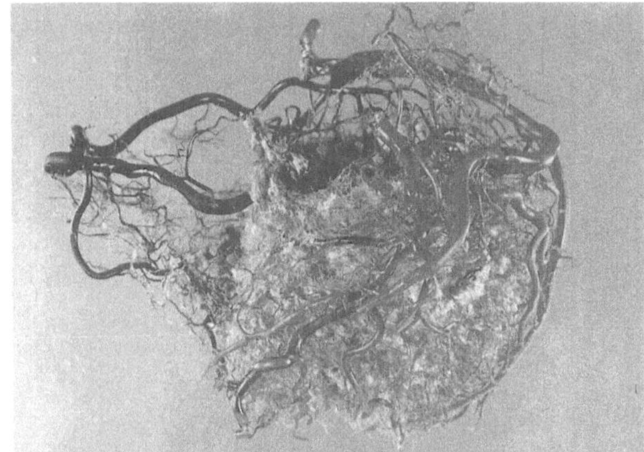
a

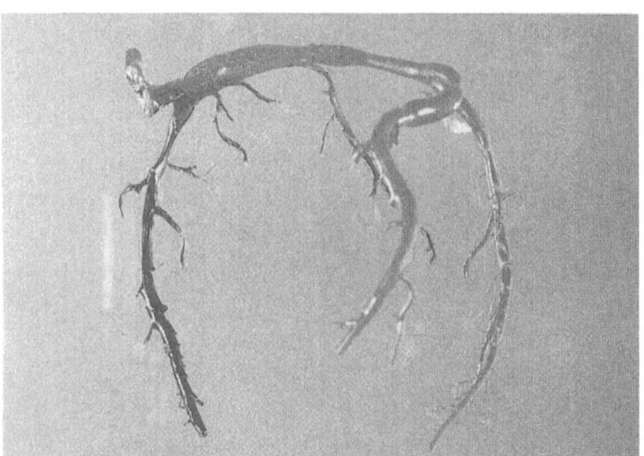
b

Fig. 7a, b. The coronary sinus system of veins shown is its normal position within the coronary network in an upright/anterior view (**a**), and in isolation (**b**). The three vertical veins are the anterior descending, obtuse marginal, and posterior descending, and the circular one is the great cardiac vein. At its wider end the latter becomes the coronary sinus which empties into the right atrium (See also Fig. 9)

vein narrows and joins the anterior descending vein. At the posterior end the great cardiac vein widens and becomes the "coronary sinus" which empties into the right atrium. Other ventricular and atrial veins empty into this system at different points along the course of the great cardiac vein, forming a network similar to that on the arterial side but with some exceptions. First, the anterior and posterior descending veins, as they run along the anterior and posterior interventricular sulci, may drain from both the left and right ventricular walls. By contrast, the corresponding arteries perfuse the left ventricular and interventricular septal walls only. Second, a separate system drains most of the right ventricular wall directly into the right atrium (Fig. 8). Finally, there are some very small so called "thebesian veins" which empty directly into the chambers of the heart. The coronary sinus system, however, accounts for the major part of venous return from the myocardium.

Fig. 8. Veins (*arrows*) from a right ventricular wall which typically do not empty into the coronary sinus system but directly into the right atrium. The large vessel which they are seen to bypass is the right coronary artery

Functional Design

The task of the coronary network is to provide blood supply to a highly nonuniform organ with a highly variable range of demands. The fluid-dynamic strategy by which the network meets this task forms the basis of its functional design.

It is important not to confuse functional design with morphological pattern. A pertinent analogy serves to illustrate the difference. The task of the heart itself is to pump blood and maintain pressure on the arterial side of the vascular system. The functional design of the heart consists of its 4 chambers, the valves, the elasticity of its walls, and the conducting system which controls its rhythmic action. These features are intrinsic to the pumping function of the heart and they do not vary from heart to heart. On the other hand, the weight of the heart and its exact shape and size vary considerably from heart to heart. These features are mere incidents of its morphological form, not part of its functional design.

Similar discussion applies to the coronary network. The course and size of individual vessels within the network vary considerably from heart to heart, which suggests clearly that these details of morphological pattern are not part of the functional design of the network. Yet much of our present knowledge about the coronary network consists of accurate mapping of the course and size of individual vessels and statistical data on variability in these morphological details. We know very little about the constant design from which this variability springs, or the basic fluid dynamic design by which the coronary network supplies the myocardium.

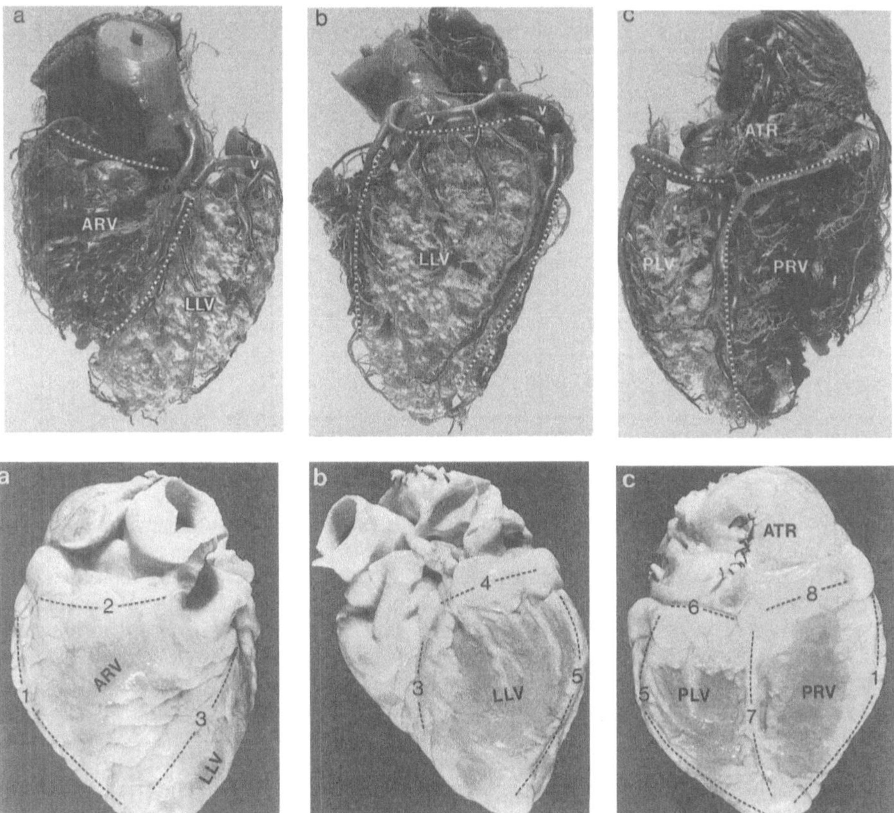

Fig. 9a–c. The human heart and coronary network in the upright position and **a** anterior, **b** left lateral, and **c** posterior views. Dotted lines define 6 zones of the myocardium which act as functional cells of the coronary network. *ARV*, anterior right ventricular zone; *LLV*, lateral left ventricular zone; *PLV*, posterior left ventricular zone; *PRV*, posterior right ventricular zone; and *ATR*, atrial zone. The interventricular septal zone (IVS) is not shown. Adapted from [6]

It was suggested recently that this design consists of dividing the myocardium into distinct zones which act as functional cells or functional units of the network [6, 17]. Each zone has a set of "distributing" vessels which bring blood supply to it. They circle it and remain on its borders but they do not enter the zone. Branches from these vessels enter the zones and implement the delivery of blood. They are the "delivering" vessels. The relationship between zones of the myocardium and their distributing and delivering vessels was found to be a constant feature of the coronary network which did not vary from heart to heart and was thus proposed as the basis of its functional design.

Six zones of the myocardium were identified under this scheme (Fig. 9): anterior right ventricular, lateral left ventricular, posterior left ventricular, interventricular septal, posterior right ventricular, and atrial. The right coronary, circumflex, acute and obtuse marginal, and anterior and posterior descending arteries were identified as distributing vessels. Diagonal, atrial, nodes and septal

	lad	lcx	lmg	lpd	rca	rmg	rpd	D/N
LLV	100/200	100/100	100/100					300/400
PLV			100/100		100/100			200/200
IVS	100/200						100/100	200/300
ARV					100/100	100/100		200/200
PRV					100/100	100/100		200/200
ATR		100/100			100/100			200/200
D/N	200/400	200/200	200/200		400/400	200/200	100/100	1300/1500

Fig. 10. Quantitative mapping of the relation between zones of the myocardium (listed in *left column*) and distributing coronary arteries (listed in *top row*). The service received by a zone from a distributing vessel along one of its borders is designated as 100 units. The total amounts received by zones are listed in the *right column* and those supplied by vessels and listed in the *bottom row*. The two values in each box represent normal (*N*) and diseased (*D*) states. In the diseased state the *lad* is assumed to provide only 50% of its service. The result in the right column indicates that service to the *LLV* zone is reduced by 25% and that to the *IVS* zone by 33%. *lad*, left anterior descending; *lcx*, left circumflex; *lmg*, left marginal; *lpd*, left posterior descending; *rca*, right coronary; *rmg*, right marginal; *rpd*, right posterior descending; *LLV*, lateral left ventricular zone; *PLV*, posterior left ventricular zone; *IVS*, interventricular septal zone; *ARV*, anterior right ventricular zone; *PRV*, posterior right ventricular zone; *ATR*, atrial zone. From [6]

branches, additional coronary arteries and other vessels were all identified as delivering vessels. It was found possible to map the relationship between zones of the myocardium and their distributing vessels. This was used to provide a quantitative assessment of the provisions for blood supply to each zone, and of the manner in which these provisions are affected by each vessel in that particular heart (Fig. 10).

In the terminology of this scheme of blood supply to the myocardium, a border defining the edge of a zone is always occupied by a distributing vessel. The identity of the vessel is variable, but its presence is constant. The border between the atrial and posterior left ventricular zones, for example, may be occupied by the circumflex artery, the right coronary artery, or partly by both. The border between the posterior left and posterior right ventricular zones may be occupied by a posterior descending branch either of the right coronary artey or of the left circumflex artery. Variability does not affect the presence of distributing vessels along these and other borders; it only affects the identity of the vessels. By contrast, variability greatly affects both the identity and presence of delivering vessels, and it is not possible to map these vessels accurately.

Such understanding of the coronary network and of its relation to the myocardium is essential in clinical cardiology. We must go beyond the purely descrip-

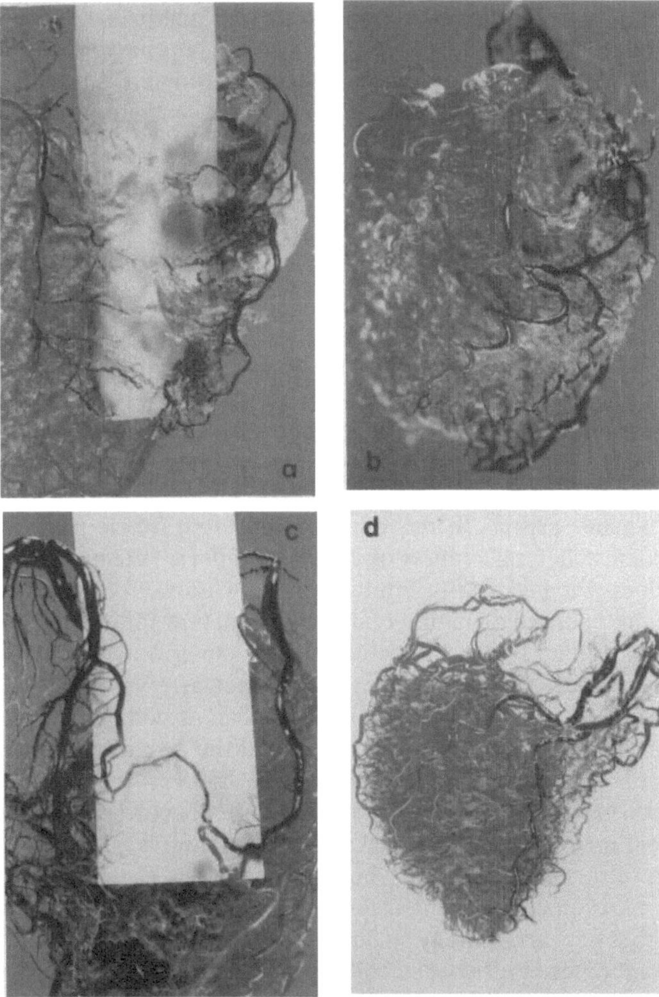

Fig. 11a–d. Human coronary networks in which the left and right coronary arteries were perfused with red and blue casting material respectively to detect collateral pathways between the two sides of the network. The figure shows a typical selection of such pathways found in the human heart: **a** numerous fine vessels extending from the bed of one coronary artery to that of another with evident mixing of the two colors; **b** a single large coronary artery from the right side of the network (*blue*) reaching a territory on the left (*red*); **c** a single medium-size collateral vessel providing a direct connection between two coronary arteries on different sides of the network; **d** several branches of a right coronary artery (*blue*) perfusing a territory on the left (*red*) side of the coronary network

tive anatomy of the coronary arteries to an understanding of their particular fluid-dynamic arrangement in each heart. The notion of "one-," "two-," or "three-vessel" disease as it is widely used in clinical cardiology is based on the assumption that the disease of one vessel in one heart can be equated to the disease of one vessel in another heart. Wide variability in the individual arrangement of the coronary arteries in different hearts makes this notion clearly untenable. The functional effect of disease in a coronary artery can only be assessed in the context of the coronary network of which it is a part, and of the particular role played by the artery in that network. A figure expressing the mere stenosis of the artery has no meaning in itself, except in this wider context. The ultimate effect to be assessed is the reduction in the provisions for blood supply to a zone of the myocardium (Fig. 10)

Collateral Circulation

There are other important questions relating to functional design aspects of the coronary network; the most prominent perhaps is that of collateral vessels and collateral circulation. The question here is essentially this: can blood supply reach a given destination in the myocardium via more than one pathway?

Much confusion and controversy surrounds this subject because the fluid-dynamic aspects of this question are often overlooked. The mere presence of some collateral connections between different elements of the coronary network does not necessarily imply multiple pathways for flow through the network under normal dynamic conditions. Fluid flow follows the path of least resistance. There may be very little or no flow along a collateral pathway that has high resistance, yet the pathway will appear as a vascular connection between two elements of the network. In particular, if collateral connections occur at the capillary level or at any point beyond the arteriolar resistance level then it is of little significance since by that stage the flow will have expended much of its driving pressure and will not be able to perfuse new territories. Under experimental conditions such collateral pathways may fill with contrast material or a tracer but the rate at which they fill is not usually known. The corresponding perfusion under the normal dynamics of the circulation may be so small that it does not result in any significant amount of blood flow. There is thus an important difference between collateral *vessels* and collateral *circulation*, which as yet has not been adequately dealt with.

Surprisingly early in the literature it was concluded that "The vessels which carry blood to the heart. .come together again and here and there communicate by anastomosis" [18], and "The human heart is potentially able to develop collaterals where they are needed" [19]. Collateral vessels were demonstrated clearly by a number of authors [7, 10, 20–25], leading some to conclude that "Anatomically the coronary arteries of man are not end arteries" [10], and "Anatomically then the coronary arterial circulation is not a circulation formed of terminal branches" [7].

These early conclusions suggested that blood supply to the myocardium is well secured by multiple pathways, which in turn implied that myocardial ischemia would be a rare phenomenon in the human heart. This view soon conflicted with the high incidence of coronary heart disease, however, where it has been established clearly that the obstruction or stenosis of a coronary artery is strongly linked to myocardial ischemia and that the phenomenon is far from rare [13, 26–30]. Attempts to explain this conflict continue today [31–41], with no conclusive explanation as yet. There is also lack of agreement on the exact mapping of collateral vessels in the coronary network; differences have been reported between species and within a species. In humans, collateral vessels can be easily demonstrated in some hearts but not in others.

The role and functional significance of collateral circulation in the human heart is thus a complex and as yet unresolved issue. In fact some fairly elementary questions on the subject require further discussion and clear definition of terms. Is the coronary network in the form of a tree with open branches, or is it in the form of an interconnected mesh? Current anatomical descriptions point clearly to the former model. The major coronary arteries and their branches are fully documented, with no permanent bridges or interconnections between them. Such communicating vessels as the palmar arches in the hand or the circle of Willis in the brain have not been identified in the human heart. That is, there are no collateral connections at the primary levels of the coronary network. Do such communications exist at higher levels of the coronary network? The answer in this case is yes, but with a rather unpredictable pattern. A selection of these are shown in Fig. 11, where the left and right sides of the coronary network were cast in different colors to detect any communications between them. The results, which are typical of those referred to earlier, indicate that collateral vessels can be found at different levels of the network but in a rather sporadic pattern. They can be found in some hearts but not in others, with considerable variation in the number reported by different authors [7]. In the absence of collaterals a distinct border between the two sides of the coronary network can be found as shown in Fig. 12. Different colors have also been used to study the distributions of individual vessels within the coronary network, with similar results [42]. In the absence of collaterals, the beds of individual vessels can in fact be separated from each other, as shown in Fig. 13.

The question of whether the coronary arteries are "end arteries" must thus be defined before it can be answered. In the sense that the coronary network is in the form of a tree with open branches that at the primary levels of the network do not anastomose with each other to provide constant collateral circulation between them, the coronary arteries are indeed end arteries. In the sense that in some hearts sporadic anastomoses can be found at some levels of the tree, however, it may be concluded that the coronary arteries are not end arteries. There is no contradiction between the two conclusions, they merely reflect two different defiinitions of the term "end artery." The first definition is more consistent with the literal sense of the term, and in that sense the coronary arteries are end arteries.

The ultimate question of course concerns the place of collateral vessels in the functional design of the coronary network, and hence their functional significance in the human heart. Here it is important to distinguish between ad hoc anastomoses and permanent vascular bridges. The absence of the latter in the human heart suggests that collateral circulation is not part of the normal functional design of the coronary network. This is fully consistent with the presence of sporadic anastomoses in some hearts and with the increase in their number in patients with coronary heart disease [39, 41]. So far there is no reason to suspect that the latter is different from a normal angiogenic response to ischemia, not

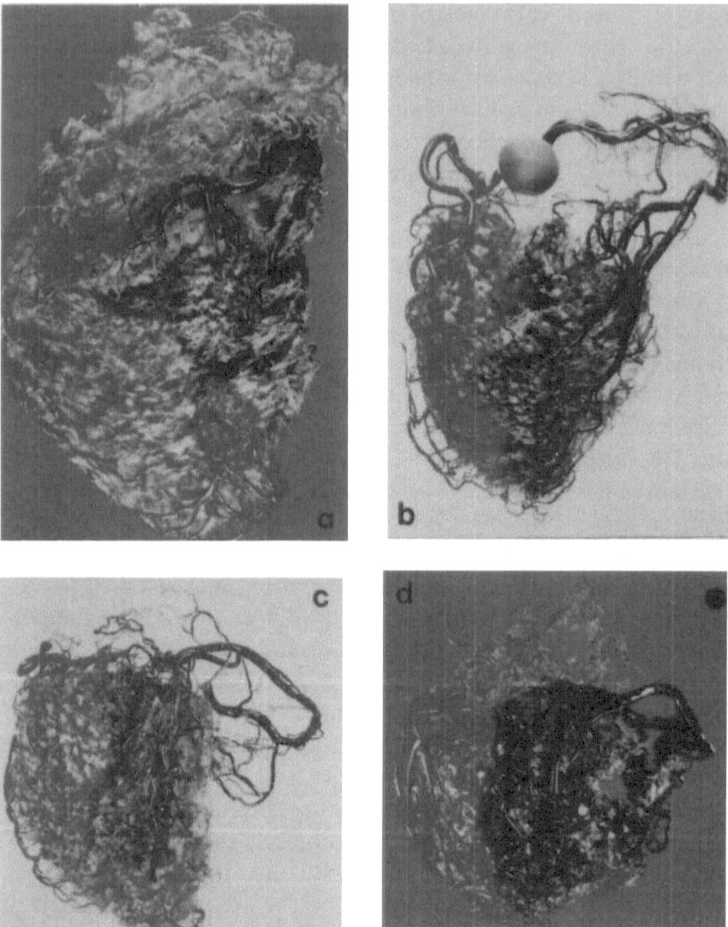

Fig. 12a–d. In the absence of collateral vessels a distinct border is observed between the two sides of the coronary network. There is no mixing of the two colors representing the perfusion territories of the left (*red*) and right (*blue*) coronary arteries. The coronary networks shown are human, in an upright posterior view. The dividing line between the two sides is rather irregular in some cases (**a**, **b**) and more straight in others (**c**, **d**)

necessarily peculiar to the heart. Indeed, the frequent failure of this mechanism to deal with the severe demands of the myocardium in the human heart suggests once again that collateral circulation is not part of the normal functional design of the coronary network.

Flow Conditions

The diameters of the main feeding vessels to an organ are usually indicative of the provisions for blood supply to that organ. According to the classical "cube law," in Poiseuille flow the vessel diameter should optimally vary as the cube

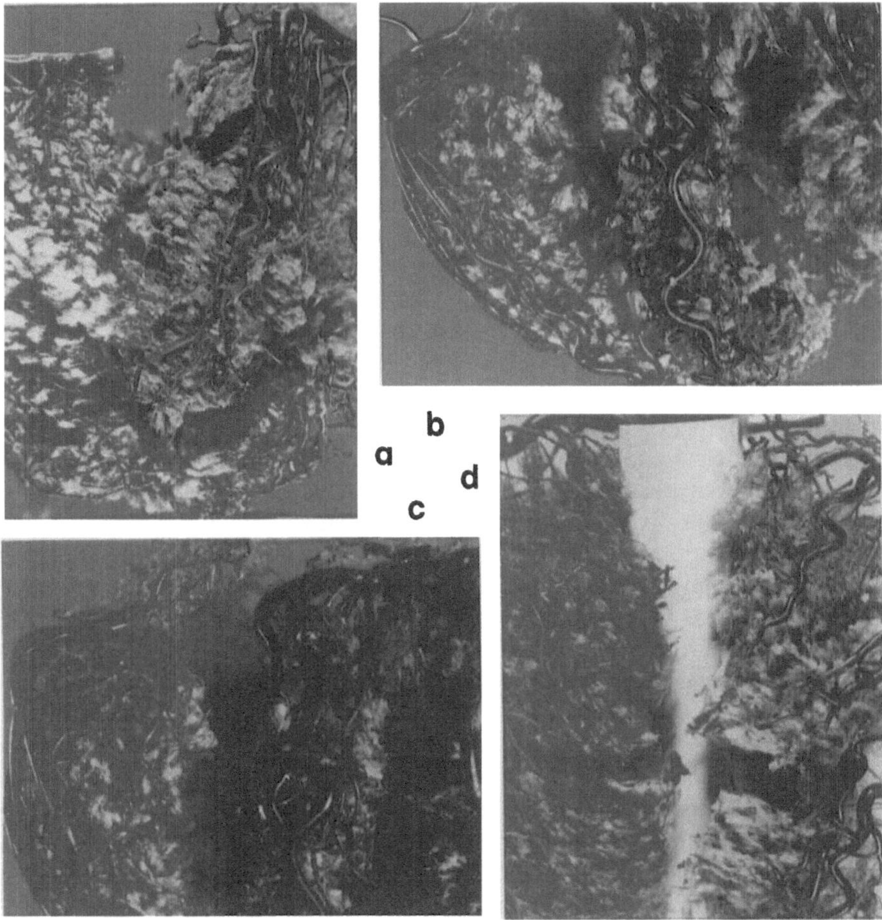

Fig. 13a–d. In the absence of collateral vessels the left (*red*) and right (*blue*) sides of the coronary network can be easily separated from each other (**a**, **d**). The beds of individual vessels on the same side of the network can also be separated from each other in some cases (**b**, **c**)

root of the flow which the vessel is destined to carry [43–45]. This result is reached as a compromise between (a) the diameter being too large, hence the amount of blood required to fill the vessel and the corresponding amount of energy required to maintain it is too large, and (b) the diameter being too small, hence the energy required to drive flow through the vessel is too large. The law can only be regarded as an approximate guide since the flow through most vessels in the arterial tree is rarely Poiseuille flow. Also, the law may be superseded by more important considerations which come into play in individual situations, such as the storage function which some of the larger vessels have, the resistance function of the arterioles, and wave reflections due to the pulsatile nature of the flow. Nevertheless the law has been found to hold in many parts of the vascular system and it may therefore serve as a useful guide in the comparison of one vascular bed with another [46–53]. In particular, in the human coronary network it has been found that the cube law holds at successive branching sites as vessels divide into smaller and smaller branches, although the results show considerable scatter (Fig. 14) [54].

Another aspect of the cube law is that the shear force acting on endothelial tissue in Poiseuille flow is proportional to the flow divided by the cube of the diameter. The ratio of flow to diameter cubed thus provides a useful yardstick, under the cube law, as we move through different levels of the arterial tree. If the law is obeyed and the ratio remains constant, the implication is that the shear stress on endothelial tissue also remains constant. Conversely, since we expect the shear stress to decrease as we move to smaller and smaller vessels on the tree, the ratio of flow to diameter cube is also expected to decrease at the same time. This test may be applied to the main feeding coronary arteries, with regard to their hierarchical position on the systemic arterial tree as a whole. A useful measure of the flow-to-diameter-cube ratio is the "bolus speed", which has a convenient physical interpretation and which is thus an appropriate measure for this test. A "bolus" is here defined as a cylindrical volume of fluid whose length and diameter are equal, and the bolus speed through a vessel is the number of such boluses (having the same diameter as the vessel) which it conveys per second. The bolus speed is thus proportional to the flow-to-diameter-cube ratio, and as we move from the aorta to the main coronary arteries both the ratio and the speed are expected to decline or at most remain constant.

Per gram of tissue, the myocardium consumes more oxygen and hence requires more blood flow than any other organ. Blood flow to the heart is commonly estimated as 5% of total blood flow. Accordingly, for an average human, if the flow through the aorta is taken as 5 l/min then the flow through the two main coronary arteries would be 250 ml/min. If the diameter of the aorta is taken as 25 mm and that of each of the two coronaries is taken as 4 mm, the bolus speed would be 6.75 boluses/s in the aorta and 41.45 boluses/s in the coronaries. If the diameters of the coronaries are taken as 3 mm each, the bolus speed there would be 98.24 boluses/s. These higher values of the bolus speed indicate that shear stress on endothelial tissue in the two main coronary arteries may be much higher than that in the aorta instead of being somewhat lower as argued above.

There are other adverse flow conditions which affect the coronary arteries but which are less amenable to quantitative analysis. The major coronary arteries are strongly tethered to, and the smaller ones completely embedded in, the myocardium. Vessels at all levels of the coronary network are therefore subject to the large and rather violent contractions of the myocardium. The major vessels are required to undergo large changes in length, hence the characteristic wrinkling of these vessels which can be observed in systole, and presumably some stretching must occur in diastole. The smaller vessels, being embedded in the myocardium, are subject to intramyocardial pressure which is believed to not only stop the flow within these vessels in systole but in fact cause some reverse flow as they empty into the larger vessels [30]. Also, the angles at which the two major coronary arteries arise from the ascending aorta are highly unusual and are not consistent with the theoretical principles of optimum branching [55]. These angles affect the flow pattern in the proximal segments of the two coronary arteries.

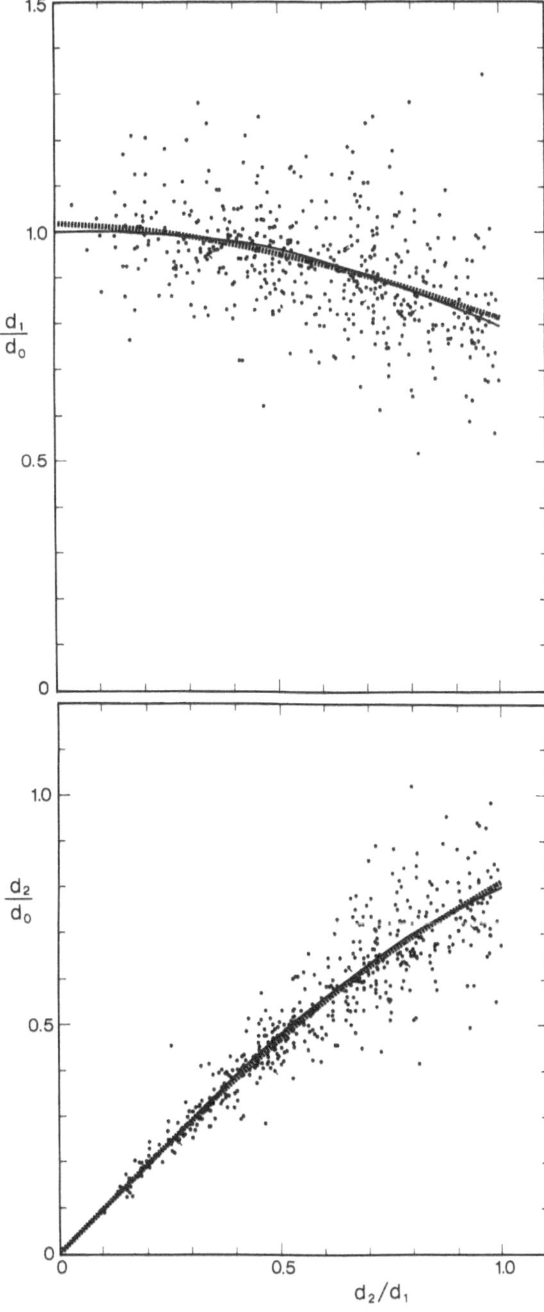

Fig. 14. The diameters of coronary arteries as measured at arterial bifurcations at various levels of the human coronary network, where d_0, d_1, d_2 are the diameters of the parent vessel, the larger branch, and the smaller branch respectively. The *solid curves* represent the theoretical relation between these diameters dictated by the "cube law" and the *dotted curves* represent statistical fits of the measured data. Adapted from [54]

Thus flow conditions within the coronary network differ in several important respects from conditions in other parts of the cardiovascular system. These differences are not adequately documented at present and very little is known about the extent to which they are incorporated into the functional design of the coronary network. A study of this design must therefore involve not only the static morphology of the coronary arteries, but the peculiar fluid dynamics of the coronary circulation. A good understanding of the functional design of this system is essential to an understanding of its malfunction in coronary heart disease.

Acknowledgments. This work was supported by the Heart and Stroke Foundation of Ontario, and in part by the Natural Sciences and Engineering Research Council of Canada.

References

1. Moberg A (1968) Anastomoses between extracardiac vessels and coronary arteries. Acta Med Scand (Suppl) 485: 1–26
2. Davies DV, Coupland RE (1967) Gray's anatomy. Longmans Green, London
3. Crouch JE (1972) Functional human anatomy. Lea and Febiger, Philadelphia
4. Walmsley R, Watson H (1978) Clinical anatomy of the heart. Churchill Livingstone, Edinburg
5. McAlpine WA (1975) Heart and coronary arteries. Springer-Verlag, New York
6. Zamir M, Silver MD (1985) Morpho-functional anatomy of the human coronary arteries with reference to myocardial ischemia. Can J Cardiol 1: 363–372
7. Baroldi G, Scomazzoni G (1967) Coronary circulation in the normal and pathologic heart. Armed Forces Institute of Pathology, Washington DC
8. James TN (1961) Anatomy of the coronary arteries. Harper and Row, Hagerstown
9. Engel HJ, Torres C, Page HL Jr (1975) Major variations in anatomical origin of the coronary arteries: Angiographic observations in 4250 patients without associated congenital heart disease. Cathet Cardiovasc Diagn 1: 157–169
10. Fulton WFM (1965) The coronary arteries. Thomas, Springfield
11. Luzsa G (1974) X-ray anatomy of the vascular system. Lippincott, Philadelphia
12. Gensini GG (1975) Coronary arteriography. Futura, Mount Kisco
13. Grayson J (1982) Functional morphology of the circulation. In: Kalsner S (ed) The coronary artery. Croom-Helm, London
14. Morris JJ, Peter RH (1971) Coronary circulation. In: Conn HL, Horowitz O (eds) Cardiac and vascular diseases, vol 1. Lea and Febiger, Philadelphia
15. Soto B, Russel RO, Moraski RE (1976) Radiographic anatomy of the coronary arteries: An atlas. Futura, New York
16. Vlodaver Z, Neufeld HN, Edwards JE (1975) Coronary arterial variations in the normal heart and in congenital heart disease. Academic Press, New York
17. Zamir M (1988) Distributing and delivering vessels of the human heart. J Gen Physiol 91: 725–735
18. Lower R (1669) Tractatus de corde. Elsevier, Amsterdam
19. Hunter J (1861) Essays and observations. Owen R (ed). Van Voorst, London, vol 1, p 126
20. Spalteholz W (1907) Die Coronararterien des Herzen. Verhandlungen der Anatomischen Gesellschaft (supplement to Anatomischer Anzeiger) 21st meeting at Wurzburg, pp 141–153
21. Schlesinger MJ (1938) An injection plus dissection study of coronary artery occlusions and anastomoses. Am Heart J 15: 528–568
22. Blumgart HL, Zoll PM, Freedberg AS, Gilligan DR (1950) The experimental

production of intercoronary arterial anastomoses and their functional significance. Circulation 1: 10–27
23. Baroldi G, Mantero O, Scomazzoni G (1956) The collaterals of the coronary arteries in normal and pathologic hearts. Cir Res 4: 223–229
24. Gensini GG, Da Costa BCB (1969) The coronary collateral circulation in living man. Am J Cardiol 24: 393–400
25. James TN (1970) The delivery and distribution of coronary collateral circulation. Chest 58: 183–203
26. Appelbaum E, Nicolson GHB (1935) Occlusive diseases of the coronary arteries. Am Heart J 10: 662–680
27. Bean WB (1937) Infarction of the heart. (A morphological and clinical appraisal of three hundred cases, part I. Predisposing and precipitating conditions). Am Heart J 14: 684–702
28. Yater WM, Traum AH, Brown WG, Fitzgerald RP, Geisler MA, Wilcox BB (1948) Coronary artery disease in men eighteen to thirty-nine years of age. Am Heart J 36: 683–722
29. Baroldi G (1983) Diseases of the coronary artery. In: Silver MD (ed) Cardiovascular pathology. University Park Press, Baltimore
30. Gregg DE (1950) Coronary circulation in health and disease. Lea and Febiger, Philadelphia
31. Schaper W (1971) The collateral circulation of the heart. New Holland, Amsterdam
32. Cohen MV (1978) The functional value of coronary collaterals in myocardial ischemia and therapeutic approach to enhance collateral flow. Am Heart J 95: 396–404
33. Sjoquist P-O, Duker G, Almgren O (1984) Distribution of the collateral blood flow at the lateral border of the ischemic myocardium after acute coronary occlusion in the pig and the dog. Basic Res Cardiol 79: 164–175
34. Gottwik MG, Puschmann S, Wüsten B, Nienaber C, Muller K-D, Hofmann M, Schaper W (1984) Myocardial protection by collateral vessels during experimental coronary ligitation: A prospective study in a canine two-infarction model. Basic Res Cardiol 79: 337–343
35. Brazzamano S, Fedor JM, Rembert JC, Greenfield JC Jr (1984) Increase in myocardial collateral blood flow during repeated brief episodes of ischemia in the awake dog. Basic Res Cardiol 79: 448–453
36. Eng C, Kirk ES (1985) Flow into ischemic myocardium and across coronary collateral vessels is modulated by a waterfall mechanism. Circ Res 56: 10–17
37. Boxt LM, Levin DC (1985) A primer of coronary and angiography. Cardiovasc Med August: 21–28
38. Anastasiou-Nana M, Nanas JN, Sutton RB, Tsagaris TJ (1985) Left main coronary artery occlusion. Cardiology 72: 208–213
39. Elayda McAA, Mathur VS, Hall RJ, Massumi GA, Garcia E, de Castro CM (1985) Collateral circulation in coronary artery disease. Am J Cardiol 55(1): 58–60
40. Hearse DJ, Muller CA, Fukanami M, Kudoh Y, Opie LH, Yellon DM (1986) Regional myocardial ischemia: Characterization of temporal, transmural and lateral flow interfaces in the porcine heart. Can J Cardiol 2(1): 48–61
41. DeWood MA, Stifter WF, Simpson CS, Spores J, Eugster GS, Judge TP, Hinnen ML (1986) Coronary arteriographic findings soon after non-q-wave myocardial infarction. N Engl J Med 315: 417–423
42. DiDio LJA, Rodrigues H (1983) Cardiac segments in the human heart. Anat Clin 5: 115–124
43. Thompson D'AW (1917) On growth and form. University Press, Cambridge
44. Murray CD (1926) The physiological principle of minimum work. I. The vascular system and the cost of blood volume. Proc Natl Acad Sci USA 12: 207–214
45. Zamir M (1977) Shear forces and blood vessel radii in the cardiovascular system. J Gen Physiol 69: 449–461
46. Rodbard S (1975) Vascular caliber. Cardiology 60: 4–49
47. Hutchins GM, Miner MM, Boitnott JK (1976) Vessel caliber and branch angle of human coronary artery branch-points. Circ Res 38: 572–576

48. Hooper G (1977) Diameters of bronchi at asymmetrical divisions. Respir Physiol 31: 291–294
49. Zamir M, Medeiros JA, Cunningham TK (1979) Arterial bifurcations in the human retina. J Gen Physiol 74: 537–548
50. Zamir M, Brown N (1982) Arterial branching in various parts of the cardiovascular system. Am J Anat 163: 295–307
51. Zamir M, Medeiros JA (1982) Arterial branching in man and monkey, J Gen Physiol 79: 353–360
52. Mayrovitz HN, Roy J (1983) Microvascular blood flow: evidence indicating a cubic dependence on arteriolar diameter. Am J Physiol 14: H1031–H1038
53. Zamir M, Phipps S, Langille BL, Wonnacott TH (1984) Branching characteristics of coronary arteries in rats. Can J Physiol Pharmacol 62: 1453–1459
54. Zamir M, Chee H (1986) Branching characteristics of human coronary arteries. Can J Physiol Pharmacol 64: 661–668
55. Zamir M, Sinclair P (1988) Roots and calibers of the human coronary arteries. Am J Anat 183: 226–234

2. Measurements of Epicardial Coronary Artery and Vein Flow Velocity Waveforms on Left and Right Ventricles

Evaluation of Coronary Blood Flow by Fiber-Optic Laser Doppler Velocimeter

FUMIHIKO KAJIYA, OSAMU HIRAMATSU, KEIICHIRO MITO, SHINICHIRO TADAOKA, YASUO OGASAWARA, and KATSUHIKO TSUJIOKA[1]

Summary. Our laser Doppler velocimeter (LDV) with an optical fiber is a powerful tool for the measurement of both coronary artery and vein flow velocities because of its excellent accessibility to the coronary vessels of a moving heart. In this paper, we briefly describe the optical arrangement of the LDV and then introduce some results of measurements of the blood velocity patterns of epicardial large coronary vessels, and epicardial small arteries and veins obtained by two different routes of access of fiber probe. We also touch upon the dual-core-fiber system which is probably promising as a Doppler catheter for clinical use

Key words: Laser Doppler velocimeter—Blood flow velocity—Coronary artery—Coronary vein—Right coronary artery

Introduction

Laser Doppler velocimetry has been considered to be a promising new technique capable of measuring blood flow velocity accurately in a small sample volume [1, 2]. Riva et al. [3] first applied a laser Doppler velocimeter (LDV) to measurement of the blood flow velocity in a 200 μm diameter tube. Blood velocity measurements by LDV subsequently were made in the retinal and skin vessels as well as in the small vessels of frog webs and the rat renal cortex [4–7]. The application of this method, however, had been restricted to measurement of the blood flow velocity in superficial fine vessels with a thin wall, because of the relatively low transparency of blood and the vessel wall to laser light. Tanaka and Benedek [8] were the first to use a thick fiber-optical catheter (500 μm outside diameter) to introduce laser light into a blood vessel. They measured the average blood flow velocity in the rabbit femoral vein by taking an autocorrelation of scattered light. Their method, however, was unable to detect instantaneous changes in pulsatile blood flow, or to differentiate reverse from forward flow.

In order to apply LDV to real-time observation of phasic arterial and venous blood flow velocity, one should be able to measure the blood flow velocity with a

[1]Department of Medical Engineering and Systems Cardiology, Kawasaki Medical School, 577, Matsushima, Kurashiki, 701-01 Japan

high temporal resolution and also to discriminate the reverse from the forward component. In this study, we developed a high-resolution LDV using an optical fiber to assess local, detailed characteristics of pulsatile blood flow. Kilpatrick independently developed an LDV with an optical fiber and demonstrated its utility for blood flow velocity measurements in the coronary vein [9]. We particularly intended to apply our method to an analysis of the blood flow velocity in the coronary vascular system. Our LDV with an optical fiber has the following advantages: (1) high spatial resolution (about 100 μm) and (2) excellent accessibility with a flexible thin fiber sensor [10–13]. This enabled us to measure local blood flow velocities point by point in a coronary artery, and blood flow velocities in the small extra- and intra-mural coronary arteries and veins.

Principle of Fiber Optic Laser Doppler Velocimeter

The basic optical system of our LDV is shown in Fig. 1. The He-Ne laser beam (632.8 nm, 5 mW) is divided by a beam splitter (BS). Half the initial light passed by the BS is focused onto the entrance of a graded-index multimode fiber (125 or 62.5 μm diameter) and transmitted through the fiber into a blood stream. A part of the light back-scattered by flowing erythrocytes is collected by the same fiber and is transmitted back to its entrance. The other half of the initial light divided at the BS is used as reference beam. A frequency shifter (40 MHz) is interposed on the path of the reference beam to differentiate the forward flow from the reverse. The optical heterodyning is made by mixing the Doppler-shift signal from the moving erythrocytes with the reference beam. The photocurrent from the photodetector (APD) is fed into a spectrum analyzer to detect the Doppler frequency.

The back-scattered light signal has a Doppler-shift frequency Δf given by

$$\Delta f = 2nV\cos\theta/\lambda$$

where V is the blood flow velocity; n is the refractive index of blood, approximately 1.33; θ is the angle between the fiber axis and the axis of the blood vessel, and λ is the laser wavelength of 632.8 nm in a free space. When θ is 60° the Doppler-shift frequency of 1 MHz corresponds approximately to blood velocity of 48 cm/s. The sample volume of our system is approximately (πx0.05^2x0.1) mm^3 and the temporal resolution is 8 ms.

Three Different Routes of Access of the Fiber Probe to Coronary Vessels

Our laser Doppler velocimeter provides excellent accessibility to coronary vessels of the moving heart. We used three different routes of access for the fiber probe according to the measuring objectives (Fig. 2), i.e., epicardial large coronary vessels, epicardial small artery and vein, and intramyocardial artery and vein [14]. First (route 1), for the blood flow velocity measurements in large and middle-sized epicardical coronary arteries and veins, we placed a cuff around the

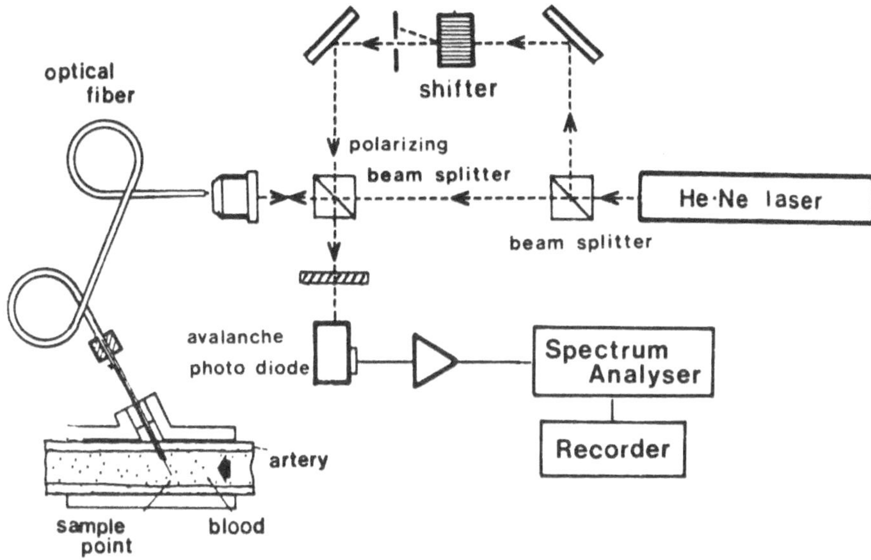

Fig. 1. Schematic diagram of the laser Doppler velocimeter with an optical fiber. From [12] with permission of Birkhauser Verlag

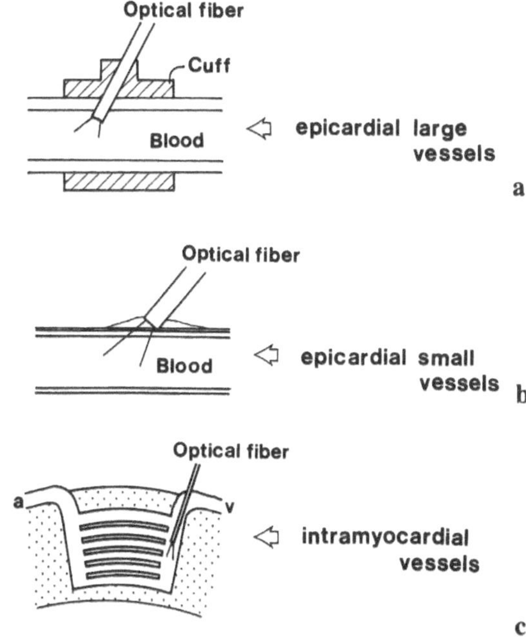

Fig. 2a–c. Three different routes of access of the fiber probe to coronary vessels. **a** Route 1 is applied for the velocity measurements in large and middle-sized epicardial coronary arteries and veins. **b** Route 2 is for the velocity measurements in small epicardial arteries and veins. **c** Route 3 is for the velocity measurements in intramyocardial arteries (A) and veins (V)

vessel and inserted the fiber into the vessel through a small hole in the vessel wall. The fiber tip was moved stepwise across the vessel to obtain the velocity profile across the vessel. Second (route 2), for the blood flow velocity measurements in small epicardial arteries and veins whose walls were thin enough to be transparent for laser light, we placed the fiber tip on the outer surface of the vessel and fixed it by a drop of cyanoacrylate. Third (route 3), for the blood flow velocity measurements in intramyocardial arteries and veins, we inserted the fiber into the vascular lumen from a position just penetrating into myocardium and introduced the fiber into a deeper portion.

Blood Flow Velocity Measurements in the Left Coronary Arteries and Veins

Eleven mongrel dogs were anesthetized with sodium pentobarbital (30 mg/kg) and ventilated with room air by a Harvard respirator pump. A thoracotomy in the left intercostal space was performed, the pericardium was opened, and a cradle was formed. The coronary blood flow velocity measurement [13, 15–18] were performed by route 1 (see Fig. 2). The left circumflex coronary artery (LCX) was isolated at its proximal and distal portions. The fiber tip was inserted into vascular lumen at an angle of 60° with the aid of a small plastic cuff, selected from several types of different diameters (0.8–3.6 mm) to fit the vessel snugly. The fiber tip was moved stepwise to traverse from the near to the far wall to measure local blood velocity at each sampling point. The position of the fiber tip on the vessel wall was determined by the position where the Doppler signals disappeared. Coronary blood flow velocity was recorded on a tape recorder (TEAC R-210) simultaneously with other tracing including pressure and electrocardiography (ECG). Direct paper recordings were also made on an 8-channel Mingograph. By keying on the R wave in ECG, the velocity profiles in the proximal and distal portions of the left circumflex coronary artery were reconstructed during one cardiac cycle.

Representative velocity profiles in the proximal and distal portion of the left circumflex coronary artery (LCX) are shown in Fig. 3. The characteristics of coronary arterial velocity are readily comprehensible using the three-dimensional display. The velocity waveform showed a diastolic-predominant pattern which is a characteristic of the coronary arterial flow. The velocity profile across the vascular lumen was flat near the axial region and declined abruptly at the vicinity of the vessel wall. The profiles were not symmetric in many cases and, in this case, skewing towards the outer wall was observed.

Compared with the velocity profiles in the proximal portion (Fig. 3a), the magnitude of blood flow velocity was smaller in the distal portion (Fig. 3b) throughout the cardiac cycle, especially during systole. Reverse flow was frequently observed during early systole. The velocity profile across the vascular lumen was more developed (parabolic) in the distal portion.

Isoproterenol administration enhanced the difference between the proximal and distal velocity waveforms (Fig. 4). In the proximal portion, the early systolic flow component increased and the reverse flow appeared in mid-systole, whereas in the distal portion, the systolic forward flow components became smaller and

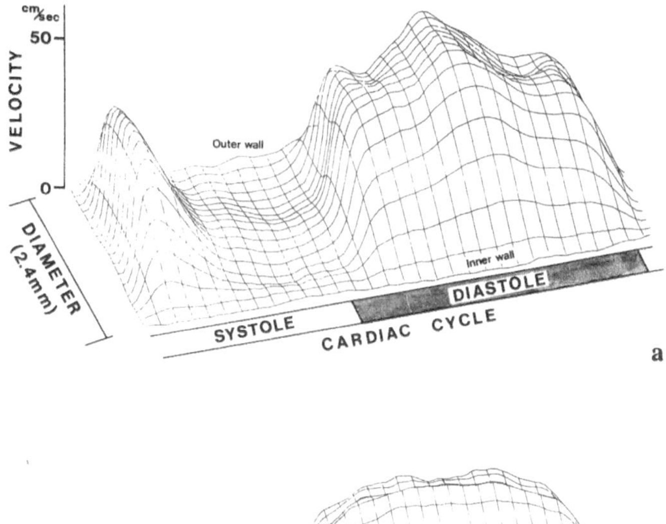

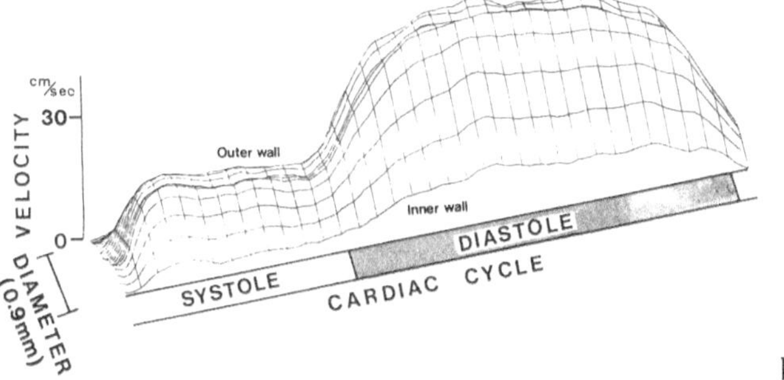

Fig. 3a, b. Three-dimensional display of the blood flow velocity in proximal (**a**) and distal (**b**) portions of the left circumflex coronary artery (LCX) of a mongrel dog. This is reconstructed from the velocity waveforms at more than 20 sampling points across the vessel by keying on the R wave in ECG. From [17] with permission of the American Society of Mechanical Engineers

the reverse flow was divided into two components, i.e., the early- and mid-systolic components. The difference in the velocity waveforms between the proximal and distal portions may be mainly caused by the compliance of the epicardial artery, as indicated by Chilian et al. [19]. Therefore, it is necessary to measure the blood flow velocity in the distal portion to evaluate the blood inflow into myocardium.

Our laser Doppler velocimeter with an optical fiber probe is a powerful tool for the measurement of the coronary venous flow, even when the vessel is easily collapsible, as is the vein. A representative tracing of blood flow velocity in the great cardiac vein (GCV) is shown in Fig. 5. The GCV velocity is always characterized by a prominent systolic flow wave. The blood velocity increased around the onset of left ventricular ejection and decreased gradually after the peak formation at mid- or late-systole. Besides this systolic flow wave, two small backflow components were found in the phase of atrial contraction and during the isovolumic contraction phase of the left ventricle in this particular case. These small wave components are observed in some other cases. Isoproterenol

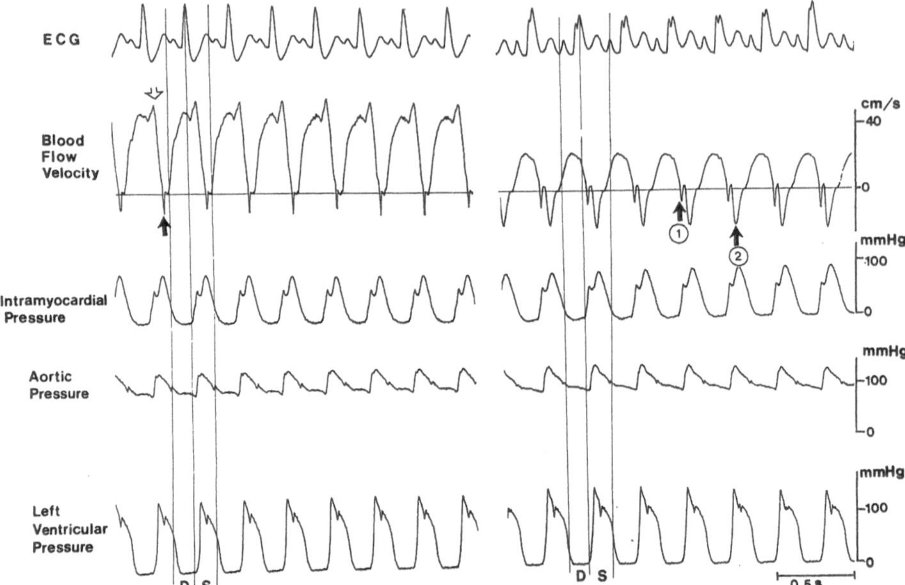

Fig. 4. Phasic blood velocity in the proximal and distal circumflex coronary artery (LCX) during systemic administration of isoproterenol. Early systolic forward flow was more prominent in the proximal portion. In the distal portion early and mid-systolic reverse flows were observed. *S*, systole; *D*, diastole. From [17] with permission of the American Society of Mechanical Engineers

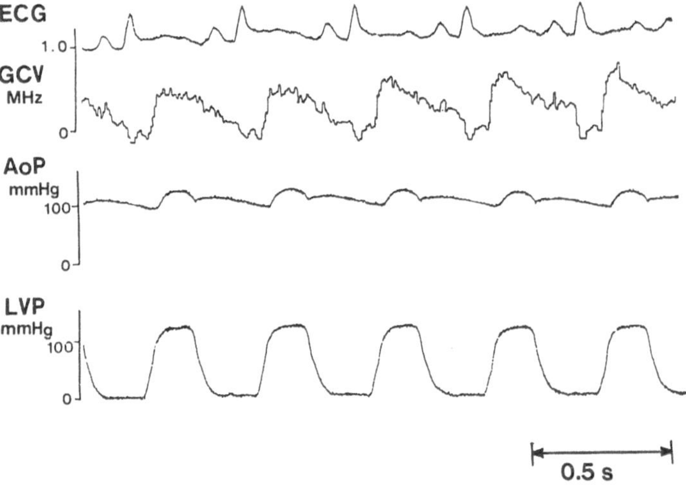

Fig. 5. Phasic blood velocity waveform in the central axial portion of the great cardiac vein. The velocity waveform was characterized by a prominent systolic flow wave. *GCV*, blood velocity in the great cardiac vein; *AoP*, aortic pressure; *LVP*, left ventricular pressure. From [15] with permission of Springer-Verlag

administration increased the peak GCV blood flow velocity by 70% without significant changes in the systolic flow, and accelerated the rise in the systolic flow velocity. Systematic administration of dipyridamole increased the blood flow velocity in the GCV. But the velocity waveform itself did not differ significantly from that under control conditions.

Hellenbrand and others have used the fibre-optic system to measure flow in the small branches of the coronary veins, during research into the capacitance of the coronary venous bed [20]. Using this technique, they have been able to determine the relationship between pressure and flow in the small vessels of the heart.

Blood Flow Velocity Measurement in the Right Coronary Small Arteries and Veins

Although it is well established that the systolic flow component is greater in the right coronary artery (RCA) than in the left coronary artery (LCA) [21, 22], the phasic patterns of the arterial inflow and venous outflow of the right ventricular myocardium are unclear because of the excessive movement of the right ventricle.

We measured the blood flow velocity in the distal right coronary artery just before its penetration into the myocardium, as well as that in a small coronary vein just after its appearance from myocardium, in the dog by route 2 (see Fig. 2b). Vessels with an outer diameter of about 150–500 μm were chosen for the measurements, as their vascular walls are transparent to laser light. The fiber tip was placed on the vessel surface and it was moved manually perpendicular to the vascular axis. When large and good-quality Doppler signals were consistently observed, the fiber tip was fixed at that position with a drop of cyanoacrylate. The blood flow velocity was measured under control conditions, during transient pulmonary stenosis.

Figure 6 shows a representative tracing of the blood flow velocities in the proximal and distal portions of the right epicardial artery under control conditions. The velocity in the proximal portion was measured by an ultrasound Doppler method. Phasic blood velocity waveforms in both proximal and distal portions were characterized by a diastolic-predominant pattern. The systolic flow commenced with a rapid increase in flow velocity after the opening of the aortic valve. But the systolic flow component in the distal portion was significantly smaller than that in the proximal. This difference of blood flow waveforms may be explained by the existence of a significant epicardial compliance flow in the proximal portion during systole. The threefold elevation of right ventricular pressure by transient pulmonary artery stenosis reduced the magnitude of the systolic velocity component, resulting in a pattern similar to the inflow pattern into the left ventricular myocardium. Thus, the systolic arterial inflow into the right myocardium is much smaller than the systolic flow in the proximal portion, but it is still larger than that of left ventricle.

The blood flow velocity waveform in the epicardial small veins of the right ventricle was always characterized by the systolic-predominant pattern. As

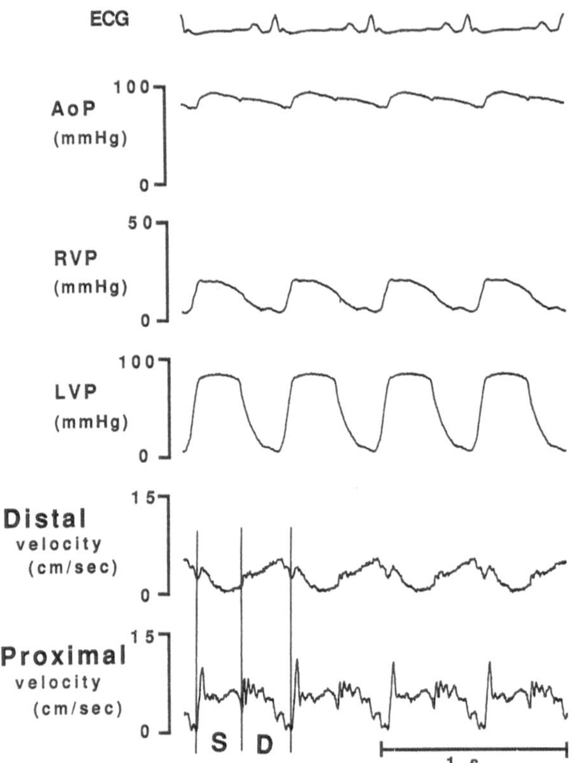

Fig. 6. The blood velocity waveform in the proximal and distal portions of the right epicardial coronary artery. Phasic blood velocity waveforms in both portions were characterized by the diastolic-predominant pattern, but the systolic flow component in the distal portion was significantly smaller than that of the proximal. *AoP*, aortic pressure; *RVP*, right ventricular pressure; *LVP*, left ventricular pressure; *S*, systole; *D*, diastole

shown in Fig. 7, the blood velocity increased with the rise of the right ventricular pressure and decreased with right ventricular relaxation. This venous systolic flow wave may be caused by the squeezing-out of the blood within intramyocardial capacitance vessels into the vein by the contraction of the right ventricular myocardium. We conclude that the velocity waveforms of arterial and venous flows of the right ventricle are fundamentally diastolic- and systolic-predominant, respectively, like their velocity waveforms for the left ventricle. Thus, the impeding effect on the arterial flow and the squeezing-out effect on the venous flow by muscle contraction seem more powerful than we have inferred.

The New Laser Doppler Velocimeter with Dual-Core-Fiber for Clinical Application

To obtain a smaller sample volume and a suitable sample position for the measurement of blood velocity, we fabricated a laser Doppler velocimeter (LDV) with a dual-core-fiber [23]. Figure 8 shows simplified illustration of the

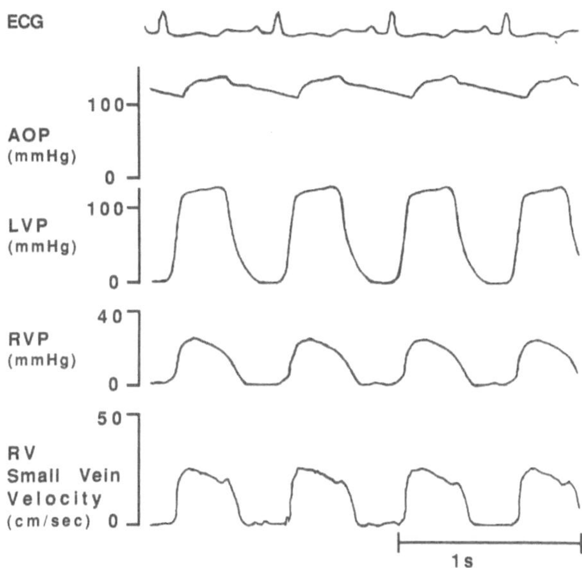

Fig. 7. The blood velocity waveform in an epicardial small vein of the right ventricle. The blood velocity waveform was characterized by the systolic-predominant pattern. *AoP*, aortic pressure; *LVP*, left ventricular pressure; *RVP*, right ventricular pressure; *RV*, right ventricle (small veins)

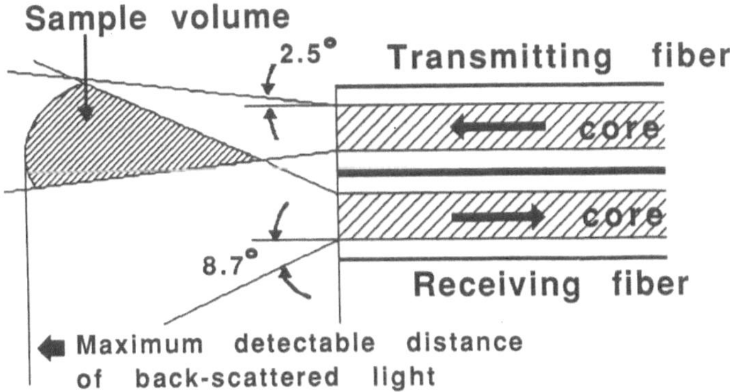

Fig. 8. The dual-core fiber pickup and simplified illustration of the sampling volume of the dual-core fiber laser Doppler velocimeter (LDV). The sample volume is approximately given by the overlapped region of the transmitted and received lights and the maximum detectable distance of back-scattered light (about 300 μm). From [23] with the permission of Pergamon Press

sampling volume of the dual-core-fiber LDV. The two fibers are placed side by side. The light is emitted from the transmitting fiber with a spreading angle of 2.5°. The back-scattered light is collected by the receiving fiber with an angle of 8.7°. Thus, the sample volume is approximately given by the overlapped region of the transmitted and received lights and the maximum detectable distance of

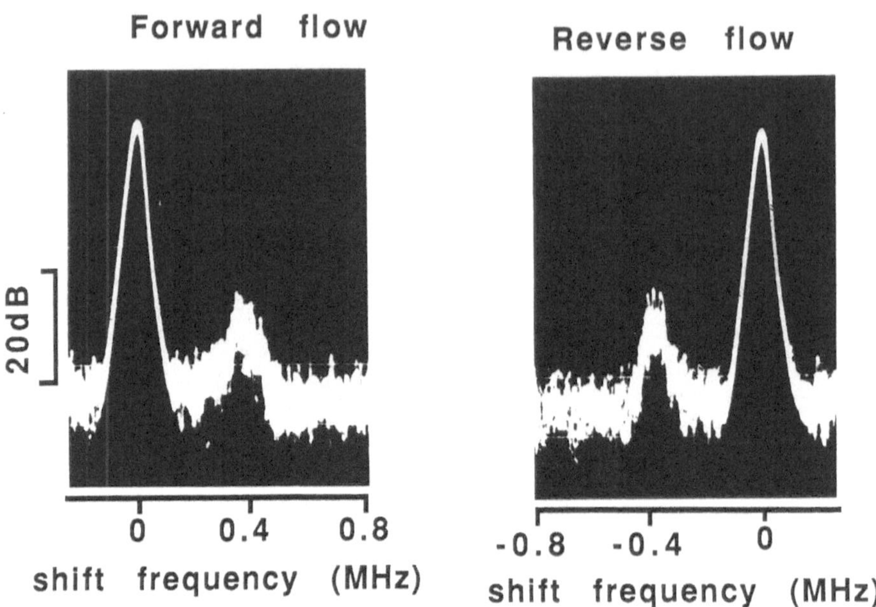

Fig. 9. The power spectra of Doppler signals for forward and reverse flows. The spectrum pattern of Doppler signals showed a sharply peaked pattern for both the forward and reverse flows. From [23] with the permission of Pergamon Press

back-scattered light (about 300 μm). Figure 9 shows typical spectra of Doppler signals for forward and reverse flows which were obtained by the combination of core = 10(clad = 125)/50(62.5) μm fibers. The spectrum pattern of Doppler signals showed a sharply peaked pattern for both forward and reverse flows, indicating that this combination is good enough to detect the Doppler signal. Also, the power spectrum of reverse flow seems better than that of the forward flow. The accuracy of blood flow velocity measurements was satisfactory, since the Doppler shift frequency showed an excellent linearity with the known blood velocities by the turntable. These results indicate that the dual-core-fiber LDV has advantage over a conventional method for coronary artery flow measurements in clinical cardiology, since we have to measure blood velocity away from the catheter tip in the latter situation and the flow may be easily disturbed.

References

1. Kajiya F, Hoki N, Tomonaga G, Saito M (1982) Engineering approaches to the evaluation of cardiac function in future. Med Prog Technol 9: 57–65
2. Roach MR (1977) Biophysical analyses of blood vessel walls and blood flow. Annu Rev Physiol 39: 51–71
3. Riva C, Ross B, Benedek GB (1972) Laser-Doppler measurements of blood flow in capillary tubes and retinal arteries. Invest Ophthalmol 11: 936–944
4. Horimoto M, Koyama T, Mishina H, Asakura T (1979) Pulsatile blood flow in arteriole of frog web. Biorheology 16: 163–170

5. Stern M (1975) In vivo evaluation of microcirculation by coherent light scattering. Nature 254: 56–58
6. Stern M, Lappe DL, Bowen PD, Chimosky JE, Holloway GA, Keiser HR, Bowman RL (1977) Continuous measurement of tissue blood flow by laser-Doppler spectroscopy. Am J Physiol 232: H441–448
7. Tanaka T, Riva C, Ben-Sira I (1974) Blood velocity measurements in human retinal vessels. Science 186: 830–831
8. Tanaka T, Benedek GB (1975) Measurement of the velocity of blood flow (in vivo) using a fiber optic catheter and optical mixing spectroscopy. Appl Optics 14: 189–196
9. Kilpatrick D, Linderer T, Sievers RE, Tyberg JV (1982) Measurement of coronary sinus blood flow by fiber-optic laser Doppler anemometry. Am J Physiol 242: H1111–H1114
10. Kajiya F (1980) Laser Doppler blood velocimetry with optical fiber. In: Digest of 2nd international conference on mechanics in medicine and biology. International seminar house of Kansai, Kobe, 16–17
11. Kajiya F, Hoki N, Tomonaga G (1981) Evaluation of blood velocity profiles in dog coronary artery by laser Doppler method. (abstract) Circulation 64: IV–40
12. Kajiya F, Hoki N, Tomonaga G, Nishihara H (1981) A laser-Doppler-velocimeter using an optical fiber and its application to local velocity measurement in the coronary artery. Experientia 37: 1171–1173
13. Tomonaga G, Mitake H, Hoki N, Kajiya F (1981) Measurement of point velocity in the canine coronary artery by laser Doppler velocimeter with optical fiber. Jpn J Surg 11: 226–231
14. Kajiya F, Tsujioka K, Ogasawara Y, Mito K, Hiramatsu O, Goto M, Wada Y, Matsuoka S (1989) Mechanical control of coronary artery inflow and vein outflow. Jpn Circ J 53: 431–439
15. Kajiya F, Tsujioka K, Goto M, Wada Y, Tadaoka S, Nakai M, Hiramatsu O, Ogasawara Y, Mito K, Hoki N, Tomonaga G (1985) Evaluation of phasic blood flow velocity in the great cardiac vein by a laser Doppler method. Heart Vessels 1: 16–23
16. Kajiya F, Mito K, Ogasawara Y, Tsujioka K, Tomonaga G (1984) Laser Doppler blood flow velocimeter with an optical fiber and its applications to detailed measurements of the coronary blood flow velocities. Proc SPIE 494: 25–31
17. Kajiya F, Tomonaga G, Tsujioka K, Ogasawara Y, Nishihara H (1985) Evaluation of local blood flow velocity in proximal and distal coronary arteries by laser Doppler method. Trans ASME J Biomech Eng 107: 10–15
18. Kajiya F, Hiramatsu O, Mito K, Ogasawara Y, Tsujioka K (1987) An optical-fiber laser Doppler velocimeter and its application to measurements of coronary blood flow velocities. Med Prog Technol 12: 77–85
19. Chilian WM, Marcus ML (1984) Coronary venous outflow persists after cessation of coronary artery inflow. Am J Physiol 247: H984–990
20. Hellenbrand WK, Klassen GA, Armour JA, Sezerman O, Paton B (1986) Automatic nervous system regulations of epicardial coronary vein systolic and diastolic blood velocity as measured by a laser Doppler velocimeter. Can J Physiol Pharmacol 64: 1463–1472
21. Gregg DE, Khouri EM, Rayford CR (1965) Systematic and coronary energetics in the resting unanesthetized dog. Circ Res 16: 102–113
22. Lowensohn HS, Khouri EM, Gregg DE, Pyle RL, Patterson RE (1976) Phasic right coronary artery blood flow in conscious dogs with normal and elevated right ventricular pressures. Circ Res 39: 760–766
23. Kajiya F, Hiramatsu O, Ogasawara Y, Mito K, Tsujioka K (1988) Dual-fiber laser Doppler velocimeter and its application to the measurements of coronary blood velocity. Biorheology 25: 227–235

Ultrasonic Measurements of Coronary Blood Flow

William M. Chilian and Susan M. Layne[1]

Summary. Measurements of coronary blood flow have been made for approximately a century, but within the last few decades ultrasonic measurements of flow, based on the Doppler Principle, have been applied towards the quantitative measurement of blood flow in the coronary circulation. The Doppler Principle maintains that as sound waves are reflected from a moving object, the frequency of the reflected wave is shifted to a higher or lower sonic frequency, depending on the direction of movement of the particle. The precise frequency shift of the sound waves is proportional to the velocity of the moving object. Modern ultrasonic flowmetric techniques use the Doppler Principle to enable quantitative measurement of coronary blood flow or coronary blood flow velocity. For accurate measurements of coronary blood flow using ultrasonic flowmetry, several details must be known: (1) the angle between the flowing red cells and the sound beam axis; (2) the position of the ultrasonic flow probe with respect to the longitudinal axis of blood flow; (3) the cross-sectional area of the blood vessel or tube; and (4) the constant position of the piezoelectric crystal relative to the blood vessel. If these details are known or satisfied, accurate quantitative measurements of coronary artery inflow or coronary venous outflow may be made. This chapter summarizes ultrasonic flowmetric techniques that have been used to document regulation of coronary blood flow during physiological and pathophysiological conditions.

Key words: Doppler—Coronary blood flow—Flowmeter—Coronary circulation

Introduction

Measurements of coronary blood flow have been performed since the late 1800s [1, 2]. To measure both mean and phasic coronary blood flow, early investigators utilized a variety of techniques, including measurement of differences between pressure curves during the cardiac cycle, movement of a foreign particle (mercury droplet) in a coronary artery, measurement of lateral pressure difference across a slight stenosis using an orifice flow meter, timed collection of coronary outflow, and measurement of the cooling rate of a heated platinum wire [3]. All these measurements were limited by insensitivity of the measurement or inadequate time resolution to record instantaneous blood flow. Within the last three decades instantaneous measurements of coronary blood flow have

[1]Department of Medical Physiology, Microcirculation Research Institute, Texas A&M University College of Medicine, College Station, TX 77843-1114, USA

been performed largely using electromagnetic flowmetry or using ultrasonic flowmetry (Doppler technique) [4, 5]. Neither the electromagnetic or ultrasonic flowmeters directly measure blood flow; rather, a signal, voltage, is generated by moving blood and is proportional to the magnitude and direction of blood flow. Using appropriate electronic signal amplification and recording techniques, the electrical analogue of coronary blood flow or blood flow velocity can be measured on-line with an oscillographic recorder.

It is the goal of this review to describe many of the studies which have been performed using ultrasonic flowmetry techniques to measure phasic coronary blood flow or blood flow velocity. For a description of the laser-Doppler methods of measuring coronary blood flow, we refer you to the report in this volume by Kayija et al. (pp. 43–53) and for a detailed description of the principles behind ultrasonic flowmetry, we refer you to the review by Baker and Daigle [6].

Principles of Ultrasonic Flowmetry

The ultimate goal of ultrasonic flowmetry when applied to the coronary circulation is the accurate, quantitative measurement of coronary blood flow in patients and experimental animals. This technique is based on the Doppler Principle, which was originally described by Johann Doppler almost 150 years ago [7]. The Doppler Principle maintains that as sound waves are reflected from a moving object, the frequency of the reflected wave is shifted to a higher or lower frequency, depending on the direction of movement. The frequency shift of the sound waves is proportional to the velocity of the moving object relative to the transmitter of the sound wave. This frequency shift is also known as the Doppler Shift, and can be quantitated as:

$$F_{P_s} = 2FV \cos\theta/c$$

where F_{P_s} is the phase shift in ultrasonic frequency, F is the frequency of the transmitted ultrasound, V is the velocity of the object from which the sound is reflected, $\cos\theta$ is the angle between the transmitted sound wave and the target, and c is the velocity of sound in the medium. Thus, ultrasonic flowmetry, using the Doppler Principle, measures blood flow velocity rather than blood volume flow. With a proper electronic device, the Doppler Shift can be transduced into an electrical output (voltage). When utilized for measurements of blood flow velocity, ultrasonic waves are generated by a piezoelectric crystal and the sound waves are reflected by moving red blood cells. As the sound "strikes" a moving object, the frequency is altered in proportion to the red cell velocity. If the difference in frequencies of the transmitted and reflected sounds are assessed and "$\cos\theta$" and "c" are known, blood flow velocity can be calculated by simple mathematical rearrangement of the above equation.

Modern ultrasonic flowmetric techniques were applied to the circulation by Franklin et al. [8, 9] in which piezoelectric crystals were used to send or receive ultrasound. The first generation Doppler flowmeter used two crystals, one to continuously generate the ultrasound, and the second to receive the shifted ultrasound (continuous wave Doppler). This device could not detect directional changes in blood flow velocity. The second generation Doppler flowmeter, termed a pulsed Doppler, uses the same crystal, both transmitting and receiving the ultrasound. Pulsed Dopplers are capable of detecting both antegrade and retrograde blood flow velocity. A third generation Doppler flowmeter has recently been developed [10]. This device involves the use of time-gated sampling of the reflected ultrasound to construct the blood flow velocity profile from 12 channels (time-gates) and provide accurate measurements of both the velocity profile and blood volume flow from a single ultrasonic piezoelectric crystal. Unfortunately, this technique has only been successfully applied to large arteries, such as the aorta, and measurements of coronary blood flow with the third generation Doppler ultrasonic flowmeter have not yet been reported. One of the distinct advantages of the ultrasonic Doppler technique is that a stable zero reference is easily obtained [6]. This fact is worth emphasizing, because electromagnetic flowmetric techniques have been characterized as suffering from excessive baseline drift in many experimental situations.

Figure 1 illustrates some of the principles behind pulsed ultrasonic flowmetric techniques. There are several important details which should be known for accurate measurements. *First*, it is essential to know the angle between the flowing red blood cells and the sound beam axis. This is critical information to calculate the absolute blood flow velocity. *Second*, it is also important to know the position (or relative position) of the ultrasonic flow probe with respect to the longitudinal axis of blood flow. Obviously, it would be inaccurate to have the ultrasonic

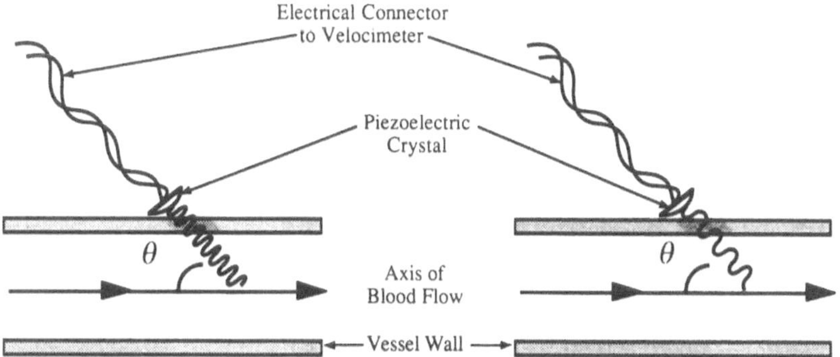

Fig. 1. An illustration of the Doppler technique as utilized for measurements of blood flow velocity. The Doppler shift is illustrated by the different frequencies of sound emitted (*left*) or received (*right*). In the "transmit" mode a pulsed piezoelectric crystal emits ultrasound at a certain frequency, and in the "receive" mode the same crystal serves as the detector for the sound reflected from the moving red blood cells. If the angle (θ) between the transmitted sound wave and the moving red blood cells is known, along with the frequencies of the transmitted and received ultrasound, and the velocity of sound in the medium, then blood flow velocity can be calculated

flow probe oriented in such a manner that blood flow velocity was not being sampled in the center line (or in proximity) of the blood vessel. *Third*, if measurements of blood volume flow are to be calculated, the cross-sectional area of the blood vessel (or tube) should be known. If relative changes in blood flow velocity are to be used to estimate relative changes in blood volume flow, it is a prerequisite that the cross-sectional area of the tube (or blood vessel) has not changed during the different experimental interventions. Such constancy in the cross-sectional area is an implicit assumption for accurate measurements of blood volume flow from blood flow velocity or accurate estimation of relative changes in blood volume flow from flow velocity measurements. *Fourth*, it is critical that the position of the piezoelectric crystal remains constant relative to the blood vessel. This is particularly apparent when using intravascular Doppler catheters, in which movement of the catheter associated with cardiac movement could cause a large artifact on the recorded blood flow velocity. Despite these many potential problems, Doppler ultrasonic flowmetry has been successfully utilized to measure blood flow velocity in a variety of experimental and clinical settings [11–14]. Blood flow velocity measurements have been used to estimate the phasic nature of intramyocardial coronary blood flow and estimate coronary capacitance from venous outflow, determine autonomic modulation of coronary blood flow, and estimate coronary vasodilator reserve in patients in the catheterization laboratory and undergoing open-heart surgery. It is without question, that when properly used, measurements obtained using ultrasonic Doppler flowmetry have contributed significantly towards our understanding of the control of coronary blood flow.

Different types of pulsed Doppler piezoelectric probes that have been utilized for clinical and basic science studies are illustrated in Fig. 2. Figure 2a shows a suction probe with the piezoelectric crystal housed in a silastic suction cup. Following application of a vacuum, the probe attaches to the heart, and if the crystal is properly positioned above a coronary vessel, relative measurements of blood flow velocity are obtained. It has been shown that changes in blood flow velocity measured with these probes are proportional to changes in blood volume flow [11, 13], although direct quantitation of blood flow velocity and blood volume flow are difficult to assess with this technique because the exact angle of the crystal with respect to the column of blood and the cross-sectional area of the vessel are not known. A distinct advantage of using this type of Doppler probe is that dissection of the coronary vessel from which flow velocity is to be assessed is not required; therefore, confounding influences such as trauma and denervation of the vessel are avoided. This, of course, is an advantage for patient studies. Another positive feature is that instantaneous estimates of coronary blood flow or flow velocity of small animals, such as rats, can be made, which are extremely difficult (if not impossible) to make with other techniques. A cuff Doppler probe, shown in Fig. 2b, can be used acutely or chronically. If used chronically, accurate measurements of blood flow velocity and blood volume flow can be accomplished because the probe fibroses securely around the vessel [15, 16]. This is reported to maintain cross-sectional area of the blood vessel relatively constant during alterations in distending pressure. A potential disadvantage is that dissection of the blood vessel could produce damage and denervation; however, recently several laboratories provide evidence that this is not a serious problem [17–19]. Figure 2c illustrates an extracorporeal Doppler flow probe in which the piezoelectric crystal is fixed in a metal or heavy plastic tube. The major advantage of this probe is that accurate measurements of blood volume flow can be made because cross-sectional area is known and is fixed. The primary disadvan-

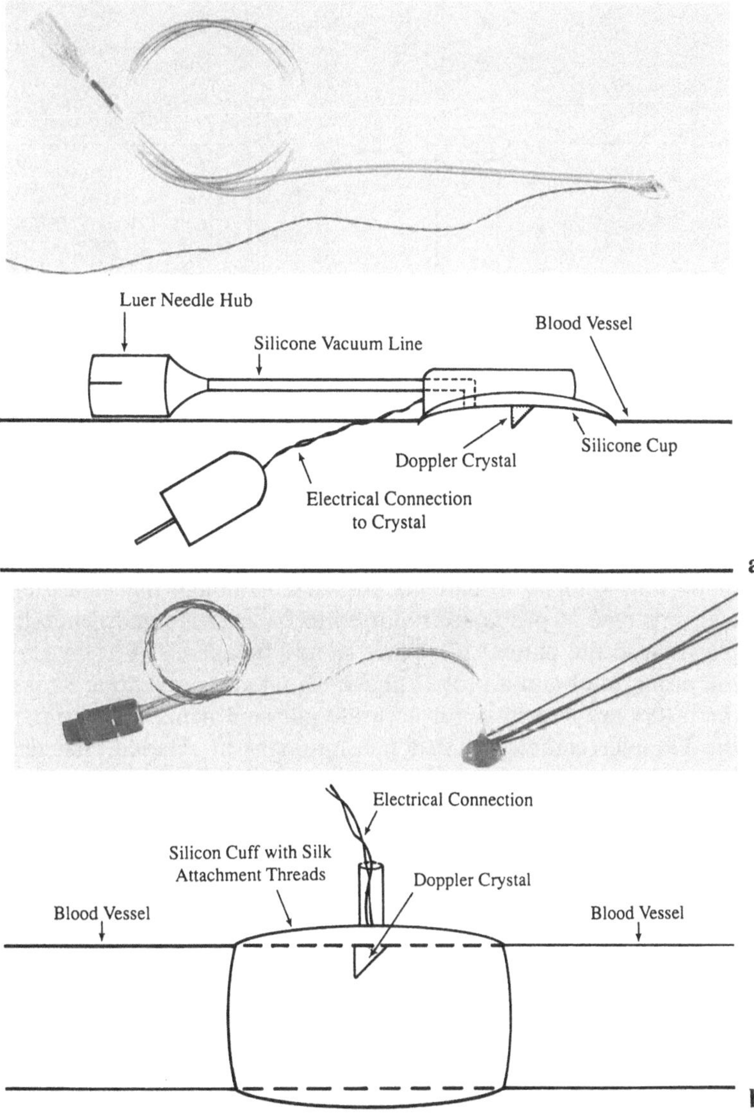

Fig. 2a–c. Three different types of ultrasonic Doppler flow probes. **a** Shows a suction-cup type probe in which a vacuum causes the probe to adhere to surface to the heart, with the piezoelectric crystal situated over a coronary blood vessel. **b** Illustrates a cuff-type Doppler probe which can be placed around a coronary artery after the vessel has been dissected free from the surrounding tissue. Doppler cuff-type probes have been utilized successfully for either acute or chronic measurements of coronary blood flow velocity

(continued)

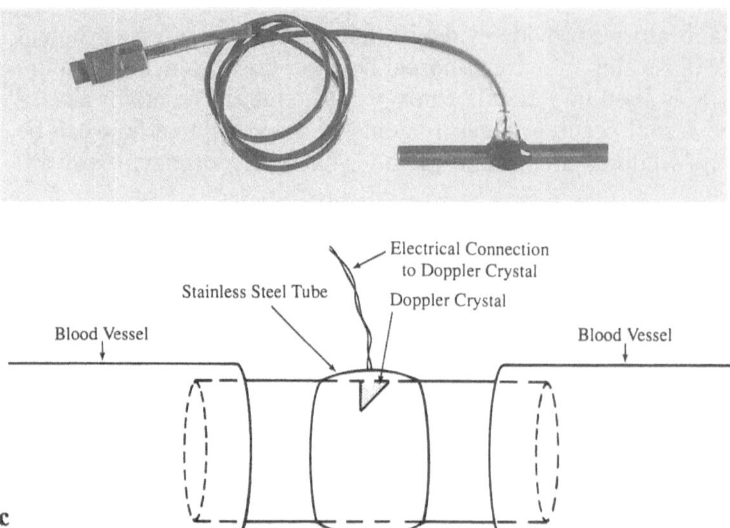

Fig. 2c. Illustrates a type of Doppler probe that is to be incorporated into an extracorporeal perfusion circuit. This latter probe type can be used to make measurements of blood volume flow

tage is that this type of probe must be used acutely and extensive dissection is required. A fourth type of Doppler probe, which is encased by a catheter suitable for coronary angiography, has also been developed for measurements of blood flow velocity in patients [20, 21]. Although the measurements obtained with this type of probe are reported to be accurate and generally without complication to the patient [21], care should be exercised in its use. Positioning of the probe to obtain a proper phasic signal and preventing its movement within the artery are necessary for accurate measurements. The primary advantage of the Doppler catheter is that measurement of relative changes in blood flow velocity can be obtained in patients undergoing cardiac catheterization; thus, functional significance of coronary lesions can be directly assessed by measuring flow velocity at rest and during administration of an intense coronary vasodilator.

In the aggregate, Doppler measurements of blood flow velocity have provided much information regarding the regulation of the coronary circulation in clinical and experimental studies. The fidelity of the phasic nature of the coronary blood flow velocity measured with ultrasonic Doppler techniques signal compares favorably to that measured with electromagnetic flowmetry (Fig. 3). Thus Doppler flowmetry can be used to assess the instantaneous nature of coronary blood flow. Generally speaking, ultrasonic flowmetry has been used to measure five different experimetal variables in the coronary circulation: (1) coronary vasodilator reserve; (2) epicardial coronary artery blood flow velocity and blood volume flow; (3) phasic intramyocardial blood flow velocity; (4) coronary venous blood flow; and (5) measurements of coronary blood flow. Many of these variables have been successfully measured in both humans and experimental animals. It is beyond the scope of this review to discuss thoroughly all of the contribu-

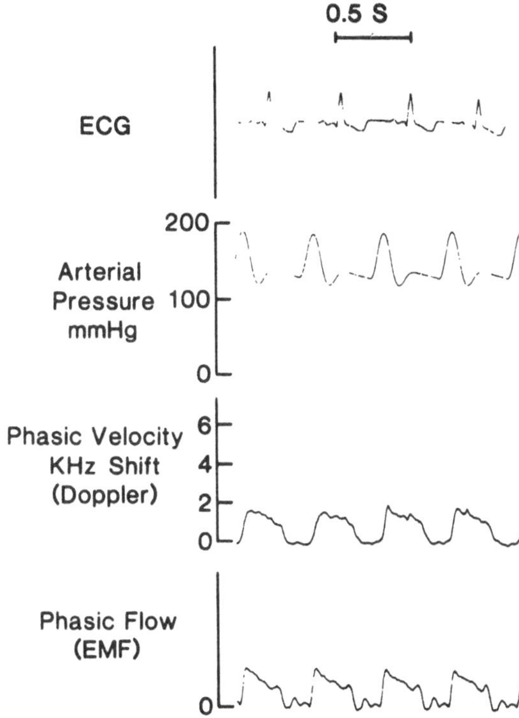

Fig. 3. Simultaneous recordings of ECG, arterial pressure, phasic left anterior descending artery blood flow velocity (measured with an epicardial suction probe) and left anterior descending artery blood volume flow (measured with an electromagnetic probe). *EMF*, electromotive force. This figure was originally presented by Marcus et al. [11] and is published by permission of the American Heart Association, Inc.

tions that have described these measurements. Rather, we will focus on some of the different variables that have been measured in the coronary circulation in clinical and basic experimental studies.

Ultrasonic Flowmetry in Patients

In patients the Doppler technique has been principally applied towards measurements of coronary vasodilator reserve or flow in bypass grafts. Coronary vasodilator reserve is defined as the ability of the coronary circulation to increase blood flow above that at rest. For instance, if blood flow is 100 ml/min at rest and is increased to 500 ml/min during maximal dilation, coronary reserve is 5. The Doppler technique has been successfully applied for measurement of coronary vasodilator reserve in both experimental and clinical studies. One manner in which the Doppler technique has been successfully utilized for measurements of coronary vasodilator reserve is using the pulsed, suction cup Doppler (Fig. 2). After positioning this type of probe, the coronary artery can be occluded transiently (15–20 s) and the characteristics of the reactive hyperemic period can be assessed following release of the occlusion. With measurements of blood flow velocity at rest and during maximal coronary vasodilation (reactive hyperemia), coronary vasodilator reserve (peak-to-resting blood flow velocity) can be calculated. The major assumption of this experimental approach is that diameter of the coronary artery does not change during reactive hyperemia from that during baseline. Many laboratories report that peak-to-resting blood flow velocity ratios are an accurate index of vasodilator reserve [11], suggesting that changes in epicardial coronary diameter during reactive hyperemia are not a major source of error.

Marcus et al. [11] characterized the phasic nature of coronary blood flow velocity and reactive hyperemic responses in humans (Figs. 4, 5). Phasic coronary blood flow velocity was primarily diastolic in the left anterior descending artery and systolic in the right coronary artery (Fig. 4), which is similar to phasic profile of coronary blood flow measured with electromagnetic flowmetric techniques. These investigators found that maximal coronary vasodilation occurred following a 5–20 occlusion (Fig. 5) and the ratio of peak blood flow velocity-to-resting blood flow velocity (coronary vasodilator reserve) was 5.8 [11]. This indicates that in the normal human, there is considerable coronary vasodilator reserve at rest, sufficient to increase blood flow almost 6 times upon demand. These same investigators have extended their studies of coronary vasodilator reserve in patients to examining the effects of pathology. It was reported that in patients with left ventricular hypertrophy secondary to aortic valvular stenosis, coronary reserve decreased from the normal value of approximately 6 to approximately 2 [12]. It is noteworthy that in many patients suffering from aortic stenosis there was almost a complete lack of coronary vasodilator reserve (coronary reserve of 1.3 or less). These results were also interpreted as providing evidence that indicated the mechanism for the occurrence of angina in patients with aortic stenosis. Coronary reserve when measured with the Doppler technique was also reported to be compromised in patients with volume-overload of the right ventricle caused by atrial-septal defects [22].

A Doppler ultrasound pencil probe has been used to estimate blood flow in coronary artery bypass grafts at the time of cardiac surgery [23]. The pencil probe is a device in which the Doppler crystal is in a plastic "pencil" and is positioned on a blood vessel directly by an investigator. It was reported that the pencil probe was remarkably insensitive to changes in the intersect angle between 17° and 42°; thus, this probe could be applied successfully at the time of open-heart surgery to estimate blood flow velocity (and blood volume flow) in both vein and artery bypass grafts. Also, this probe was more readily applied to measuring graft flow velocity than electromagnetic flow probes, where circumferential contact is necessary and often slight constriction of the graft is a prerequisite to induce proper contact. The most significant error in the calculation of blood volume flow from blood flow velocity was estimation of the cross-sectional area of the blood vessel. If external diameter measurements were used, there was significant overestimation of the absolute blood volume flow, especially in artery bypass grafts. This technique, if used in concert with high-frequency imaging transducers that accurately measure internal diameter of coronary vessels

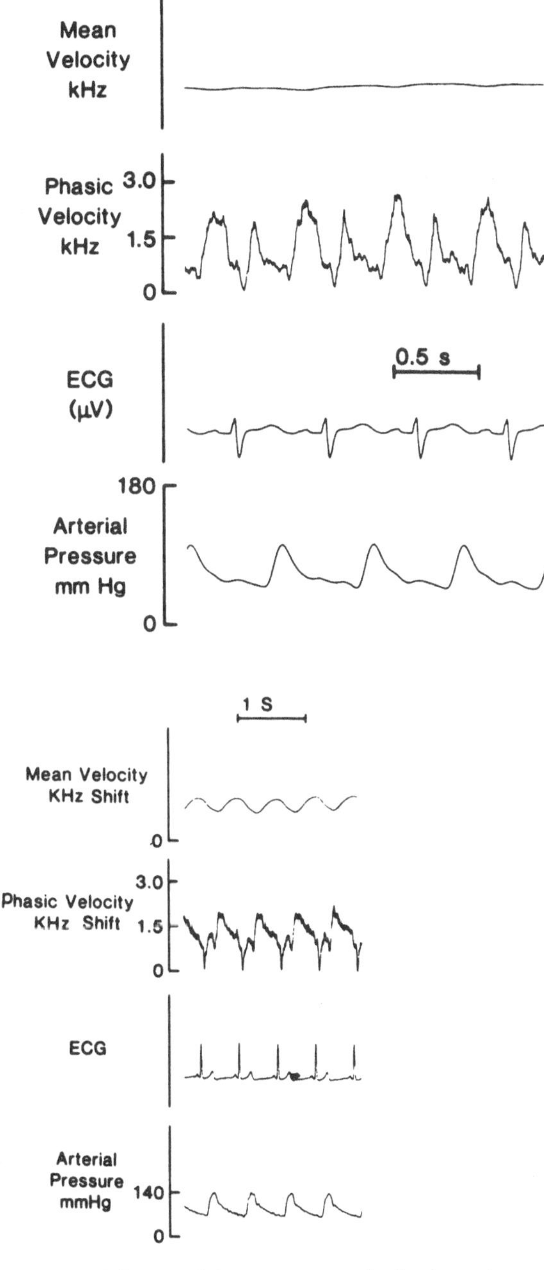

Fig. 4. Simultaneous recordings of the ECG, arterial pressure, and phasic and mean coronary blood flow velocity from the left anterior descending and right coronary arteries from patients. Note that the left anterior descending (bottom) blood flow velocity appears to be primarily diastolic in nature, whereas right coronary (top) blood flow velocity is predominately systolic. These data were originally presented by Marcus et al. [11] and are published by permission of the American Heart Association, Inc.

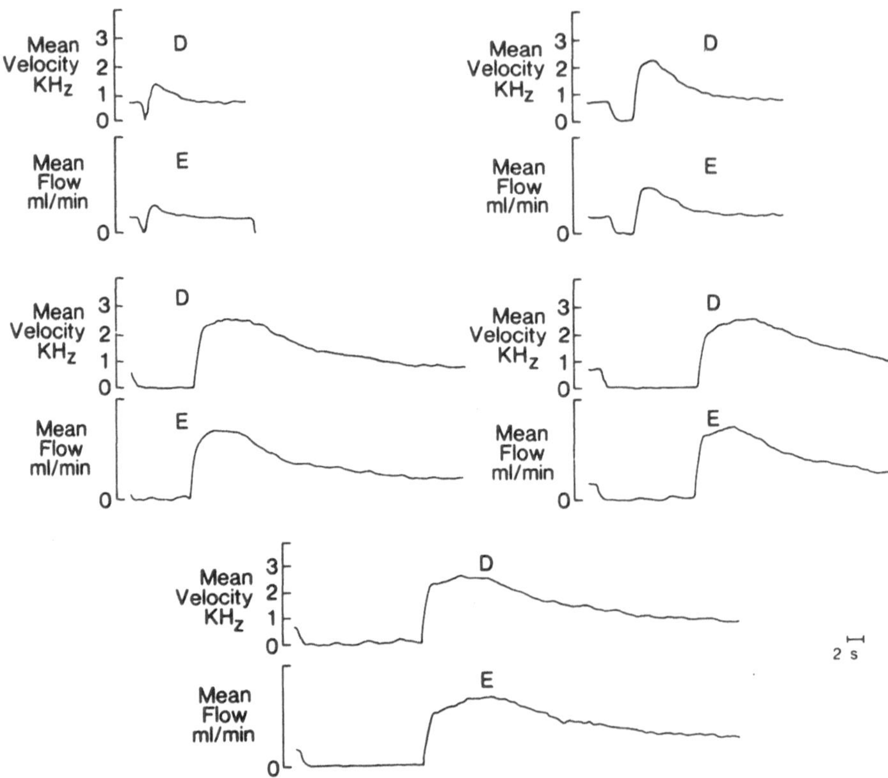

Fig. 5. Reactive hyperemic responses in the normal human. The durations of the occlusions vary from 1–20 s. Note that maximal coronary blood flow velocity is attained after about a 15 s occlusion, and at the maximal blood flow velocity during the peak, reactive hyperemic response is approximately 6-fold greater than that at rest. *D*, Doppler; *E*, EMF. These data were originally presented by Marcus et al. [11] and are published by permission of the American Heart Association, Inc.

[24] may prove to be very useful for measurement of absolute blood volume flow in the bypass grafts; thus, it may prove useful in the evaluation of a successful grafting procedure.

Recently, Wilson et al. refined a technique to measure coronary blood flow velocity in patients during transluminal cardiac catheterization with a Doppler catheter to measure phasic coronary blood flow velocity [21]. These investigators validated the catheter system against timed-venous collections of coronary sinus blood flow and against relative changes in coronary blood flow velocity measured with an epicardial Doppler. The phasic coronary blood flow velocity recordings in both the left anterior descending artery and right coronary artery of a patient are shown in Fig. 6, and are remarkably similar to those reported by Marcus et al. (Fig. 4). Also, these investigators found in normal patients (those without coronary artery disease but suffering from atypical chest pain) that coronary blood flow reserve ranged between 3.8 and 7.0 with an average of 4.9. This method has an important potential for the functional assessment of the

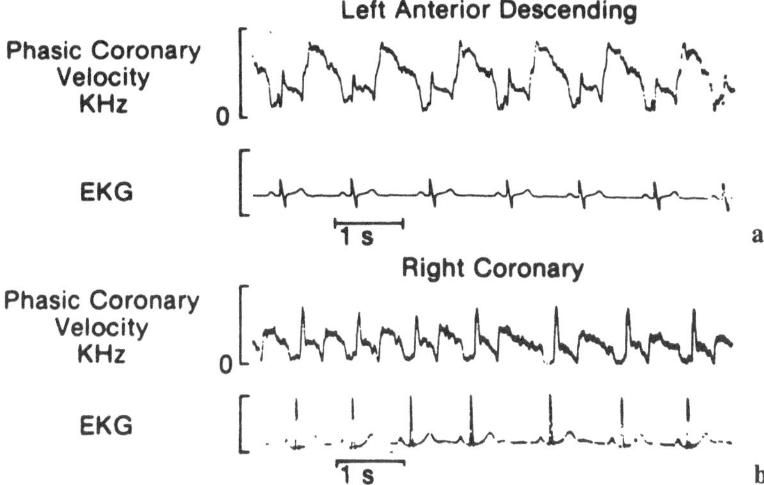

Fig. 6a, b. Phasic coronary blood flow velocity and the electrocardiogram in the left anterior descending artery (**a**) and the right coronary artery (**b**) obtained with the ultrasonic Doppler catheter. Note the marked similarity of the phasic nature of coronary blood flow velocity to that shown in Fig. 3. This figure was originally presented by Wilson et al. [21] and has been published by permission of the American Heart Association, Inc.

deleterious nature of coronary artery lesions in patients. Moreover, this method should prove to be very useful for the understanding of factors which regulate coronary blood flow in the conscious human. In this regard, Wilson et al. have recently found that pulmonary inflation does not produce reflex coronary vasodilation in patients, which is in marked contrast to the reflex vasodilation found in canines [25].

Ultrasonic Flowmetry in Basic Experimental Studies

The Doppler technique has been successfully applied using many different experimental approaches to procure measurements of coronary blood flow and coronary blood flow velocity in conscious and anesthetized animals. Investigators have been able to make measurements of coronary blood volume flow, coronary vasodilator reserve, intramyocardial blood flow velocity, and coronary collateral blood flow in experimental animals. This information has greatly advanced our understanding of the control of the coronary circulation under physiological and pathophysiological conditions.

The Doppler technique has also been used to evaluate the effect of pathophysiological disturbances of left ventricular hypertrophy on coronary vasodilator reserve. In the normal rat (WKY) coronary reserve usually approximates 2.2 [26]. The explanation for this relatively low coronary reserve when compared to humans (or other larger animal species) relates to the fact that the rat is characterized by extremely high myocardial oxygen demands and high coronary

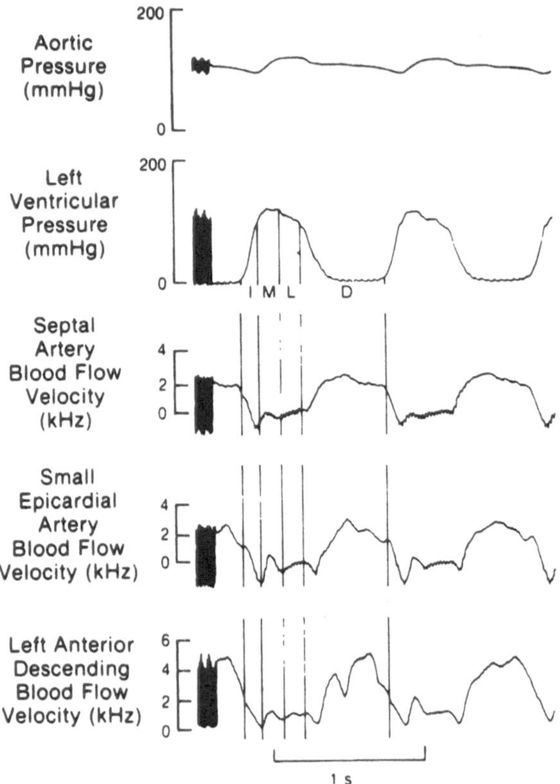

Fig. 7. Simultaneous recordings of aortic pressure, left ventricular pressure, and blood flow velocity in the septal artery, a small epicardial coronary artery, and the left anterior descending artery. The blood flow velocity is divided into different portions of the cardiac cycle: *I*, isovolumic systole; *M*, mid-systole; *L*, late-systole; *D*, diastole. This illustration was originally presented by Chilian and Marcus [28] and is reprinted with permission from the American Physiological Society

blood flow at rest. Left ventricular hypertrophy as a result of hypertension (spontaneously hypertensive rat) was characterized by diminished coronary vasodilator reserve, usually in the range of approximately 1.5–1.7. In contrast, ventricular hypertrophy caused by thyrotoxicosis did not impair coronary vasodilator reserve [27]. These results have been interpreted as suggesting that the stimulus that produces left ventricular hypertrophy has profound influences on the interaction between the hypertrophying myocardium and the coronary vasculature. As with humans, the Doppler technique has proved to be useful in the evaluation of pathology on coronary reserve in experimental animal models of disease states.

Doppler measurements of intramural coronary blood flow velocity have also been performed to evaluate the extent of epicardial capacitance and estimate phasic intramural perfusion of the myocardium [28, 29]. Figure 7 illustrates aortic pressure, left ventricular pressure, and coronary blood flow velocity in a large epicardial artery, a small epicardial artery immediately before penetration into

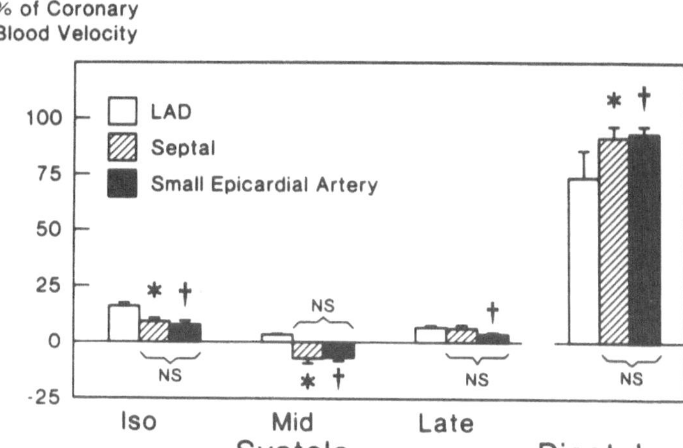

Fig. 8. The percentage of the total coronary blood flow that occurs during a specific portion of the cardiac cycle: *Iso*, isovolumic systole; *Mid*, mid-systole; *Late*, late systole in the left anterior descending (*LAD*) artery, septal artery, and small epicardial coronary arteries. Note that blood flow was negative in the small epicardial artery during mid-systole and diastolic flow was proportionally greater in these two arteries when compared to that in the left anterior descending artery. *NS*, $P>0.1$; *, $P<0.05$ (septal vs LAD); †, $P<0.05$ (small epicardial artery vs LAD). This figure was originally presented by Chilian and Marcus [29] and is reprinted by permission of the American Heart Association, Inc.

the myocardium, and septal artery. Flow velocities in the epicardial arteries were measured with cuff probes, and that in the septal artery with a small suction probe. Note the marked differences in the phasic blood flow velocity pattern during the course of the cardiac cycle; namely, the septal artery and small epicardial artery have negative blood flow velocity during a portion of systole, whereas that in the large epicardial coronary is exclusively antegrade. Figure 8 summarizes the percent of total blood flow that occurs during the different portions of the cardiac cycle. Mid-systolic flow was negative in the septal and small epicardial arteries, whereas that in the left anterior descending (LAD) artery was positive. During diastole the proportion of blood flow was greater in the small epicardial artery and in the septal artery than that in the LAD artery. These data are compatible with the concept that epicardial capacitance obscures the actual phasic perfusion of the myocardium. The septal artery, being encased by muscle, and the small epicardial artery, downstream from the majority of epicardial capacitance, more accurately reflect phasic myocardial perfusion. In contrast, the pattern of blood flow in the large epicardial artery is markedly obscured by the capacitance function. This interpretation can be highlighted by emphasizing the observation that when there is negative blood flow in the small epicardial artery, there continues to be forward flow in the large epicardial artery. The only way for antegrade blood flow and retrograde blood flow to occur simultaneously in these interconnected blood vessels is to have an intermediate site for blood volume storage. It is also worth noting that the phasic profile of these ultrasonic measurements of coronary blood flow velocity in the septal and small epicardial artery have been found to be markedly similar to intravital microscopic measurements of red cell velocities in small epicardial coronary arterioles [30]. Thus, the pulsed Doppler technique has proved instrumental in detecting retrograde blood flow velocity in the coronary circulation.

In addition to measurements of epicardial capacitance, the ultrasonic Doppler technique has been used to measure coronary venous outflow to estimate intramyocardial capacitance [31]. Since veins are very compliant and small changes in pressure may produce large alterations in cross-sectional area, an extracorporeal Doppler probe was situated in a perfusion circuit between the coronary sinus and the right atrium to facilitate quantitation of blood volume flow. During long diastoles, produced by overdrive suppression in dogs with atrioventricular nodal blockade, coronary venous outflow was observed to persist for several seconds after the cessation of coronary artery inflow (Fig. 9). Coronary venous outflow occurred after cessation of arterial inflow for about 3 s with vasomotor tone intact and for about 6 s during intense coronary vasodilation produced by asphyxia. The importance of this measurement is that it demonstrated that intramyocardial capacitance is of sufficient magnitude to discharge for several seconds after the cessation of arterial inflow. Capacitance estimates from these data concurred with those using other experimental approaches [32, 33].

The ultrasonic Doppler technique has been recently utilized to provide measurements of instantaneous collateral blood flow in conscious dogs [34]. These investigators have estimated instantaneous collateral flow from the step changes in blood flow in an epicardial artery during occlusion and subsequent release of the other major epicardial artery. The instantaneous reduction of left circumflex artery blood flow (LCCA) associated with release of the left anterior descending occlusion is interpreted as the magnitude of collateral blood flow (Fig. 10). Both mean and phasic collateral blood flow can be derived from subracting circumflex artery blood flow after release of the occlusion from that during the occlusion. Moreover, these investigators have reported that certain pharmacological agents, such as nitroglycerin, are capable of increasing collateral blood flow. This application of the Doppler technique certainly appears to have the potential to resolve many of the dilemmas regarding development and control of the coronary collateral circulation. For details describing this procedure, please refer to Franklin et al. (pp. 267–278).

Concluding Remarks

Unfortunately, due to the concise nature of this review it is impossible to discuss all of the studies which have utilized the Doppler technique to make measurements of coronary blood flow or blood flow velocity. The ultrasonic Doppler technique has proved invaluable towards advancing our understanding of the regulation of coronary blood flow in both humans and in experimental animal models. It would appear that one of the major future directions of ultrasonic flowmetry techniques in the coronary circulation would be development and application of the third generation Doppler probe to measurements of coronary blood flow in humans and in animal studies. This application has the potential to make measurements of absolute coronary artery blood flow, because the velocity profile of coronary artery blood flow is measured across the entire wall of the

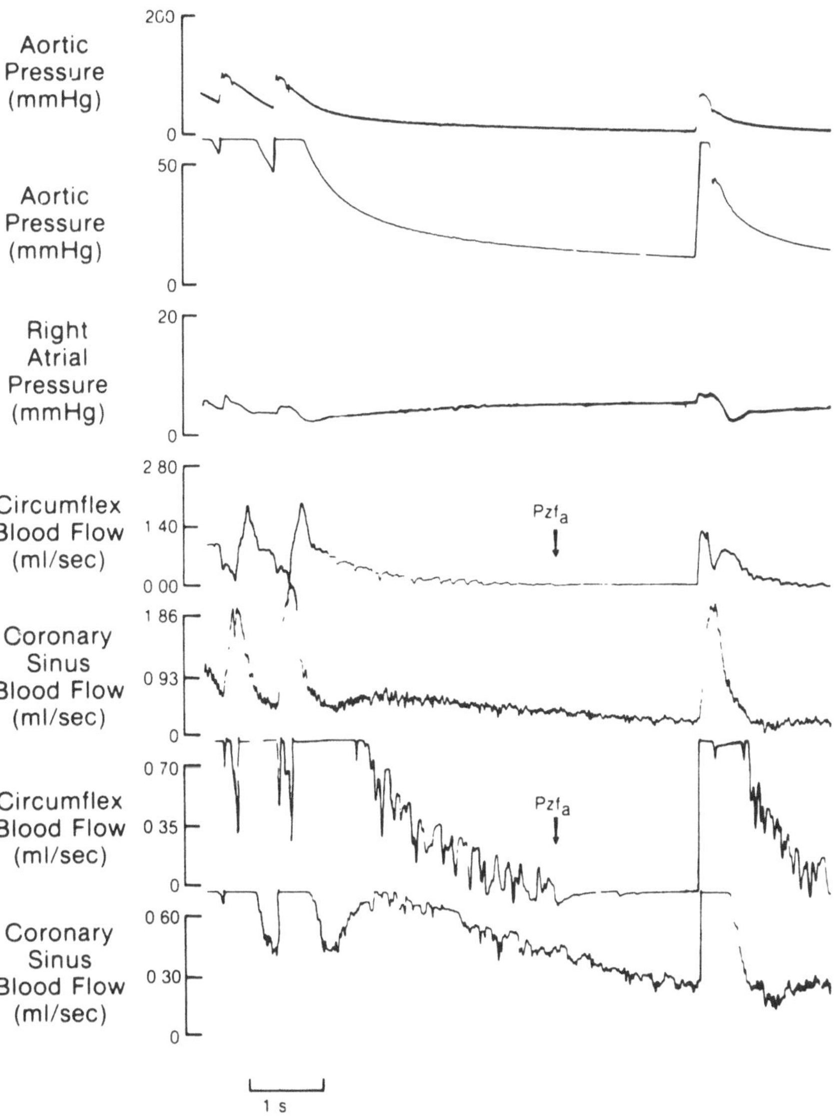

Fig. 9. Recordings of aortic pressure (at two different sensitivities), right atrial pressure, and circumflex and coronary sinus blood flow (at two different sensitivities) during a long diastole produced by overdrive suppression. The pressure at zero flow in the artery (Pzf_a) is at approximately 20 mmHg. Note that at Pzf_a (cessation of arterial inflow) coronary sinus blood flow continues until the subsequent escape beat. This figure was originally presented by Chilian and Marcus [30] and is reprinted with permission from the American Physiological Society

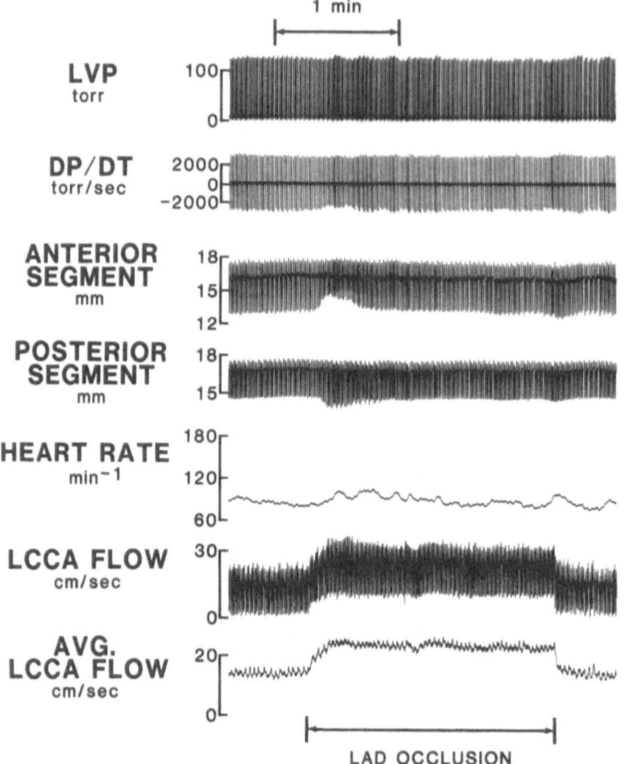

Fig. 10. Simultaneous recordings of left ventricular pressure (*LVP*), left ventricular dP/dt, segmental dimensions in anterior and posterior regions, and both average and phasic left circumflex coronary artery blood flow (*LCCA*). Note the step changes in left circumflex artery blood flow upon LAD occlusion and following the release. The decrement in LCCA blood flow upon release of occlusion represents collateral flow from the left circumflex to the anterior descending region. This figure was provided to the authors by Professor Dean Frankin of the University of Missouri. For further details of the Doppler method for measuring collateral blood flow, please refer to the contribution by Franklin and colleagues (pp 267–278)

vessel. Because the absolute velocity profile is known, this will enable calculations of vascular shear stresses. These measurements could prove to be beneficial towards advancing our understanding of the role of shear stresses in the atherogenic process in coronary artery disease.

Acknowledgments. The authors thank Susan Gard and Nita Blackwell for their patient preparation of this manuscript and acknowledge the support of the following grants from the National Heart, Lung, and Blood Institute of the U.S. Public Health Service, HL32788 and HL01570, that provided partial support for the authors' studies.

References

1. Langendorff O (1895) Untersuchungen am überlebenden Säugethierherzen. Pflügers Arch ges in Physiol 61: 291–325
2. Porter WT (1898) The influence of the heart beat on the flow of blood through the walls of the heart. Am J Physiol 1: 145–173
3. Gregg DE, Fisher LC (1963) Blood supply to the heart. In: Hamilton WF (ed) Handbook of physiology, sec. 2: Circulation. American Physiological Society, Washington DC, vol 2, pp 1517–1584
4. Kolin A (1960) Circulatory system: Methods, blood flow determination by electromagnetic method. In: Glasser O (ed) Medical physics. Yearbook Publishers, Chicago, 3: 141–156
5. Vatner SF, Franklin D, Van Citters RL (1970) Simultaneous comparison and calibration of the Doppler and electromagnetic flow meters. J Appl Physiol 29: 907–910
6. Baker DW, Daigle RE (1977) Non-invasive ultrasonic flowmetry. In: Hwang NHC, Norman NA (eds) Cardiovascular flow dynamics and measurements. University Park Press, Boston, pp 151–187
7. Doppler CJ (1842) Uber das farbiqe licht der dopplesterne. Abhandlungen der konig-lishen bohmischen gesellschaft der wissenschaften 2: 465
8. Franklin DL, Watson NW, Van Citters RL (1964) Blood velocity telemetered from untethered animals. Nature 203: 528–530
9. Franklin DL, Schlegel W, Rushmer RF (1961) Blood flow measured by Doppler frequency shift of back-scattered ultrasound. Science 134: 564–565
10. Rumberger JA, Fastenow CF, Laughlin DL, Marcus ML (1984) Validation of a third-generation Doppler system for studies of detailed aortic flow. Am J Physiol 247 (Heart Circ Physiol 16): H847–H856
11. Marcus ML, Wright C, Doty D, Eastham C, Laughlin D, Krumm P, Fastenow C, Brody M (1981) Measurements of coronary velocity and reactive hyperemia in the coronary circulation of humans. Cir Res 49: 877–891
12. Marcus ML, Doty DB, Hiratzka LF, Wright CB, Eastham CL (1982) Decreased coronary reserve. A mechanism for angina pectoris in patients with stenosis in normal coronary arteries. New Engl J Med 307: 1362–1366
13. Wangler RD, Peters KG, Laughlin DE, Tomanek RJ, Marcus ML (1981) A method for continuously assessing coronary blood flow velocity in the rat. Am J Physiol 241: H816–H820
14. Hartley CJ, Cole JS (1974) An ultrasonic pulsed Doppler system for measuring blood flow in small vessels. J Appl Physiol 37: 626–629
15. Vatner SF, Higgins CB, Braunwald E (1974) Effects of norepinephrine on circulation and left ventricular dynamics in the conscious dog. Circ Res 34: 812–823
16. Stone HL, Stegall HF, Kardon MB, Payne RM (1971) Changes in aortic, coronary, and carotid flows during $+G_x$ acceleration. J Appl Physiol 30: 21–26
17. Knight DR, Thomas JX Jr, Randall WC, Vatner SF (1987) Effects of left circumflex coronary flow transducer implantation on posterior wall innervation. Am J Physiol 252 (Heart Circ Physiol 21): H536–H539
18. Heusch G, Guth BD, Roth DM, Seitelberger R, Ross J Jr (1987) Contractile responses to sympathetic activation after coronary instrumentation. Am J Physiol 252 (Heart Circ Physiol 21): 1059–1069
19. Roth DM, White FC, Costello-Mathieu O, Guth BD, Heusch G, Bloor CM, Longhurst JC (1987) Effects of left circumflex ameroid constrictor placement on adrenergic innervation of myocardium. Am J Physiol 253 (Heart Circ Physiol 22): H1425–H1434
20. Cole JS, Hartley CJ (1977) The pulsed Doppler coronary artery catheter: Preliminary report of a new technique for measuring rapid changes in coronary artery flow velocity in man. Circulation 56: 18–25

21. Wilson RF, Laughlin DE, Ackell PH, Chilian WM, Holida MD, Harley CJ, Armstrong ML, Marcus ML, White CW (1985) Transluminal, subselective measurement of coronary artery blood flow velocity and vasodilator reserve in man. Circulation 72: 82–92
22. Doty DB, Wright CB, Hiratzka LF, Eastham CL, Marcus ML (1984) Coronary reserve in volume-induced right ventricular hypertrophy. Am J Cardiol 54: 1059–1067
23. Simpson IA, Spyt TJ, Wheatley DJ, Cobbe SM (1988) Assessment of coronary artery bypass graft flow by intraoperative Doppler ultrasound technique. Cardiovasc Res 22: 484–488
24. McPherson DD, Armstrong M, Rose E, Marcus ML, Kerber RE (1986) High frequency epicardial echocardiography for coronary artery evaluation: in vitro and in vivo validation of arterial lumen and wall thickness measurements. J Am Coll Cardiol 8: 600–606
25. Wilson RF, Marcus ML, Laughlin DE, White CW (1988) The pulmonary inflation reflex: Its physiologic significance in man. Am J Physiol 255: H866–H871
26. Peters KG, Wangler RD, Tomanek RJ, Marcus ML (1984) Effects of long-term cardiac hypertrophy on coronary vasodilator reserve in HSR rats. Am J Cardiol 54: 1342–1348
27. Chilian WM, Wangler RD, Peters KG, Tomanek RJ, Marcus ML (1985) Thyroxin-induced left ventricular hypertrophy in the rat: Anatomical and physiological evidence for angiogenesis. Circ Res 57: 591–598
28. Chilian WM, Marcus ML (1982) Phasic coronary blood flow velocity in intramural and epicardial coronary arteries. Circ Res 50: 775–781
29. Chilian WM, Marcus ML (1985) Effects of coronary and extravascular pressure on intramyocardial and epicardial blood velocity. Am J Physiol 248 (Heart Circ Physiol 17): H170–H178
30. Ashikawa K, Kanatsuka H, Suzuki T (1984) A new microscope system for the continuous observation of the coronary microcirculation in the beating canine left ventricle. Microvasc Res 28: 387–394
31. Chilian WM, Marcus ML (1984) Coronary venous outflow persists after cessation of coronary artery inflow. Am J Physiol 247 (Heart Circ Physiol 16): H984–H990
32. Scharf SM, Bromberger-Barnea B (1973) Influence of coronary flow and pressure and cardiac function and coronary vascular volume. Am J Physiol 224: 918–925
33. Spaan JAE, Breuls NO, Laird JD (1981) Diastolic-systolic coronary flow differences are caused by intramyocardial pump action in the anesthetized dog. Circ Res 49: 584–593
34. Fujita M, McKown DP, McKown MD, Franklin D (1988) Effects of glyceryl trinitrate on functionally regressed newly developed collateral vessels in conscious dogs. Cardiovasc Res 22: 639–647

3. Mechanical Properties of Coronary Circulation and Its Contribution to Coronary Hemodynamics

Coronary Circulation Mechanics

Jenny Dankelman[1], Henk G. Stassen[1], and Jos A. E. Spaan[1,2]

Summary. The use of linear models is very common in studying the mechanical events of the coronary circulation. In this chapter some of the these linear models are discussed. However, special attention is devoted to a nonlinear model. This nonlinear model consists of an arteriolar, capillary, and a venular compartment, each composed of transmural pressure-dependent resistances and compliances. With this model, the influence of pressure-dependent resistances and compliances on the arterial signals is analyzed. This analysis shows that the phasic arterial signals mainly depend on the pressure dependency of the arteriolar compartment. Changes in the parameters of capillary and venular compartments hardly affected the arterial phasic signals. Furthermore, it can be concluded that the interpretation of results obtained by application of linear system theory is highly questionable and could easily lead to misleading conclusions on the magnitude of coronary compliance.

Key words: Mechanical models—Pressure dependency—Distribution—Coronary flow—Nonlinear models

Introduction

Flow and pressure in coronary arteries and veins are normally time-variant. The relationships between those variables find their basis in mechanical events occurring in the microvessels submerged into the myocardial wall. Apart from the very subepicardium, flow and/or pressure in these vessels can not be measured yet. Hence, conclusions regarding the mechanical events at the microcirculatory level have to be inferred from measurements at the level of large and small epicardial arteries and veins by the use of models.

A model is a representation, for the most part simplified, of reality. In a physical model reality is mimicked by a material system. For example, it is possible to study the fluid dynamic behavior of elastic blood vessels in tissue by studying the dynamics of flow through elastic tubes surrounded by air or another fluid.

[1]Laboratory for Measurement and Control, Faculty of Mechanical Engineering and Marine Technology, Delft University of Technology, Mekelweg 2, 2628 CD Delft, The Netherlands

[2]Department of Medical Physics, University of Amsterdam, AMC, Meibergdreef 15, 1105 AZ Amsterdam, The Netherlands

The similarity in fluid dynamic behavior between the model and the real system of course depends on how well the vessel wall mechanics and suspension in tissue is mimicked by the model system.

A mathematical model attempts to describe the properties of the variables by mathematical equations. One may distinguish between several different mathematical models on basis of their use. A model can be constructed in order to (1): reduce data, (2) recognize analogies between different systems, and (3) do predictions, often by means of simulations. In this chapter only the last type, predictive models, are of interest.

This chapter deals with different models and the problems related to their validation. As the central issue, the pressure dependency of compliance and resistance of the coronary vessels will be considered, as well as their possible effect on the interpretation of the pressure and flow variables. First the major models will be reviewed briefly.

Models of the Coronary Circulation

Extravascular Resistance Model

In the model it is assumed that during systole the resistance is increased because of the extravascular forces acting on the vessels within the myocardium (e.g., [1]). A cause for this increased resistance could be the suggested kinking and throttling effects due to deformation of the heart [2]. As shown below, in more recent models the extravascular resistance effect is assumed to be caused by alteration of the lumens of microvessels due to compression related to contraction of the myocardium.

According to the extravascular resistance model, coronary arterial flow in systole is lower than in diastole because of increased resistance. Very early, it was already noted that the model was too simple because it predicts that venous outflow would also be lower in systole than in diastole, whereas the opposite is the case [3].

Waterfall Model

In analogy to effects observed with perfusion of the lungs, Downey and Kirk [4] formulated the waterfall model to describe the impeding effect of cardiac contraction on coronary flow. It was assumed that when tissue pressure exceeds vascular blood pressure at a certain location the vessel would then collapse locally. In the case where this pressure is lower than arterial pressure, the blood pressure just proximal to the point of collapse would then equal the tissue pressure and form the local back pressure for flow. When the tissue pressure would exceed arterial pressure then flow would stop completely.

The waterfall model predicts the coronary arterial flow in systole to be lower than in diastole because of the increased back pressure. The advantage of the waterfall model is that it explains the decrease in time-averaged flow due to contraction of the heart. However, it falls short in predicting the magnitude of the systolic-diastolic swing in coronary arterial flow [5]. Systolic coronary arterial flow can become zero or even retrograde when the coronaries are perfused at constant arterial pressure [6]. This can not be explained by the waterfall model since it is also assumed that in the outer layers of the heart the tissue pressure is small and only slightly hampered by cardiac contraction. Hence, the waterfall model always predicts significant forward flow, even when the flow at the subendocardium stagnates completely. Moreover, the waterfall model in its original form does predict a coronary venous outflow signal similar to the arterial inflow signal, which is not in agreement with experimental observations.

Linear Intramyocardial Pump Model

In the linear intramyocardial pump model [5] it is assumed that compression of the coronary microvessels results in squeezing out of blood, and that this displacement of volume contributes transiently to the coronary arterial and venous flows. In systole, the former is then decreased and the latter increased. For this mechanism to be operative it is required that the time constants related to the volume changes are longer than a heart beat.

The advantage of the linear pump model is that it explains easily the arterial and venous swings in flow, the latter in phase and the former out of phase with left ventricular pressure. Retrograde arterial flow is easily explained by this model as well [7]. The disadvantage of this model is that it does not explain the reduction in time-averaged coronary flow due to contraction of the heart.

Nonlinear Intramyocardial Pump Model

This model [8–10], although formulated earlier, can be considered as an extension of the linear intramyocardial pump model. The difference is that now the vascular volume changes are assumed to be so large as to effect resistance of the vascular bed. Resistance in this model depends in a nonlinear way on vascular volume (law of Poiseuille). Furthermore, volume is related to transmural pressure via a sigmoid transmural pressure-volume curve. According to this model cardiac contraction affects coronary arterial and venous flow in two ways. The transient effect due to displacement of volume from the vascular space remains, as with the linear model. The second effect is the increased resistance due to the decreased vascular volume. This secondary effect contributes to the phasic behavior of coronary flow as long as resistance is changing, but also predicts a steady-state impeding effect of cardiac contraction because of the increase in time-averaged vascular resistance.

The advantage of the nonlinear intramyocardial pump model is obvious since it has none of the disadvantages of the previous three models described. However, the effect of cardiac contraction is, as with the waterfall model, still related to tissue pressure, which in itself is a controversial concept.

Time-Varying Elastance Model

This model was recently proposed by Krams et al. [11, 12] from the group of Westerhof. Strictly speaking, the model is an intramyocardial pump model, but the driving force for the pump action is not the varying tissue pressure but the varying elastance. The intramyocardial blood space is considered as an additional chamber of the heart but without valves. The pump action of this chamber is described by analogy with the action of the left ventricle, as delineated by Suga et al. [13]. The pressure-volume relation of the left ventricle was described by a time-varying elastance. With the time-varying elastance model for the coronary circulation, the increase in blood pressure and/or displacement of blood are explained on the analogy of the well-known pressure-volume relations of the left ventricle. In the model, the displacement of blood is hampered by the vascular resistance, as within the pump models. The model also accounts for the steady effects of contraction on time-averaged flow by relating a decreased vascular volume to an increased resistance, as in the nonlinear pump model.

Time-varying elastance is a sweet and simple concept to explain the impeding effect of cardiac contraction on coronary flow. However, the model still has not been formulated mathematically in such a way that it can be validated more quantitatively. Moreover, some basic questions still have to be answered. It is not clear yet how the transmural difference in impeding effect on coronary flow can be explained. It is also known that pulsation of the right coronary artery flow is much less than in the left coronary artery, while the interaction on the microscopic level between myocytes and microvessels is probably much the same. Nevertheless, it is a promising concept that allows one to quantify a direct effect of contractility on coronary flow which is not possible by the four other models discussed before.

Possible Interacting Effects of Compliance and Changing Resistance on the Arterial Coronary Flow

Nowadays, it has generally been accepted that both resistance and compliance of the vascular bed are pressure dependent [10, 14]. In this section we will analyze in what way this might effect the phasic coronary arterial flow signal. This will be done by using elements from the nonlinear intramyocardial pump model. In its original form the nonlinear pump model divides the heart into a number of concentric layers, with the tissue pressure varying linearly with layer depth between instantaneous left ventricular pressure at the subendocardium to atmospheric level at the subepicardium. The total coronary arterial flow is then the sum of flows into all the layers. This complicates the interpretation of inflow signal in terms of pressure-dependent compliance and resistance, since these effects depend on the level of tissue pressure, which is different for the different layers. In order to focus on these pressure-dependent effects we will study here only one layer with tissue pressure varying between 0 mmHg (diastole) and 50 mmHg (systole). This model layer can be considered to represent the mid-myocardium when left ventricular pressure varies between 0 and 100 mmHg. The model layer consist of three compartments; these are supposed to represent the arteriolar, capillary, and venular spaces, respectively.

The three compartmental nonlinear model is illustrated in Fig. 1. Each compartment is compliant and can alter volume by a change either in luminal blood pressure or in external pressure. The distributed compliance of a part of the

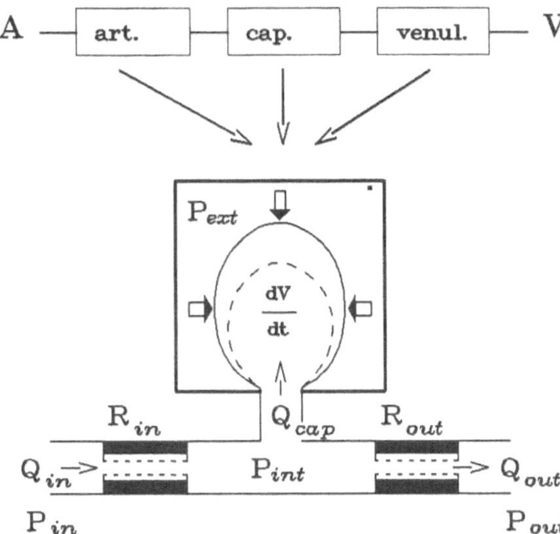

Fig. 1. Schematic representation of one layer of the non-linear intramyocardial pump model. The volume of a compartment, V, changes when blood pressure, P_{int}, or external pressure, P_{ext}, changes. Furthermore, the resistances, R_{in} and R_{out}, depend on the volume of the compartment. A, artery; *art.*, arteriolar; *cap.*, capillary; *venul.*, venular; V, vein; Q_{in}, inflow; Q_{out}, outflow; Q_{cap}, capacitive flow; P_{in}, inflow blood pressure; P_{out}, outflow blood pressure; dV/dt,change in volume with time

vascular bed represented by the compartment is lumped into a single compliance which is illustrated by the balloon in the box. When volume of the compartment is decreased, resistance is increased, which is symbolized by the narrowing of the inflow and outflow resistances. These resistances, which are assumed to be equal, represent the distributed resistance of the vascular compartment. The resistances are assumed to be coupled with the volume of the compartment according to:

$$R_i = \frac{K_i}{V_i^2} \tag{1}$$

where $i = 1$, 2, or 3 and refers to the arteriolar, capillary and venular compartment, respectively; K_i = constant; V_i = volume of compartment i; and R_i = resistance.

Inflow and outflow of each compartment is related to the first derivative of volume to time,

$$\frac{dV_i}{dt} = q_{\text{in},i} - q_{\text{out},i} \tag{2}$$

where $q_{\text{in},i}$ and $q_{\text{out},i}$ are the flows into and out of compartment i.

The volume of each compartment is related to transmural pressure via either its compliance or, more generally, its pressure-volume relation. When compliances are defined, the relation between pressure and volume is given by:

$$dV_i = C_i \cdot dP_{trans,i} \qquad (3)$$

where C_i = the compliance of compartment i; and $P_{trans,i}$ = the transmural pressure of compartment i.

Furthermore, it holds that

$$P_{trans,i} = P_{b,i} - P_{im} \qquad (4)$$

where $P_{b,i}$ = the blood pressure of compartment i; and P_{im} = the intramyocardial pressure.

When pressure-volume relations are defined, Eq. 3 is replaced by the pressure-volume relation symbolized by:

$$V_i = f_i(P_{trans,i}) \qquad (5)$$

where f_i indicates that V_i is a function of $P_{trans,i}$.

From a reference condition in which flow and distribution of V_i and $P_{trans,i}$ are defined, the distribution of R_i and the constants K_i can be determined. The model is then defined, except for the boundary conditions, which for this study are pressures at the inlet and outlet and tissue pressure. For the simulation, the parameters and variables shown in Table 1 were used.

Difference Between Linear and Nonlinear Intramyocardial Pump Models

To illustrate the characteristic behavior of the nonlinear intramyocardial pump model with reference to the linear models, the following model simulations were done. With the nonlinear pump model, flow was simulated using sigmoidal pressure-volume curves for the compartments as explained in Bruinsma et al. [9]. Hence, in the reference condition the distribution of volume and transmural pressure over the three compartments was defined. Then, all variables were allowed to vary according to the model due to the alterations in tissue pressure. This implies that the transmural pressure and volume of each compartment shift according to its respective pressure-volume curve, thereby altering compliance and resistance according to the formulas given. With the linear model, arterial flow was simulated, keeping resistances and compliances of the different compartments at their reference value. In order to accentuate the characteristics of the models, the transition to a permanent diastole was simulated as well.

The simulation of the arterial flow with the linear model is depicted in Fig. 2b. It exhibits the classical characteristics of a linear model with compliances and resistances. The time-averaged flow, both in the beating condition and with arrest, is constant and determined by the ratio between the pressure drop over the layer and the sum over all the resistances in series. The compliances only

Table 1. Values of variables and parameters in the reference condition

	Arterial	Arteriolar	Capillary	Venular	Venous
Pressure (mmHg)	100	56	9.5	6	5
Volume (ml/100 g)	—	1.9	3.5	3.3	—
Flow (ml/s per 100 g)	1.5	1.5	1.5	1.5	1.5

affect the transients in the signal. When tissue pressure is suddenly kept constant, flow decays to this time-averaged value. From the decay of the flow signal, a characteristic time for the model can be estimated.

The simulation of the arterial flow with the nonlinear model is depicted in Fig. 2c. In this panel also the ratio between pressure drop over the layer and the sum of the resistances in series is depicted with the dotted line. This ratio provides an indication how the sum of the resistances in the model does vary with time. Note that in this case the time-averaged flow in the beating state is lower than the time-averaged flow in the arrested state. When tissue pressure is suddenly kept constant flow decays to a time-averaged value higher than that in the beating state. This obviously is due to the resistances which are changing as long as compartmental volumes are changing. From the decay of the flow signal, a characteristic time for the flow decay can be estimated which, however, is not representative for the other physical events in the compartments. Figure 2d depicts the variations in the resistances of the three compartments in the nonlinear pump model.

As is clear from Fig. 2d, resistance variation is most pronounced in the venu-

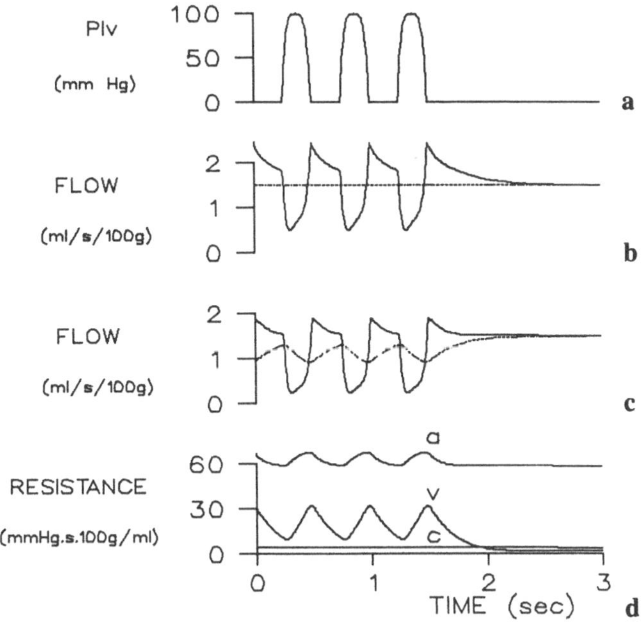

Fig. 2a–d. Instantaneous arterial inflow and resistive flow as predicted by the linear and nonlinear pump model. In these simulations only one layer is used. After three heart beats a long diastole is simulated. **a** Left ventricular presure as used in the simulations. **b** Simulations using the linear pump model with constant compliances and resistances in all three compartments. The resistive flow is thus constant. In this situation it takes a longer time before a steady state flow is reached. **c** Arterial and resistive flow as predicted by the nonlinear intramyocardial pump model with transmural pressure-dependent resistances and compliances. **d** The resistance variation in the three compartments of the nonlinear pump model. *a*, *c*, and *v* represent the arteriolar, capillary, and venular resistance, respectively

lar compartment. The increase of coronary inflow due to cardiac arrest is mainly due to the decrease of venular resistance. Hence, the effect of cardiac contraction on the time-averaged flow is mainly determined by the venous compartment. A similar conclusion is reached using the waterfall model. The difference between the two models, however, is that the waterfall model predicts an increased impediment of flow in systole and no reduction in diastole, whereas the nonlinear pump model predicts an increased venular resistance over the complete cycle, although resistance is somewhat larger in systole than in diastole.

Distribution of Compliance and Sensitivity of Resistance to Volume: Effect on the Arterial Flow Signal

In order to study the effect of compliance and resistance variability on the arterial flow, simulations as presented in Fig. 2c were repeated, but now the pressure-volume curve was linearized around a working point. This was done for all three compartments. First, the result of alterations in the arteriolar compartment will be discussed. This is illustrated in Fig. 3. Compliance at a working point can easily be altered by changing the slope of the pressure-volume relation. The sensitivity of resistance to volume is altered by choosing a different volume for the reference condition. Note that resistance in the reference condition is unaltered, since this follows from the flow and the pressure distribution over the compartments in the reference condition, which was not altered. However, from the altered reference volume a different value for K_i follows from Eq. 1. One may write for the sensitivity of resistance, R_i, on volume, V_i,

$$\frac{dR_i}{dV_i} = -\frac{2K_i}{V_i^3} = -\frac{2R_i}{V_i} \tag{6}$$

Hence, at lower initial volume the resistance sensitivity due to a volume change is higher.

Arterial flow was simulated for the compliance and initial volume values as presented in Fig. 3b. The results are illustrated in Fig. 4. A comparison of the profiles in Fig. 4 demonstrates that when the compliance and initial volume are decreased together (from *left upper panel* to *right lower panel*) the flow profile is hardly affected. On the other hand when volume is decreased and compliance is increased, large changes in diastolic flow signal were obtained (*lower left panel* to *upper right panel*). In the *upper right panel* the resistance variation is more important than the compliance effect, which is apparent from the increasing diastolic flow.

Similar variations of compliance and resistance variability were studied for the capillary and venular compartment. The capillary compliance was changed from 0.002 to 1.0 ml/mmHg per 100 g at a capillary volume of 4.0 ml/100 g. The venular compliance was modified from 0.01 to 2.0 ml/mmHg per 100 g at a venular volume of 3.6 ml/100 g. The capillary and the venular volumes in the reference condition were changed from 1.0 to 10 ml/100 g; the capillary and venular compliance were 0.08 and 0.15 ml/mmHg per 100 g, respectively.

Notwithstanding that these parameters varied over more than one order of magnitude, the changes hardly influenced the arterial diastolic flow profile. Larger compliance and lower volume values both resulted in a slight decrease in systolic flow, leading to a slight reduction of the mean flow. With larger compliances, it took a little longer time to obtain steady inflow after the onset of a long diastole. Compared with the effects of the arteriolar compartment on the arteriolar flow profile, the influences of the capillary and venular compartments are only very small.

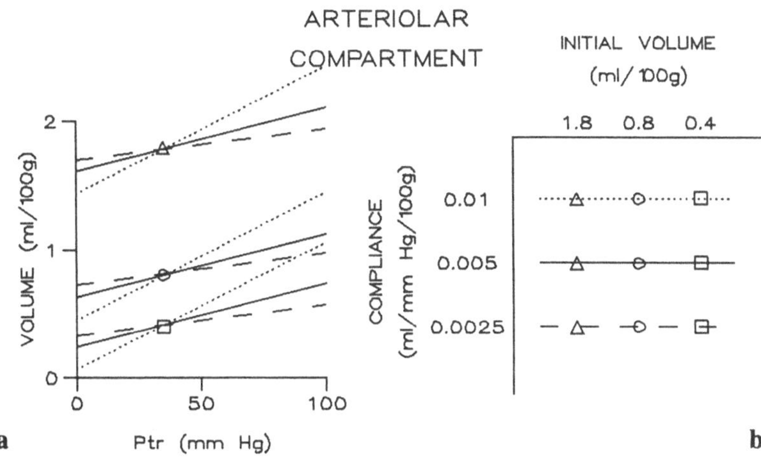

Fig. 3a, b. Linear transmural pressure-volume relations for the arteriolar compartment. The transmural pressure-volume relations (**a**) are obtained by using different slopes and different initial values in the reference condition. The different values related to the transmural pressure-volume curves are given in (**b**). *Ptr*, transmural pressure

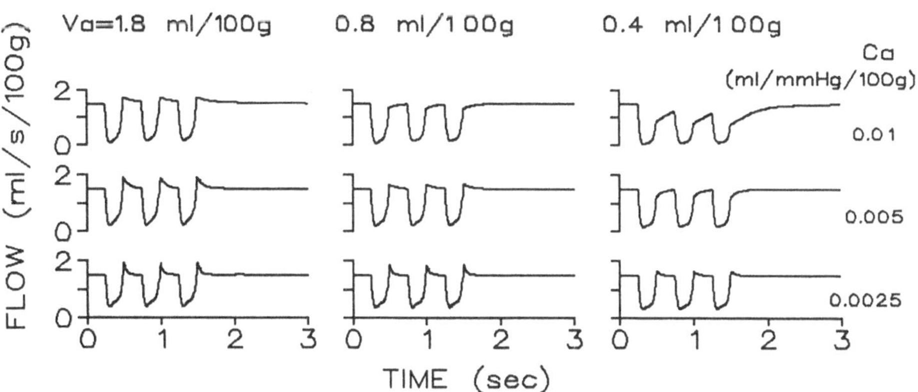

Fig. 4. Instantaneous arterial inflow predicted by the nonlinear pump model with different linear arteriolar transmural pressure-volume relations. *Left*, *middle*, and *right panels* correspond with arteriolar initial volumes (V_a) of 1.8, 0.8, and 0.4 ml/100 g, respectively; whereas *upper*, *middle*, and *lower panels* correspond with arteriolar compliances (C_a) of 0.01, 0.005, and 0.0025 ml/mmHg per 100 g. After three heart beats a long diastole was simulated

Arterial Pressure Dependency of Pulses in Arterial Pressure and/or Flow: Multiple Interpretations

There are at least two ways by which coronary arterial pressure can affect the waveform of the coronary arterial flow and pressure signals. The first one is due to autoregulation. At constant pressure perfusion, lowering arterial pressure reduces arteriolar resistance resulting in less damping of the flow wave generated by the pressurization of the distal intramyocardial blood vessels [5, 7]. The second effect follows from the dependency of inlet compliance on pressure. Obviously, the former effect is only relevant when local control is not exhausted or abolished. The latter effect always plays a role. Lee et al. [15] found experimentally that at constant flow perfusion the pulsations in coronary arterial pressure decreased in magnitude with decreasing flow. This was interpreted by the authors as evidence for a waterfall model extended with a proximal compartment, similar to the linear intramyocardial pump model. This model is illustrated in Fig. 5a. For the justification of this analysis the reader is referred to the paper of Lee et al. [15]. However, the observation can also easily be explained by the pressure dependency of a proximal compliance in the coronary system.

Figure 5b illustrates a one-compartmental linear pump model for the coronary circulation extended with a pressure-dependent epicardial compliance with a proximal resistance. The proximal resistance was assumed to amount 10% of the total coronary resistance but has no physical significance when the system is perfused with a flow source. A flow source has in theory an infinite source resistance and is in series with the proximal resistance. The compliance was assumed to be pressure dependent, as deduced from input impedance measurements performed by Canty et al. [14] and provided in Fig. 5c. The continuous curve was used as instantaneous relation between arterial pressure and compliance. The simulated pressure pulsations on the coronary arterial pressure with constant flow perfusion are now dependent on the level of arterial pressure, as is shown in Fig. 5d. This is consistent with the experimental finding of Lee et al. [15].

Validation of Models on the Coronary Circulation

It seems appropriate to start this section with a reminder that there is no such thing as a "true model." However, even if this were possible, the model might be so complex that for specific applications a more restricted model would be preferred. In first instance, the waterfall model was designed to interpret the impeding effect of cardiac contraction on time- and space-averaged coronary flow through the myocardial wall. The waterfall model has been very constructive for the interpretation of this effect but fell short for the interpretation of the magnitude of phasic flow. The linear intramyocardial pump model was designed for the interpretation of the phasic coronary arterial flow signal but fell short in explaining the time-averaged flow. None of the models except the time-varying elastance model provide the possibility for relating contractility of the heart muscle to coronary flow, a shortcoming which does not make the other models worthless.

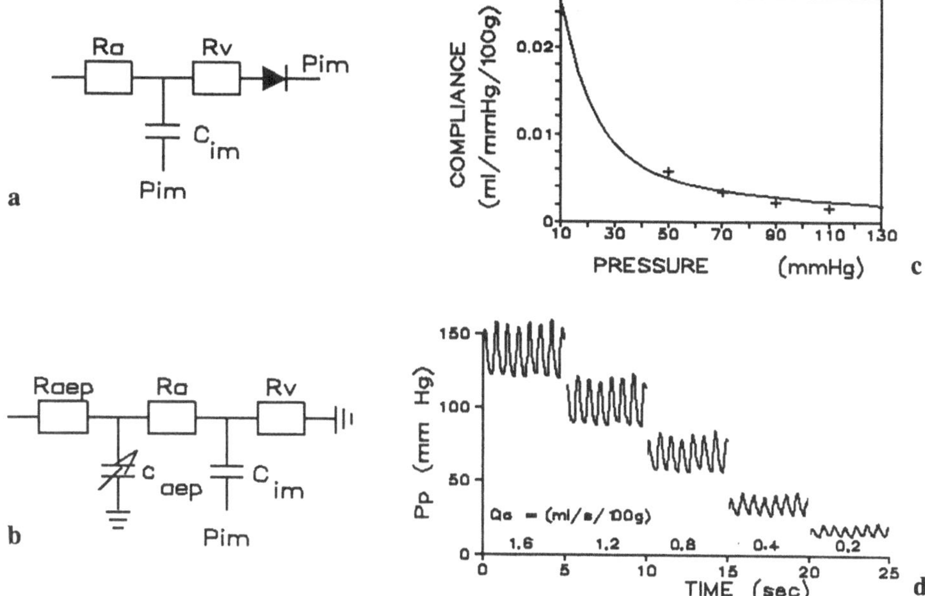

Fig. 5. a electrical analog of the extended waterfall model as proposed by Lee et al. [15]. **b** electrical analog of the one-compartmental linear pump model, extended with a pressure-dependent epicardial compliance. **c** Transmural pressure versus arterial compliance, as given by Canty et al. [14] (+) and as used in the model of **b** (*solid line*). **d** Simulated coronary phasic perfusion pressure as a result of decreasing mean flow. For this simulation the model of **b** is used with a pressure-dependent epicardial compliance. R_a, arteriolar resistance; R_v, venular resistance; P_{im}, intramyocardial pressure; C_{im}, intramyocardial compliance; R_{aep}, epicardial resistance; C_{aep}, epicardial compliance; P_p, coronary phasic perfusion pressure; Q_a, arterial flow

Apart from the restrictions on the applicability of the different models, the issue of uniqueness should be considered. It might very well be that a set of data can be explained by more than one model. This was illustrated in the previous section by the pressure pulsations occurring at constant flow perfusion. Hence, when presenting a new model or a modification of an already-existing one, one should carefully analyze whether the observations chosen are critical in that they can be explained exclusively by the model under study. This again does not imply that, when observations can be explained by a certain model, the evaluation of a different model on that point would be of no use. It is very important to know the limitations and predictive power of the different models, especially in the position where microvascular data on the myocardial circulation is so hard to obtain.

In this chapter we leant heavilyon the nonlinear intramyocardial pump model. This is because for the time being, the nonlinear intramyocardial pump model forms an attractive model as a reference for studies on coronary flow mechanics, both experimental as well as theoretical. This is especially so since at present a central issue in the field of coronary flow mechanics is the pressure dependency of vascular resistance and compliance at all levels of the circulation. These pressure dependencies are the crux of this model. Moreover, it allows the evalution of the effect of distribution of these factors over the vascular bed.

Conclusions

One may formulate two important conclusions from this analysis. *First*, pressure dependency of resistance and compliance of the arterial compartment has a significant influence on the coronary arterial flow signal, but these factors may vary widely at the capillary and venular level without affecting the coronary arterial flow signal. *Second*, the direct consequence of the pressure dependency of resistance and compliance makes the interpretation of results obtained by the application of linear system theory very difficult. Conclusion Two cannot be emphasized enough. The coronary input may be described by, e.g., a linear combination of constant resistances and compliances despite its nonlinear behavior. From apparent linearity, one may underestimate the real coronary compliance by a considerable magnitude.

References

1. Snyder R, Downey JM, Kirk ES (1975) The active and passive components of extravascular coronary resistance. Cardiovasc Res 9: 161–166
2. Sabiston DC, Gregg DE (1957) Effect of cardiac contraction on coronary blood flow. Circulation 15: 14–20
3. Wiggers CJ (1954) The interplay of coronary vascular resistance and myocardial compression in regulating coronary flow. Circ Res 2: 271–278
4. Downey JM, Kirk ES (1975) Inhibition of coronary blood flow by a vascular waterfall mechanism. Circ Res 36: 753–760
5. Spaan JAE, Breuls NPW, Laird JD (1981) Diastolic-systolic coronary flow differences are caused by intramyocardial pump action in the anesthetized dog. Circ Res 49: 584–593
6. Chilian WM, Marcus ML (1982) Phasic coronary blood flow velocity in intramural and epicardial coronary arteries. Circ Res 50: 775–781
7. Spaan JAE, Breuls NPW, Laird JD (1981) Forward coronary flow normally seen in systole is the result of both forward and concealed back flow. Basic Res Cardiol 76: 582–586
8. Arts MGJ (1978) A mathematical model of the dynamics of the left ventricle and the coronary circulation, PhD thesis, R.U. Limburg, Maastricht, The Netherlands
9. Bruinsma P, Arts T, Dankelman J, Spaan JAE (1988) Model of the coronary circulation based on pressure dependence of coronary resistance and compliance. Basic Res Card 83: 510–524
10. Spaan JAE, Bruinsma P, Vergroesen I, Dankelman J, Stassen HG (1987) Distensibility of microvasculature and its consequence on coronary arterial and venous flow. In: Sideman S, Beyar R (eds) Activation, metabolism and perfusion of the heart. Martinus Nijhoff, Dordrecht pp 398–407
11. Krams R, Sipkema P, Westerhof N (1989) Varing elastance concept may explain coronary systolic flow impediment. Am J Physiol 257: H1471–H1479

12. Krams R, Sipkema P, Westerhof N (1989) Can coronary systolic-diastolic flow difference be predicted by left ventricular pressure of time varying intramyocardial elastance. Basic Res Cardiol 84: 149–159
13. Suga H, Sagawa K, Shoukas AA (1973) Load independence of the instantaneous pressure-volume ratio of the canine left ventricle and effects of epinephrine and heart rate on the ratio. Circ Res 32: 314–322
14. Canty JM, Klocke FJ, Mates RE (1985) Pressure and tone dependence of coronary diastolic input impedance and capacitance. Am J Physiol 248 (Heart Circ Physiol 17): H700–H711
15. Lee J, Chambers DE, Akizuki S, Downey JM (1984) The role of vascular capacitance in the coronary arteries. Circ Res 55: 751–762

Functional Characteristics of Intramyocardial Capacitance Vessels and Their Effects on Coronary Arterial Inflow and Venous Outflow

Katsuhiko Tsujioka, Masami Goto, Osamu Hiramatsu, Yoshifumi Wada, Yasuo Ogasawara, and Fumihiko Kajiya[1]

Summary. The intramyocardial capacitance vessels have two functional components, unstressed volume and ordinary capacitance. The estimated value of unstressed volume was about 5% of the left ventricular mass, and the time constant in relation to ordinary capacitance was about 1 s, although both are pressure-dependent. The coronary venous outflow was closely related to the total displaceable blood volume stored in the intramyocardial capacitance vessels, i.e., the more intramyocardial blood volume, the more coronary venous outflow. On the other hand, the diastolic coronary arterial inflow was decreased when the blood volume in the intramyocardial vessels increased above the unstressed volume. This may be important as a mechanical control of the coronary circulation system.

Key words: Intramyocardial capacitance vessel—Unstressed volume—Coronary venous outflow—Coronary arterial inflow—Coronary artery pressure-flow relationship

Introduction

The phasic flow in the left coronary artery is diastolic-predominant. Scaramucci (1689, cited by Porter [1]) hypothesized that the deeper coronary vessels are squeezed by the contraction of the muscle fibers around them, which displaces the intramyocardial blood into coronary veins, and the vessels are refilled from the aorta during diastole. To prove this hypothesis, it is necessary to investigate coronary arterial inflow and venous outflow of the myocardium. Since Anrep et al. [2] investigated the circulation in the coronary artery and vein more than fifty years ago by making blood flow measurements using a hot wire method, there have been few reports [3–5] describing arterial inflow and venous outflow simultaneously. This is partly because the measurement of coronary venous flow using conventional methods including the electromagnetic flowmeter has until recently been difficult, and also because the vein was regarded as the only conduit of coronary venous outflow.

Our laser Doppler velocimeter (LDV) with an optical fiber is a powerful new tool for the measurement of both coronary arterial flow and venous flow. The

[1] Department of Medical Engineering and Systems Cardiology, Kawasaki Medical School, 577 Matsushima, Kurashiki, 701-01 Japan

most important advantage of the LDV method over conventional velocimeters is its excellent accessibility to the vessel, even when the vessel is easily collapsible, as is a vein. In this section, we report some results on the following topics: (1) the functional characteristics of intramyocardial capacitance vessels [6], (2) the effect of the capacitance vessels on coronary arterial inflow [7], and (3) their effect on coronary venous outflow [8].

The Functional Characteristics of Intramyocardial Capacitance Vessels

Mongrel dogs were anesthetized with sodium pentobarbital. After intubation, the animals were ventilated by a Harvard respirator pump with room air, which was supplemented with 100% oxygen at a rate sufficient to maintain arterial oxygen tension at a physiological level. A left thoracotomy was performed at the 4th or 5th intercostal space. The heart was exposed and suspended in a pericardial cradle.

The peripheral portion of the great cardiac vein (GCV) was isolated and the optical fiber tip was inserted into the vessel. Both the left main coronary artery (LM) and the left anterior descending coronary artery (LAD) were cannulated and connected to a reservoir to regulate perfusion pressure. In our earlier study, only the LAD was cannulated and connected to the reservoir. Coronary inflow and perfusion pressure were measured at the peripheral portion of the cannula inserted into the LAD. To induce long diastole, the atrioventricular node was destroyed by the injection of 40% formalin. Pacing electrodes were sewn onto the right ventricular wall. The experimental procedure is shown in Fig. 1. During continuous infusion of adenosine into the coronary artery, the cannulae were occluded to shut off the LAD flow. The blood velocity in the GCV decreased and reached a minimal steady value within 15 s. Then a long diastole was induced by the cessation of pacing. Two seconds after the cessation of pacing, the cannulae were reopened and the perfusion pressure was increased stepwise to a preset target pressure. The time course of the GCV flow velocity was analyzed after the initiation of reperfusion.

The time course of the coronary hemodynamic data is displayed in Fig. 2. After occlusion of the coronary inflow, the GCV decreased and reached a minimal steady value. Then it fell to zero with the cessation of pacing. After reopening the inflow, it was still absent for a few seconds (dead time). Then it reappeared and increased with a first-order time delay. The presence of the dead time indicates the existence of an unstressed volume in the intramyocardial vascular compartments, which is defined as the volume of the blood in a vessel at zero transmural pressure. The time constant of the first-order delay relates to the product of resistance and capacitance of the diastolic coronary circulation with minimal vasomotor tone.

Thus, the mechanical lumped model illustrated in Fig. 3 was adopted as the simple and optimal one for explaining the results of these animal experiments. The model consists of a combination of the unstressed volume, the resistance, and the capacitance. The unstressed volume (UV) was estimated from the coronary arterial inflow during the dead time, and resulted in a value of 5.2 ml/100 g LV on an average. The value of capacitance was obtained by dividing the time constant by the resistance. The mean value of capacitance was 0.08 ml/mmHg per 100 grams LV. With vessels embedded in tissue as intramyocardial vessels, transmural pressure at a volume less than the UV may be negative [9]. To return the vessels from a collapsed to a cylindrical configuration requires

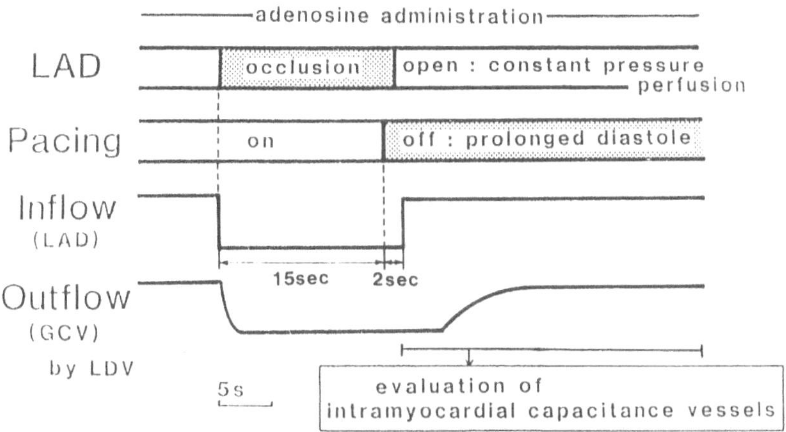

Fig. 1. Schematic diagram of the experimental procedure. After coronary inflow was shut off, the great cardiac venous (*GCV*) flow decreased gradually. Fifteen seconds after inflow cannula occlusion, a long diastole was induced. Two seconds later, the cannula was reopened and the perfusion pressure was increased stepwise. The response of the venous outflow was analyzed. *LAD*, left anterior descending coronary artery; *LDV*, laser Doppler velocimeter. From [6] by permission of the American Heart Association, Inc.

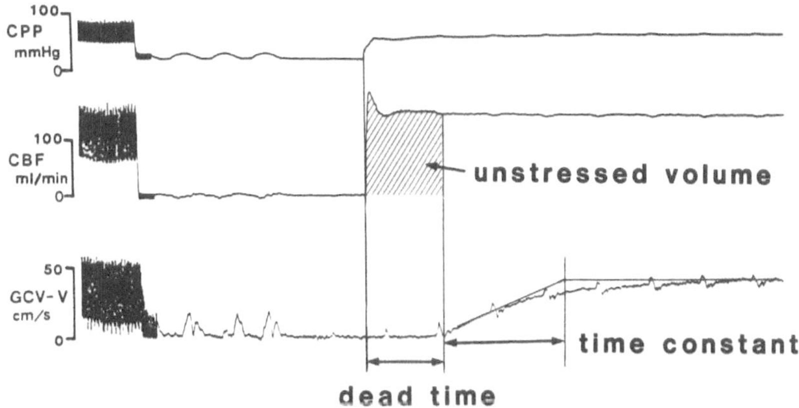

Fig. 2. A representative tracing of variables recorded during a trial. The coronary venous flow was absent for about 1 s (dead time) and increased with a first-order time delay after reperfusion. *CPP*, coronary perfusion pressure; *CBF*, coronary arterial blood flow; *GCV-V*, great cardiac venous velocity

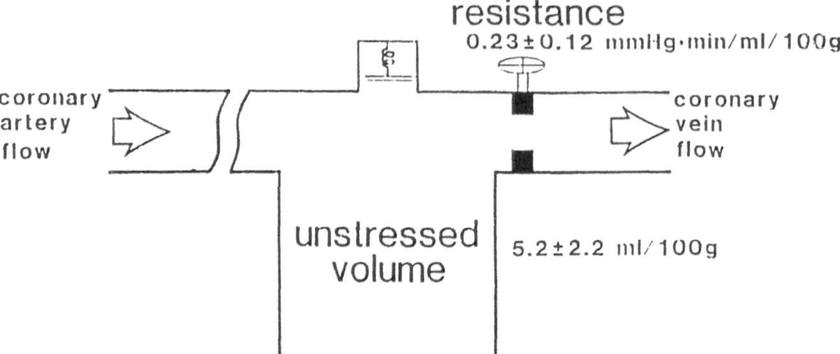

Fig. 3. Mechanical lumped model representing the characteristics of the intramyocardial capacitance vessels and estimated values of the unstressed volume, the capacitance, and the resistance. From [6] by permission of the American Heart Association, Inc.

only a small change in the intraluminal pressure. This characteristic of the vessels may contribute to the UV. This volume is distributed continuously in the coronary circulation system, but not homogeneously. The UV may be mostly distributed in veins and capillaries because of their distensibility. Therefore, some inflow into the UV may both refill the intramyocardial capacitance vessels and allow nutrient exchange. The UV is approximately 5% of the myocardium and the time constant in relation to ordinary capacitance is about 1 s, although they are both pressure-dependent.

Effect on Coronary Arterial Inflow of the Blood Filling Condition in Intramyocardial Vessels

To evaluate the effect of the intramyocardial capacitance vessels on coronary arterial inflow, the last part of the protocol in Fig. 1 was used. The target pressure after the reopening of the cannulae was changed at 7 different levels. The pressure flow (P/F) relation was analyzed for different conditions in the intramyocardial capacitance vessels, i.e., with the UV unfilled vs filled. Figure 4 shows an example of the GCV responses following stepwise increase in the perfusion pressures. Recall that the intramyocardial capacitance vessels functionally consist of unstressed volume and ordinary capacitance. The delayed onset of the GCV flow indicates an initially unfilled condition of the unstressed volume (UV-unfilled), followed by the resumption of the GCV flow (UV-filled). The LAD flow after the reperfusion reached an initial quasi-static level following transient overshoot and undershoot. Then, it decreased with reappearance of the GCV flow and converged to a final steady level. Recently, Katz and Feigl [10] also observed the declining asystolic flow in most cases during constant perfusion in the dog. We analyzed the pressure-flow relationship at two points; first, about 1 s after reperfusion (the quasi-steady-state under UV-unfilled conditions) and,

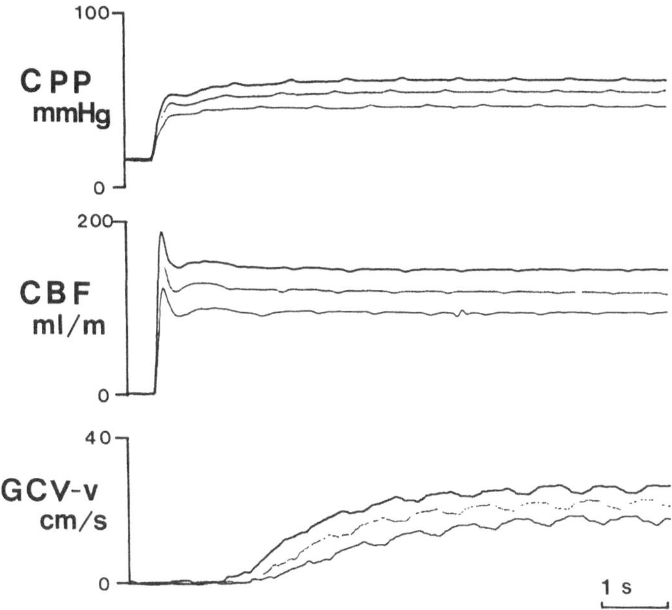

Fig. 4. Responses of the great cardiac venous flow for different perfusion pressures. The left anterior descending coronary arterial flow appears to be greater before the appearance of the great cardiac venous flow. *CCP*, coronary perfusion pressure; *CBF*, coronary arterial blood flow; *GCV-V*, great cardiac venous velocity

second, when the pressure and flow reached a steady state at about 3–5 s after the GCV flow resumed (UV-filled).

Figure 5 shows a typical example of the P/F relations before and after resumption of the GCV flow. The LAD velocities for given coronary arterial pressures were always higher in the UV-unfilled phase than those in the UV-filled. Notice that when the UV is not saturated, its intraluminal pressure is less than or equal to the GCV pressure and the flow remains at zero level, but when it is filled with blood, the intravascular pressure exceeds the outflow pressure and the GCV flow resumes. Since the P/F relation was linear ($r = 0.97$–0.99), we simply used the inverse of its slope as an index of the impeding effect, and estimated the zero-flow intercept by linear regression. The inverse of the slope was higher for the UV-filled phase than that for UV-filled ($P < 0.01$), but the respective zero-flow pressure intercepts were not different. However, it cannot be concluded that the zero-flow pressure is not the determinant of the difference in the coronary inflow accompanying the change in the state of UV, since we did not examine either the linearity of the P/F relation for very low flow regions in detail, or the stop-flow pressure after clamping the perfusion line. The higher impeding effect for the UV-filled phase is important, since it acts as a negative feedback to the coronary arterial inflow, i.e., the increase in the blood volume in the intramyocardial vessels decreases the coronary arterial inflow.

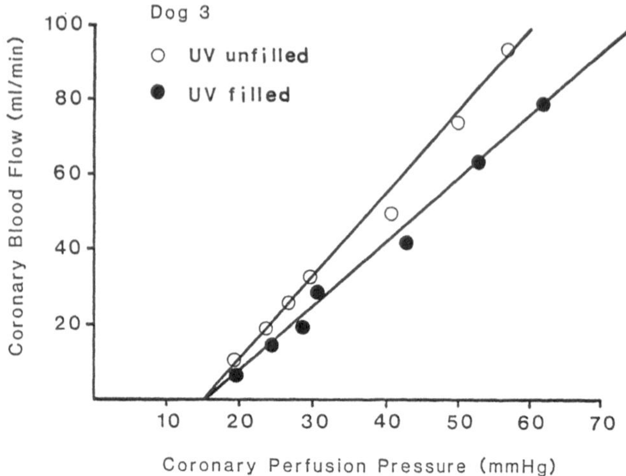

Fig. 5. A typical example of the pressure/flow relations before and after resumption of the great cardiac venous flow. The left anterior descending coronary arterial flow was smaller when the GCV flow reappeared. *UV-unfilled*, unstressed volume not filled; *UV-filled*, unstressed volume filled

Effect of Intramyocardial Capacitance Vessels on Coronary Venous Outflow

To investigate the effect of the intramyocardial capacitance vessels on coronary venous outflow, the initial part of the protocol shown in Fig. 1 was used: the time course of the GCV flow velocity was analyzed after occlusion of the cannulae. Before the occlusion, the GCV velocities were altered to various values by changing the perfusion pressure and vasomotor tone, to determine the relation between the GCV velocities before occlusion and during the squeezing out of the blood after occlusion.

Figure 6 shows the time course of decay of the GCV flow velocity after coronary inflow occlusion. It should be noted that the GCV flow decreased exponentially. Thus, the time course can be expressed as

$$V_{GCV}(t) = V_{GCV}(0)e^{-\frac{t}{\tau}} \tag{1}$$

where $V_{GCV}(t)$ is the decaying velocity and t is time after the inflow occlusion, $V_{GCV}(0)$ is the initial velocity before occlusion and τ is the time constant. Integration of Eq. (1) with respect to time t gives

$$\int_0^{\infty} V_{GCV}(0)e^{-\frac{t}{\tau}}dt = V_{GCV}(0)\cdot\tau \tag{2}$$

The value of $V_{GCV}(0)\cdot\tau$ implies the total displaceable volume V_D stored in the intramyocardial capacitance vessel before occlusion:

$$V_D = V_{GCV}(0)\cdot\tau \tag{3}$$

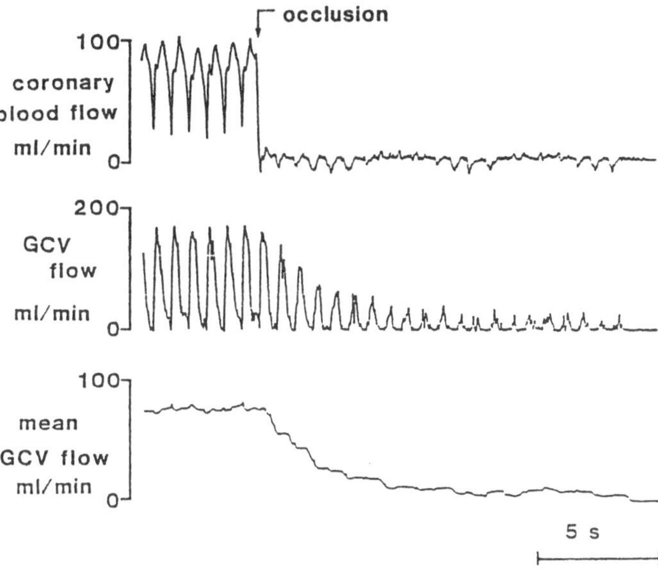

Fig. 6. A typical example of the time course of decay of the coronary venous flow velocity after coronary arterial inflow occlusion. The great cardiac venous (*GCV*) flow decreased exponentially

It follows that

$$V_{GCV}(0) = V_D/\tau \quad (4)$$

Then, the problem is "which parameter is more sensitive (contributable) to the change in $V_{GCV}(0)$, τ or V_D?." Figure 7 shows the relations between $V_{GCV}(0)$ and V_D, and between $V_{GCV}(0)$ and τ. The correlation coefficient between the total displaceable volume V_D and the coronary venous flow $V_{GCV}(0)$ was significantly high ($P < 0.01$), whereas the correlation between the time constant τ and the venous flow $V_{GCV}(0)$ was not statistically significant. This indicates that the coronary venous flow is mainly dependent on the total displaceable blood volume stored in the intramyocardial capacitance vessels, i.e., "the more intramyocardial blood volume, the more coronary venous flow." Although the real mechanism for the close correlation between the intramyocardial blood volume and the coronary venous flow is not clear, the relationship is of interest especially by analogy with the "Starling law" of the heart—the larger the preload, the more the cardiac output. The Gregg effect [11] may be attributable to this relationship.

Concluding Remarks

The contents of this paper are summarized as follows. (1) The intramyocardial capacitance vessels during diastole have two functional components, unstressed volume and ordinary capacitance. (2) The systolic coronary venous outflow is

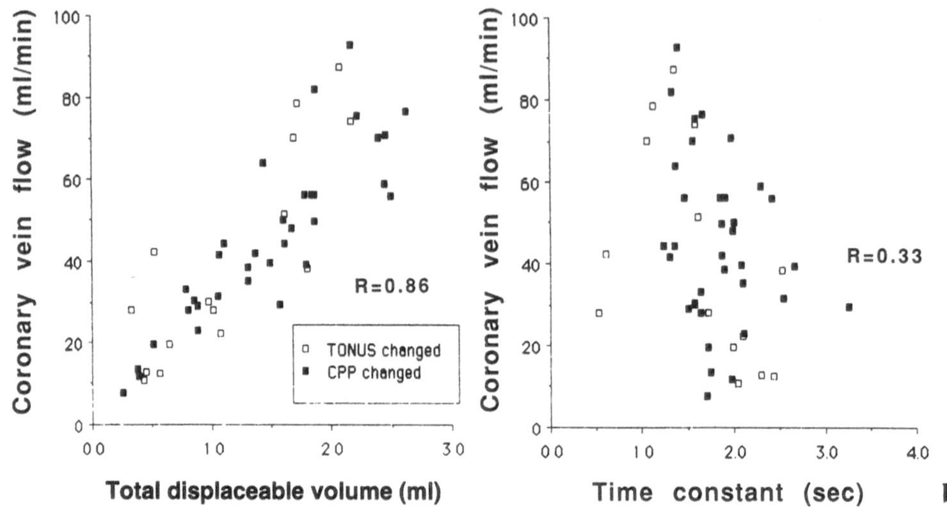

Fig. 7a, b. The relationship **a** between the great cardiac vein flow and the total displaceable blood volume in the intramyocardial vessels, and **b** between the venous flow and the time constant. The former relationship showed a significant correlation, but the latter did not. *CPP*, coronary perfusion pressure

closely related to the total displaceable blood volume stored in the intramyocardial capacitance vessels. (3) When the unstressed volume was saturated, the diastolic coronary inflow was decreased significantly compared with that for the unsaturated condition. Thus, the increase in the intramyocardial blood volume decreases the coronary arterial inflow, whereas it enhances the coronary venous outflow. This is important as a mechanical control of the coronary circulation system.

References

1. Porter WT (1898) The influence of the heart-beat on the flow of blood through the walls of the heart. Am J Physiol 1: 145–163
2. Anrep GV, Cruickshank EWH, Downing AC, Sabba RA (1927) The coronary circulation in relation to the cardiac cycle. Heart 14: 111–133
3. Chilian WM, Marcus ML (1984) Coronary venous outflow persists after cessation of coronary arterial inflow. Am J Physiol 247: H984–H990
4. Spaan JAE (1982) Intramyocardial compliance studies by venous outflow at arterial occlusion (abstract). Circulation 66 (Suppl II): 42
5. Vergroesen I, Noble MIM, Spaan JAE (1987) Intramyocardial blood volume change in first moments of cardiac arrest in anesthetized goats. Am J Physiol 253: H307–H316
6. Kajiya F, Tsujioka K, Goto M, Wada Y, Chen X-L, Nakai M, Tadaoka S, Hiramatsu O, Ogasawara Y, Mito K, Tomonaga G (1986) Functional characteristics of intramyocardial capacitance vessels during diastole in the dog. Circ Res 58: 476–485
7. Goto M, Tsujioka K, Ogasawara Y, Wada Y, Tadaoka S, Hiramatsu O, Yanaka M, Kajiya F (1990) Effect of blood filling in intramyocardial vessels on coronary arterial inflow. Am J Physiol 258: H1042–H1048

8. Kajiya F, Tsujioka K, Ogasawara Y, Mito K, Hiramatsu O, Goto M, Wada Y, Matsuoka S (1989) Mechanical control of coronary artery inflow and vein outflow. Jpn Circ J 53: 431–439
9. Rothe CF (1983) Venous system: Physiology of the capacitance vessels. In: Shepherd JT, Abboud FM, Geiger SR (eds) The cardiovascular system. Handbook of physiology, vol 3, part 1. American Physiological Society, Washington DC, pp 397–452
10. Katz AS, Feigl EO (1988) Systole has little effect on diastolic coronary blood flow. Circ Res 62: 443–451
11. Gregg DE (1963) Effect of coronary perfusion pressure or coronary flow on oxygen usage of the myocardium. Circ Res 13: 497–500

Input Impedance of the Canine Coronary Arterial Tree

NICO WESTERHOF and PIETER SIPKEMA[1]

Summary. The input impedance of the canine left circumflex coronary arterial tree was determined in systole and in diastole using an impulse response technique. The input impedances in systole and diastole are not different; the impedances at zero Hz are 2.4 ± 0.5 and $3.1 \pm 1.1 \cdot 10^9$ Pa.s.m^{-3}, respectively. Characteristic impedances in systole and diastole are 1.0 ± 0.2 and $1.1 \pm 0.1 \cdot 10^9$ Pa.s.m^{-3}, respectively. The zero-Hz component of these impedances is comparable with the slope of the instantaneous pressure-flow relation, first reported by Bellamy [1]. The finding that impedance does not vary over the cardiac cycle implies that the effects of cardiac contraction on the coronary vasculature are to be found in the periphery which is poorly characterized by impedance. The input impedance of the left circumflex artery exhibits similar characteristics to that of the femoral bed, indicating similar properties of the respective proximal artery (i.e., circumflex and femoral artery, as previously reported by Gow and Hadfield [2]) and similar total compliance of the beds.

Key words: Coronary model—Systole—Diastole—Impulse response—Coronary and femoral bed

Introduction

For the characterization of an arterial system it has been customary to use the concept of input impedance. "Input impedance" describes an arterial bed not only in terms of mean pressure and mean flow, as in using "peripheral resistance," but also takes into account the oscillatory aspects of the pressure and flow. Knowledge of impedance has led to greater understanding of the function of the arterial system. For instance, it was from impedance characterization of the entire systemic arterial tree that the three-element windkessel was derived [3]. Not only did this simple model describe the impedance characteristics, but the three elements could be shown to have their basis in properties of vessels (diameter and compliance) and blood (viscosity and mass) [3].

Input impedance commonly has been derived from the measured pressure and flow wave in the steady state. After Fourier analysis the (sinusoidal) harmonics,

[1] Laboratory for Physiology, Free University of Amsterdam, v.d. Boechorststraat 7, 1081 BT Amsterdam, The Netherlands

found at multiples of the heart rate, of pressure and flow are related. The amplitude ratio of the pressure and flow harmonics gives the impedance modulus. The phase difference of the pressure and flow harmonics gives the phase angle of the impedance. The full characterization of an arterial bed by means of impedance determination is thus obtained at frequencies that are multiples of the heart rate [4]. It would be advantageous to characterize the coronary arterial tree in a similar fashion. This could lead to greater understanding of its function and could enable comparison with other peripheral beds. However, the derivation of input impedance is only permitted when the system under study is linear and invariant with time (i.e., not changing its properties during the time of determination). Since the coronary arterial tree may exhibit nonlinear behaviour, and since its properties may vary during the heart beat, the commonly used derivation by means of a Fourier analysis cannot be used. In other words, the application of Fourier analysis to the coronary arterial pressure and flow may be performed, but the subsequent division of the Fourier components of pressure by flow does not lead to sensible results. We must therefore use another approach to obtain the input impedance of the coronary arterial tree. By using the impulse response technique [5] we could determine input impedance in the diastolic phase and in the systolic phase of the heart assuming that the properties of the system remained unchanged over the period of determination (about 250 ms). The results show that the input impedance of the coronary arterial tree is similar in systole and diastole. Input impedance can be modelled with the three-element windkessel, and the coronary bed is therefore not very different from other arterial trees. We can conclude that the effect of cardiac contraction is not manifest in the impedance.

Methods

The details of the method have been described elsewhere [6]. The left circumflex coronary artery of the anesthetized dog ($n = 5$) was cannulated. An impulse of flow was generated by means of a fast pneumatic device that partially occluded a short elastic tube which was mounted on the otherwise stiff cannulating system. The mean flow was provided by means of an extracorporeal circuit from the femoral artery and included a pump, heat exchanger, and bubble catcher. When the device displaced the wall of the compliant section, a volume of blood ($50.2 \pm 2.0 \mu l$) was propelled toward the arterial system, since on the upstream end of the perfusion system a large (i.e., much larger than coronary resistance) resistor was mounted. The impulses generated in this way had a duration of 14.1 ± 2.8 ms. As a test for linearity an impulse with smaller amplitude was delivered with a volume displacement of 32 μl.

The pressure resulting from the impulse of inflow is the impulse response, and was measured with a catheter tip manometer mounted in the cannula (3F Millar). Pressures are given in kPa with 1 kPa = 7.5 mmHg. The impulse response lasted, on the average, 250 ms. After analog-to-digital conversion with 5000 samples per second, Fourier analysis was performed on the impulse response. Since, for the Fourier series of the impulse itself, the amplitudes of the harmonics were the same up to 50 Hz (i.e., the impulse was sufficiently short in duration), the input can be considered ideal. Fourier analysis of the impulse response thus directly gave input impedance after calibration with the volume of blood injected. The impulses were triggered from the electrocardiogram, and given in mid-systole and mid-diastole assuming a time-invariant system over the period of determination. The method was tested in vitro and in vivo; for details see Van Huis et al. [6].

The data will be reported here in terms of modulus and phase of the input impedance plotted as a function of frequency for the range between zero and 15 Hz.

The input impedance of the left circumflex coronary bed of the dog was also determined by Canty et al. [7] using sinusoidal forcing during long diastoles. The input impedance in the control situation and during maximal vasodilation (at perfusion pressure 12 kPa = 90 mmHg) as published by Canty et al. [7] were used for comparison with our findings. Canty's data were converted to the units used by us.

Input impedance of the femoral bed was used for comparison. Data on this bed were taken from Cox [8] and Milnor and Nichols [9]. These impedance data were derived from pressure and flow signals in the steady state using Fourier series.

The impedance data were fitted with the three-element windkessel model [10]. The moduli and phase data were converted into real and imaginary parts of the impedance and a complex nonlinear Marquardt fit was applied. The zero-Hertz term of the impedance was not included in the fitted data. The drawn lines in the impedance graphs are based on the fit of the three-element windkessel.

Results

We performed the following test for linearity. When impulses with different amplitudes were given in mid-diastole (50 μl and 32μl, respectively) and the impulse responses measured, the amplitudes of the first peak of the responses were 6.7 ± 0.1 and 4.7 ± 0.3 kPa, respectively (average of 4 impulses). When the larger response was scaled by a factor 32/50 and plotted in the same graph as the smaller response the result of Fig. 1 was obtained. It can be concluded that the impulse response is not dependent on the magnitude of the impulse given and the system is thus linear.

An example of the input impedance of the coronary arterial tree in the dog in mid-diastole and mid-systole is presented in Fig. 2. It may be seen that the modulus of the impedance is high for zero frequency and precipitously decreases for increasing frequencies to reach a constant level. The phase of the impedance is, by definition, zero for zero Hz; it decreases to negative values for low frequencies, then slowly approaches zero values again for the higher frequencies. The average values of the impedance modulus at zero Hz in mid-diastole and in mid-systole were 3.1 ± 1.1 and $2.4 \pm 0.5 \cdot 10^9$ Pa.s.m^{-3}, respectively. These values were not plotted in Fig. 2 as will be discussed below. The average levels to which the impedance decreased at high frequencies in mid-diastole and in mid-

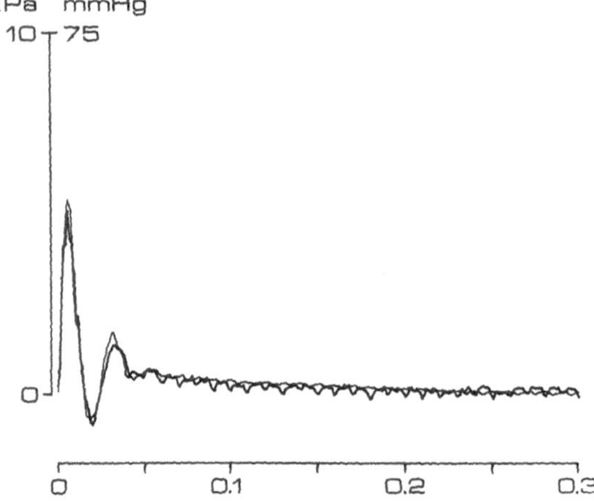

Fig. 1. Impulse responses in diastole for two impulses of different magnitude after scaling. Note the almost perfect superposition. The impulse responses (pressure as a function of time) are the result of rapid volume injections of 50 μl and 32 μl, respectively. The larger response is scaled by a factor 32/50

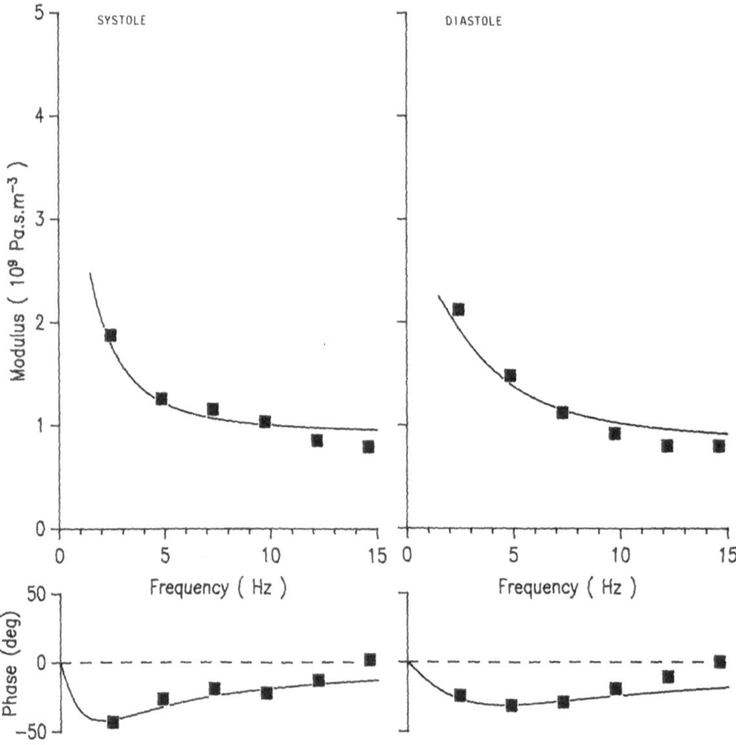

Fig. 2. Example of canine left circumflex coronary artery input impedances calculated from impulse responses determined in systole and diastole. The drawn lines are the fits of

systole were 1.1 ± 0.1 and $1.0 \pm 0.2 \cdot 10^9$ Pa.s.m^{-3}, respectively. As can be seen from Fig. 2 the differences in impedance in the systolic and the diastolic phase are small. The lines through the data points follow from the fit.

Figure 3 shows the input impedance of the left circumflex coronary arterial tree as determined using the impulse response technique (averaged data of Fig. 2); these data are compared with the input impedance of this bed during long diastoles using sinusoidal excitation, determined by Canty et al. [7]. The impedance data during control and during the dilated state are shown. It may be seen that the impedance derived by us is in between the two conditions studied by Canty et al. [7].

Figure 4 shows the input impedance of the circumflex coronary bed of the dog (averaged data from Fig. 2) and the input impedance of the femoral bed of the dog as obtained from the literature [8, 9]. It may be seen that the impedance of coronary system and femoral artery bed are similar both qualitatively and quantitatively.

The results of fitting the input impedances with the three-element windkessel model are given in Table 1. The fit of the coronary impedance resulted in values for the left circumflex coronary artery characteristic impedance, and for circumflex total arterial compliance. The femoral impedance fit gave femoral characteristic impedance and femoral arterial compliance. As can be seen from Fig. 2 the impedance approaches characteristic impedance less rapidly in the diastolic phase as compared to the systolic phase. This observation is in agreement with the result on total arterial compliance presented in Table 1, where compliance is smaller in diastole than in systole. The compliance estimated from the data of Canty et al. is, in the control situation, about 25% larger than found by us. Compliance of the femoral bed is close to circumflex arterial compliance. It may also be seen that the compliance during coronary vasodilation is larger than under control conditions.

Discussion

The input impedance of the coronary arterial tree (left circumflex bed) is not strongly dependent on the phase of cardiac contraction and is similar to the input impedance of the femoral bed. The input impedance is here derived using the impulse response technique, which is suitable for measuring impedance in a system which possibly varies with time. The impulse response and input impedance are Fourier pairs: they contain the same information but the information is presented in different form [5].

Before going into the discussion of the similarities and differences in impedances we will first discuss why the steady term of the impedance is not given in the impedance plots. It is customary to designate the ratio of mean pressure over mean flow ("peripheral resistance") as the zero-Hertz impedance value. However, when the input impedance is determined either by Fourier analysis or by means of the impulse response method, one works with perturbations superimposed on the steady state values of pressure and flow. As has been shown by Bellamy [1] for diastole and by Van Huis et al. [11] for the entire cardiac cycle,

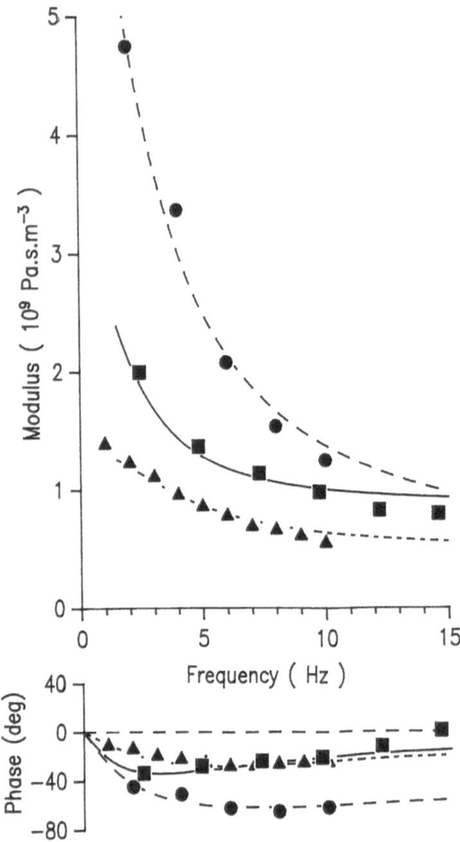

Fig. 3. Circumflex coronary input impedance determined by the impulse response method (averaged impedance of Fig. 2, ■) compared with input impedances determined in long diastoles using sinusoidal forcing by Canty et al. [7]. The two impedances taken from Canty et al. [7] were determined during control (●) and vasodilation (▲); the average perfusion pressure level was 90 mmHg (12 kPa). Drawn lines (solid and dashed) follow from three-element windkessel fit

the pressure-flow relation for perturbations deviates from peripheral resistance. The slope of this instantaneous pressure-flow relation is a better measure of the resistance at zero Hertz since it is determined under the same conditions as the input impedance. The data on the resistance given here are derived from the ratio of the area under the impulse response and the volume injected (area under the impulse), and are, as has been discussed by Van Huis [6], equal to the instantaneous resistance (slope of the instantaneous pressure-flow relation). The instantaneous resistance, although on the average slightly lower in diastole than in systole, was not significantly different between these two phases. This finding implies that the instantaneous pressure-flow relationship shifts in a parallel fashion between systole and diastole. Van Huis et al. [11] found that the slope of the instantaneous pressure-flow relation as determined over the cardiac cycle was not different from the slope determined in diastole by Bellamy [1] which also suggests the same slope in diastole and systole.

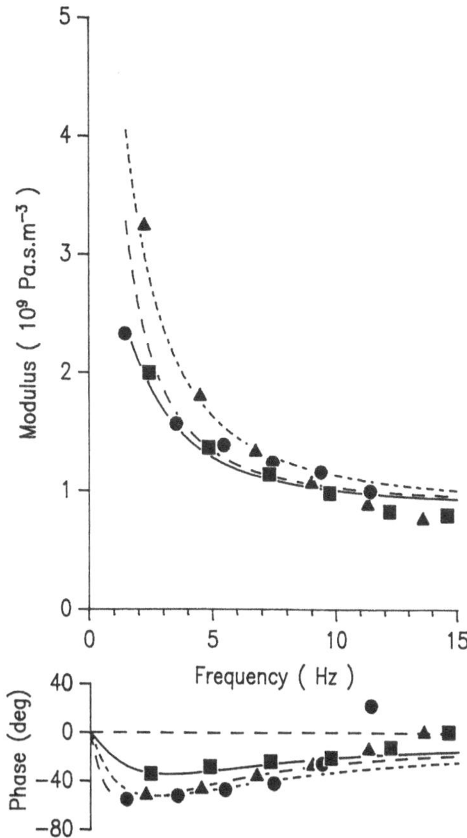

Fig. 4. Circumflex coronary input impedance (■) compared with femoral artery input impedance of the dog as reported by Cox [8] (▲) and Milnor and Nichols [9] (●). The lines through the points are obtained via the fit with the three-element windkessel model. The characteristics of both vascular beds are similar

The coronary bed appears to behave linearly since the impulse responses resulting from impulses of different amplitudes were proportional. This finding can be explained from the reasoning given later that the input impedance and impulse response give little information on distal parts of the bed, together with the fact that the larger arteries behave rather linearly over the pressure variations studied, as was already known for other peripheral beds [4]. The similarity of the input impedances in systole and diastole is explained by the fact that the impedance gives little information about the distal bed. In diastole, arterial compliance is smaller than in systole (Table 1), which is in agreement with the finding by Van Huis et al. (see Fig. 2 of [6]) that the decay time of the impulse response is faster in diastole than in systole. This smaller epicardial compliance in diastole could perhaps result from stretch of epicardial vessels in the diastolically large heart.

Table. 1. Characteristic impedance and total arterial compliance of left circumflex coronary and femoral beds of the dog

	Present study			Canty et al. [7]	
Coronary	Systole	Diastole	Average	Control	Dilated
Characteristic impedance*	0.925	0.882	0.876	0.450	0.489
Total arterial compliance*	0.049	0.036	0.043	0.013	0.053
Femoral		Cox [8]		Milnor & Nichols [9]	
Characteristic impedance*		0.980		0.902	
Total arterial compliance*		0.03		0.035	

* Characteristic impedance in 10^9 Pa.s.m^{-3}, compliance in m^3 Pa^{-1}.

The input impedance of the coronary arterial tree exhibits similar characteristics to the input impedance of the femoral artery [8, 9] and is qualitatively similar to that of other vascular beds [4]. Gow and Hadfield [2] have shown that the dynamic elastic modulus values of the left circumflex artery and the femoral artery in the dog are similar. With diameters of the circumflex coronary artery and femoral artery also similar, the characteristic impedances (equal to the square root of blood density divided by the product of cross-sectional area and compliance [4]) are about the same as well.

The characteristics of the input impedance can be explained on the basis of vessel and blood properties. The zero-Hertz value is high and is the slope of the instantaneous pressure-flow relation. This slope is determined by the vasoactive state of the smooth muscle (vasomotor tone, [12]). For low and intermediate frequencies (i.e., the range from zero to about 10 Hz) the input impedance is mainly determined by the compliance properties of the arterial bed: the modulus decreases with frequency and the phase angle is negative. For high frequencies the input impedance approaches a constant level which is close to the characteristic impedance of the proximal circumflex artery. The value of characteristic impedance found by us is about twice as high as the value reported by Canty et al. [7]. When the pulse wave velocity in the circumflex artery is taken as 7.7 m/s [13] and the radius is taken as 1.5 mm, a characteristic impedance of $1.1 \cdot 10^9$ Pa.s.m^{-3} is obtained using cross-sectional area as the relation between characteristic impedance and phase velocity [6]. This value is close to our value of $0.9 \cdot 10^9$ Pa.s.m^{-3}. The input impedance of the coronary arterial system can therefore also be modeled with the three-element windkessel as shown in Fig. 5. The characteristic impedance is represented by the first resistor, total arterial compliance by a capacitor, and the instantaneous resistance represented by a second resistor.

Coronary input impedance is similar to impedance of other vascular beds [3, 4]. This means that the basics of the oscillatory pressure-flow relations are the same for different beds. The input impedance characteristics can be explained as follows. For low frequencies the wavelength of the travelling pressure and flow wave is long, i.e., long with respect to the length of the arterial tree. The pressure and flow and thus input impedance are prescribed by the "periphery"—the smallest vessels—which are for the coronary bed intramyocardial vessels. For intermediate frequencies the impedance is characterized by more proximal structures: the large vessels. These vessels mainly contribute to compliance characteristics. For high frequencies, when the wavelengths are short, the waves generated at the entrance of the system are returning with randomly distributed phases. The system approaches the characteristics of a tube (the left circumflex artery) that seems infinitely long, and the input impedance approaches the characteristic impedance of the circumflex artery. In other words the higher the frequency, the more proximal the vascular properties are that are described by the input impedance.

The input impedance of the coronary bed can therefore also be described by the three-element windkessel. Since the first two elements (characteristic impedance and arterial compliance) are mainly determined by the large vessels that run epicardially, cardiac contraction has little affect on these elements. The periphery is subject to cardiac contraction. We have recently shown [14, 15] that it is the properties of the heart muscle surrounding the vessels that affect (small) vessel diameter and (small) vessel compliance. In systole, resistance is larger and compliance smaller than in diastole, so that coronary flow decreases in systole. This effect of cardiac contraction can be presented with a (lumped) model of the periphery, so that the entire model of the coronary system is like the one given in Fig. 5. The effects of cardiac contraction are, except at very low frequencies, filtered by the arterial compliance so that differences between systole and diastole hardly play a role in the impedance, while the coronary model has to contain elements that account for the effects of cardiac contraction.

This reasoning leads to the conclusion that the input impedance does not give information about the far periphery. It is concluded that, in the far periphery which is not characterized by the input impedance, the differences between systole and diastole, i.e., the effects of cardiac contraction on the coronary vasculature, are to be found.

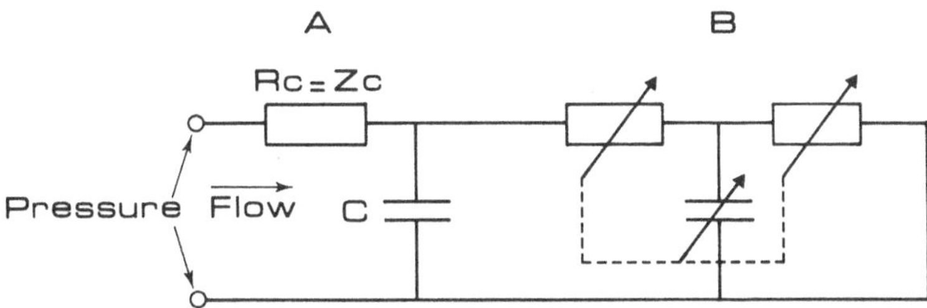

Fig 5. Proposed (lumped) model of the coronary circulation. *Part A* is mainly determined by epicardial vessels, *part B* represents the intramyocardial vessels. Characteristic impedance (Z_c or R_c) and (large) artery compliance (C) are constant. The effect of cardiac contraction is manifested in a changing vessel resistance and compliance. The changes in the periphery are filtered by arterial compliance and therefore not observed in the input impedance

References

1. Bellamy RF (1978) Diastolic coronary artery pressure-flow relations in the dog. Circ Res 43: 92–101
2. Gow BS, Hadfield CD (1979) The elasticity of canine and human coronary arteries with reference to postmortem changes. Circ Res 45: 588–594
3. Westerhof N (1968) Analog studies of human arterial hemodynamics. Ph D dissertation, University of Pennsylvania. University Microfilms, Ann Arbor, 69–5676
4. Westerhof N, Sipkema P, Elzinga G, Murgo JP, Giolma JP (1979) Arterial impedance. In: Hwang NHC, Gross DR, Patel DJ (eds) Quantitative cardiovascular studies. University Park Press, Baltimore
5. Sipkema P, Westerhof N, Randall OS (1980) The arterial system characterized in the time domain. Cardiovasc Res 14: 270–279
6. Van Huis GA, Sipkema P, Westerhof N (1987) Coronary input impedance during the cardiac cycle as determined by impulse response method. Am J Physiol 253: H317–H324
7. Canty JM, Klocke FJ, Mates RE (1985) Pressure and tone dependence of coronary diastolic input impedance and capacitance. Am J Physiol 248: H700–H711
8. Cox RH (1971) Determination of the true phase velocity of arterial pressure waves in vivo. Circ Res 29: 407–418
9. Milnor WR, Nichols WW (1975) A new method of measuring propagation coefficients and characteristic impedance in blood vessels. Circ Res 36: 631–639
10. Westerhof N, Elzinga G, Sipkema P (1971) An artificial arterial system for pumping hearts. J Appl Physiol 32: 776–781
11. Van Huis GA, Sipkema P, Westerhof N (1985) Instantaneous and steady-state pressure-flow relations of the coronary system in the canine beating heart. Cardiovasc Res 19: 121–131
12. Braakman R, Sipkema P, Westerhof N (1983) Steady state and instantaneous pressure-flow relationships: characterization of the canine abdominal periphery. Cardiovasc Res 10: 577–588
13. Arts T, Kruger RTI, Van Gerven W, Lambregts JAC, Reneman RS (1979) Propagation velocity and reflection of pressure waves in the canine coronary artery. Am J Physiol 237: H469–H474
14. Krams R, Sipkema P, Westerhof N (1989) Can coronary systolic-diastolic flow difference be predicted by left ventricular pressure or time varying intramyocardial elastance? Basic Res Cardiol
15. Krams R, Sipkema P, Westerhof N (1989) The varying elastance concept may explain coronary systolic flow impediment. Am J Physiol 257: H1471–H1479

Pressure-Flow Relationships of the Coronary Arteries

JULIEN I. E. HOFFMAN[1]

Summary. During a long diastole, coronary flow decreases as pressure decreases, and zero flow may be about 40–50 mmHg during autoregulation and about 10–20 mmHg when vessels are maximally dilated. These high zero-flow pressures have been used as evidence for diastolic intramyocardial waterfalls, but other explanations are possible. The high zero-flow pressures are partly due to the capacitance of the extramural coronary arteries. When capacitive effects are eliminated by steady-state, partial coronary occlusions, high zero-flow pressures may be due in part to collateral flow from other branches. Even in the absence of such causes of high zero-flow pressures, those pressures reflect only the last portion of muscle to be perfused, and zero-flow pressures are probably higher elsewhere in the myocardium. Alternative explanations for the pressure-flow behavior emphasize the large intramyocardial blood volume with its long time constants; according to this hypothesis, the zero-flow pressures in the extramural arteries represent the input pressures to the intramyocardial compartment. In the beating heart, it is not possible to reach equilibrium in all the vessels in one cycle, and not yet possible to determine if indeed there are intramyocardial waterfalls in diastole. In systole, recent studies have shown that blood is pumped retrogradely from the subendocardium to the subepicardium, which thus manifests forward systolic flow, even though little or no blood enters the myocardium in systole from the extramural coronary arteries. Therefore, in systole there is no evidence for typical waterfalls, but more evidence in favor of regional changes in intramyocardial compliance to explain coronary pressure-flow relationships.

Key words: Diastolic pressure-flow relations—Systolic pressure-flow relations—Vascular waterfall—Intramyocardial compliance—Zero-flow pressures

Introduction

The relationships of pressure and flow in the coronary arteries have for many years been used to understand the mechanisms involved in regulating blood flow to the myocardium. Initial measurements were hampered by inadequate instrumentation, and it was only when Gregg and his colleagues developed better methods that good data began to be collected. Pressures in the coronary arteries were found to be similar to those in the aorta, but flow patterns were unique. There was a predominance of diastolic coronary flow in the canine left coronary

[1] Department of Pediatrics and Cardiovascular Research Institute, 1403 HSE, University of California, San Francisco, CA 94143, USA

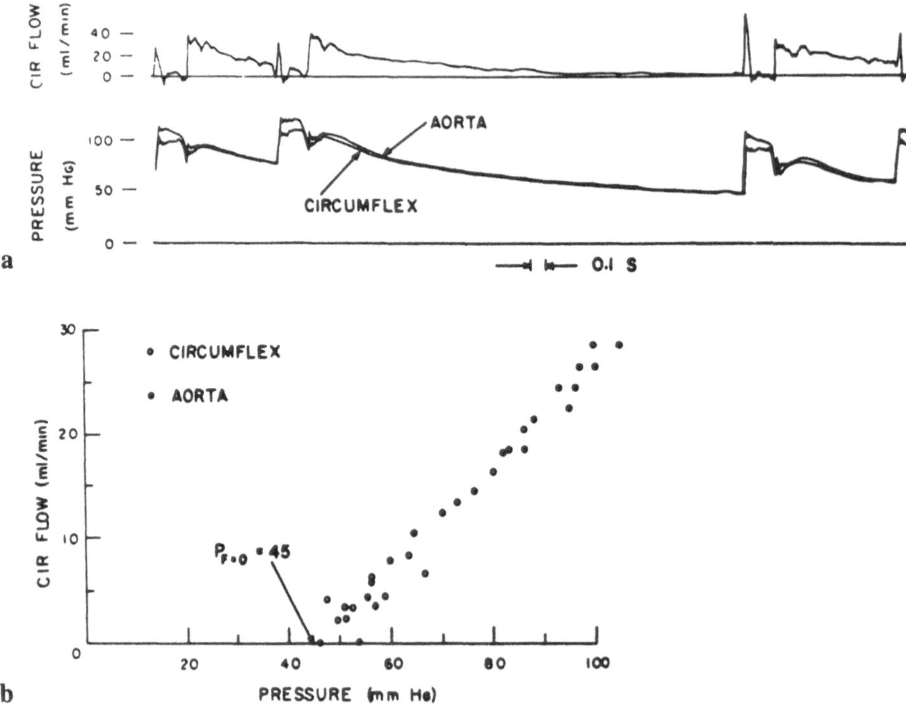

Fig. 1. a Left circumflex arterial flow (*CIR*) and aortic and circumflex pressures in a conscious instrumented dog. During a long diastole due to sinus arrhythmia, flow decreases to zero when coronary pressure is about 45 mmHg. **b** Diastolic pressures and flows measured every 0.1 s during the long diastole give a linear pressure-flow relationship with a zero-flow intercept ($P_{F=0}$) of 45 mmHg. From [1], with permission

artery and similarity of systolic and diastolic flows in the canine right coronary artery, so that flow and pressure were in phase in the right coronary artery but in antiphase in the left coronary artery. These pressure-flow patterns were interpreted as reflecting the different opposition to flow induced by the different intramyocardial pressures of the two ventricles, and indicated the dominant role of physical forces in regulating coronary arterial blood flow. Subsequently, the discovery that during diastole there was a relatively linear relationship between pressure and flow and that flow ceased at a pressure considerably above right atrial pressure (Fig. 1) lead to an extensive exploration of the interaction between wall forces and coronary flow.

It soon became clear that interpreting the results of measurements of pressure and flow in the extramural coronary arteries would be difficult because the coronary system was very complex. Firstly, the extramural coronary arteries have considerable capacitance, so that the input to the major coronary arteries was not identical to the input to the intramyocardial arteries. Secondly, opposition to forward flow in the myocardium could be due to strains as well as, or in place of, stresses in the wall. Thirdly, flow in the left ventricular wall cannot be regarded as a lumped system, and it is necessary to take account of different flow patterns and mechanisms in different layers of the left ventricular wall. Fourthly, differences between coronary venous and right atrial pressures had to be taken into account. Finally, the relatively large volume of blood in the myocardium has an important modulating effect on myocardial blood flow, and the fact that substantial systolic backflow can take place has a major affect on the interpretation of the pressure-flow relations in the major coronary arteries.

Diastolic Pressure-Volume Relations

The study of coronary pressure-flow relations is important for understanding the mechanical determinants of coronary blood flow. In the last decade, initiated by Bellamy [1], the emphasis has been on diastolic pressure-flow relations, in particular on their shape and their intercepts with the pressure axis. The intercept of the curves with the pressure axis at zero flow defines the zero-flow pressure (symbolized by P_{zf} or $P_{f=0}$) which may give information about static or dynamic back pressures to flow. These back pressures are due to vascular waterfalls, coronary vascular compliance, or to both mechanisms combined. The shape of the pressure-flow curves is important in defining coronary vascular resistance and determining the zero-flow pressure when this estimated by extrapolating the pressure-flow line to the pressure axis. Finally, a crucial issue is to what extent the shape of the pressure-flow curve and the value of zero-flow pressure depend on unsteady states, due to extramural and intramyocardial compliance on one hand and vasomotor changes on the other.

Zero-Flow Pressure and Tissue Pressures

The simplest coronary pressure-flow relationship is found when coronary vessels are maximally dilated and there is cardiac arrest or a prolonged diastole in which tissue pressures are low and unchanging. The pressure-flow relationship with flow plotted on the vertical axis is very steep when calculated per 100 g left ventricle, reflecting the low resistance to flow. The relationship is linear or almost so above perfusion pressures of about 40 mmHg, but at lower perfusion pressures may remain linear or may be concave towards the flow axis (Fig. 2). The pressure at which flow reaches zero is above right atrial pressure, and often about 18–25 mmHg in dogs and cats, but about 9–15 mmHg in pigs.

The high zero-flow pressure has been interpreted as a back pressure to flow due to a vascular waterfall, so that on this hypothesis it represents a high tissue pressure applied to the outside of a collapsible vessel in diastole. If this explanation is correct, however, it indicates a tissue pressure in diastole that is much higher than had been believed to occur. Furthermore, if diastolic tissue pressures were not equal across the left ventricular wall, then zero-flow pressure would indicate the last portion of muscle to be perfused and would therefore indicate the minimal diastolic tissue pressure; this would imply that diastolic tissue pressures in other layers would be even higher. That regional waterfall pressures in the ventricle may differ was indicated by Rouleau and White [2], who raised diastolic coronary sinus pressure to 18–28 mmHg at a constant coron-

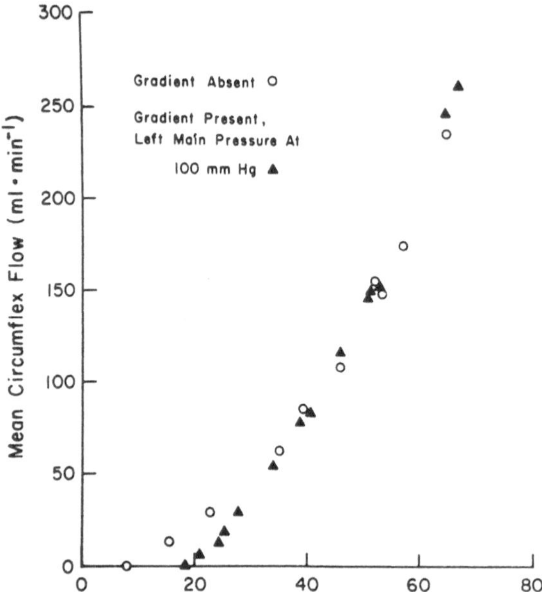

Fig. 2. Pressure and flow in the maximally dilated circumflex coronary artery of the dog when pressures are lowered simultaneously throughout the left coronary arterial system (*open circles*) and when pressure is held at 100 mmHg in the left anterior descending and septal arteries. Note that the zero-flow pressure is lower and there is more curvature at the low end of the pressure-flow relation when collateral flow is absent (*open circles*) than when it is present (*closed circles*). From [15], with permission

ary inflow pressure, and observed that flow decreased in the subepicardium but not in the subendocardium; they argued that the higher venous pressure exceeded tissue pressure only in the subepicardium. Direct confirmation that the zero-flow pressure reflects the lowest intramyocardial pressure has recently been provided. Satoh et al. [3] compared zero-flow pressures in an isolated arrested heart with intramyocardial pressures measured in the deep and superficial muscle by small solid-state transducers. As left ventricular pressure was increased from zero to 30 mmHg, the zero-flow pressure increased and was similar to the superficial intramyocardial pressures, which were much lower than the deep intramyocardial pressures. It is true that several groups of investigators have described tissue pressures during diastole or cardiac arrest that are higher in the subepicardium than the subendocardium, and may be considerably above left ventricular diastolic pressure in all layers. However, the most recent and probably most accurate direct studies of tissue pressures indicate that in diastole these pressures are near atmospheric beneath the epicardium, and about equal to left ventricular cavity pressure beneath the endocardium [4]; the higher pressures described by others might be artifacts of the measuring techniques, and failure to separate surface pressures from liquid pressures. The contradiction between low diastolic tissue pressures and high zero-flow pressures now seems to be explained by a number of factors that lead to a high value for zero-flow pressure.

Factors Affecting Zero-Flow Pressure

Linear extrapolation. Although the pressure-flow relation is often concave upwards towards the flow axis, many investigators have obtained zero-flow pressure by linearly extrapolating the pressure-flow line to the pressure axis, so that there will be overestimation of the zero-flow pressure if there is any curvature of the relationship. This problem is made worse if the lowest pressures and flows examined are not very low, for then the curvature that is often seen at very low flows and pressures will be missed (Fig. 2).

Coronary vascular capacitance. The capacitance of the extramural coronary arteries can affect zero-flow pressures. During a prolonged diastole both pressure and flow decrease in the coronary arteries. Because blood is stored in the extramural arteries, blood is discharged into the myocardium in diastole because of change in volume of the extramural arteries. As a result, a flowmeter at the origin of the left coronary artery or its large branches will record no flow when there is still flow continuing downstream, and thus the true zero-flow pressure at the *micro-vascular* level will be overestimated. There is agreement that capacitive effects have to be considered in these measurements, but no agreement about the magnitude of these effects. Some experiments to eliminate input capacitance effects showed that the zero-flow pressure could be lowered almost to the level of right atrial pressure [5, 6] (Fig. 3), and the investigators therefore believed that the high zero-flow pressures reported in other experiments were due to the effects of capacitance. Kajiya et al. [7] also observed large differences between instantaneous and steady-state zero-flow pressures. On the other hand, Klocke and his colleagues [8] pointed out that at the end of a long diastole the rate of change of pressure is very slow and thus incapable of causing large capacitive effects. In fact, a recent study of theirs [9] did indicate relatively high zero-flow pressures in some steady-state pressure-flow curves in which extremely low flows were attained.

The issue about the presence or absence of a steady state at low perfusion pressures is crucial in making inferences from experiments about coronary pressure-flow relations. Chilian and Marcus [10] showed that coronary venous outflow persists after cessation of coronary arterial flow at high zero-flow pressure (Fig. 4), and concluded that zero-flow pressure was due to a substantial intramyocardial compliance. This basically was also the argument of Spaan [11]. However, Klocke and co-workers [8] argued that a compliance distal to the vascular waterfall need not be involved in inflow regulation and that persistence of venous outflow after inflow has ceased does not exclude a vascular waterfall at the precapillary level; hence, zero-flow pressure would still be the back pressure to arterial inflow. There are two objections to this argument. Firstly, with a large intramyocardial compliance, a high zero-flow pressure can be explained without the need for an arteriolar waterfall. Secondly, the basic assumption for a vascular waterfall to operate is that the occluding pressure should be higher than the distal pressure. If the distal pressure is right atrial pressure, there is no dispute. However, if the intramyocardial compliance is distal to the vascular waterfall, the intravascular pressure of this compliance should be defined, in order to jus-

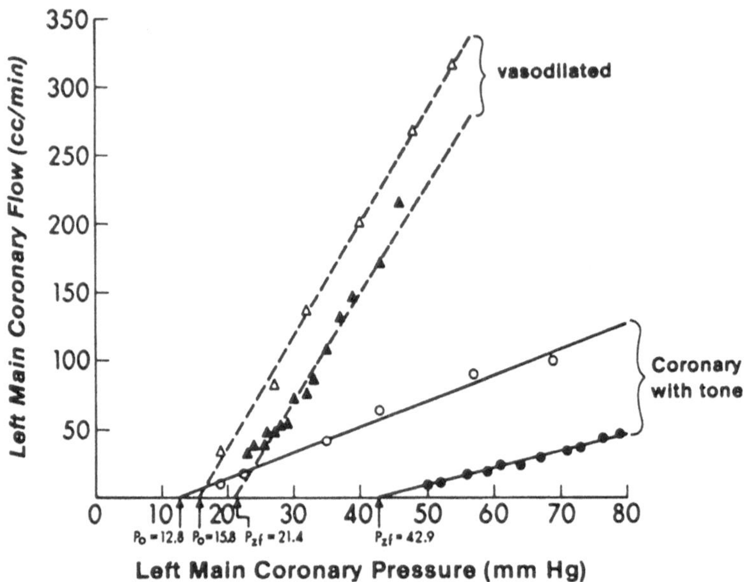

Fig. 3. Static and dynamic pressure-flow relations in the canine left main coronary artery with and without tone. With vasodilatation the slope increases, and with static perfusion the zero-flow pressure is much lower than dynamic perfusion obtained by allowing aortic pressure to decay. △ Vasodilated, reservoir perfusion ($y = 8.25\ (x - 15.8)$; $r = 0.999$); ▲ vasodilated, declining aortic pressure perfusion ($y = 7.63\ (x - 21.4)$; $r = 0.980$); ○ with tone, reservoir perfusion ($y = 1.90\ (x - 12.8)$; $r = 0.991$); ● with tone, declining aortic pressure perfusion ($y = 1.24\ (x - 42.9)$; $r = 0.993$); P_{zf}, P_o, zero-flow pressure. From [5], with permission

tify the conclusion that zero-flow pressure is higher than the distal pressure. Because the value of this intravascular pressure is uncertain, it is difficult to assert that a vascular waterfall mechanism is responsible for the zero-flow pressure. Klocke et al. have also pointed out that a waterfall mechanism explains the uncoupling of the rapid arterial and slow coronary venous responses to changes in arterial pressure. However, because it is possible to explain the cited experimental findings by the nonlinear nature of the coronary vascular bed, it is clear that additional experiments will be needed to distinguish among the various potential mechanisms.

If there is a vascular waterfall in intramyocardial vessels, it is appropriate to ask which microvessels collapse. Direct observation of the subepicardial vessels has not yet shown a point of collapse [12], though negative findings cannot be as strong as positive ones. From vascular properties, Chadwick et al. [13] concluded that the venules have the smallest buckling pressure and thus are likely to be the sites of any collapse, whereas Oddou and his colleagues [14], using different models, concluded that the greatest reduction in luminal cross-sectional area occurred in the arterioles; they did not comment on complete collapse. Obviously, if it is the venules that collapse, then the problem of the time constant of the intramyocardial blood volume becomes crucial in evaluating zero-

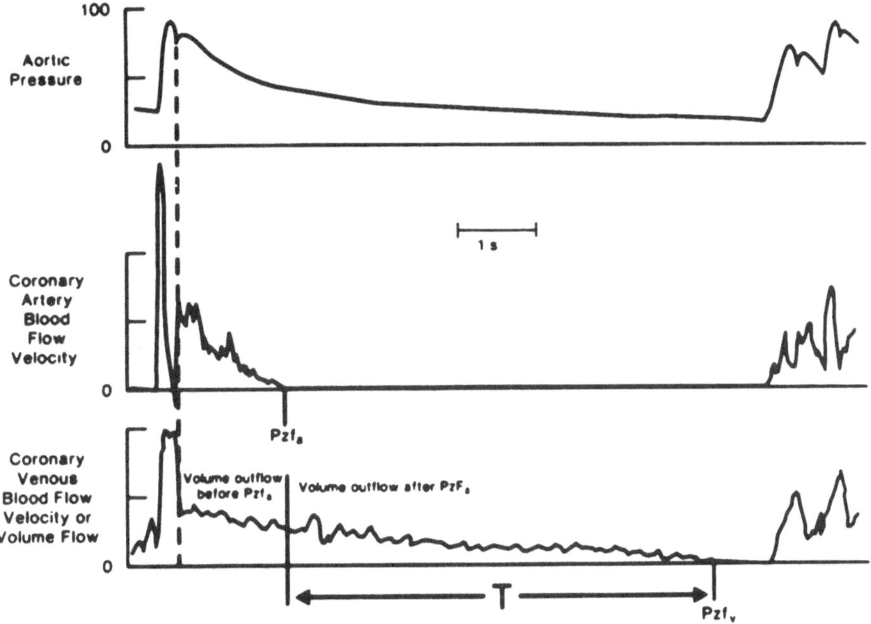

Fig. 4. Aortic pressure and flow velocities in coronary artery and vein during a long diastole. After coronary flow velocity reaches zero, there still is substantial venous outflow for several seconds. Pzf_a, arterial pressure at zero flow; Pzf_v, venous pressure at zero flow. From [10], with permission

flow pressures, whereas this time constant may not be nearly as important if the vascular waterfall is in the arterioles. Spaan has pointed out [11] that if the observed zero-flow pressures in the left coronary artery are indeed due to waterfalls, then the relatively high zero-flow pressure in the autoregulating heart implies that the waterfall is probably arteriolar, whereas the relatively low zero-flow pressure in the heart with maximally dilated vessels implies that the waterfall is probably venous. Further experiments will be needed to determine whether zero-flow pressures are determined by arteriolar, capillary, or venous waterfalls, or are due entirely to the time constants in the large intramyocardial blood pool.

To avoid capacitive effects, many investigators turned to steady state measurements. A hydraulic occluder was placed on the left circumflex or left anterior descending coronary artery and inflated to different degrees, thus producing different decreases in distal pressures. Flows were measured when pressures were stable, thereby eliminating capacitive effects in the extramural arteries. These studies usually also showed high zero-flow pressures. One explanation for this finding was proposed by Messina et al. [15] who attributed it to the effect of collateral flow from the other arteries. When vessels are maximally vasodilated, flow through them is a function of pressure. If, at a very low intravascular pressure, some flow enters, for example, the distal left circumflex coronary artery via collaterals, then the inflow into the proximal circumflex coronary artery will be reduced by the amount of collateral flow, so that total flow through the circumflex capillary bed is what it would have been in the absence of collateral flow. Thus flow measured by a flowmeter at the origin of the circumflex coronary artery will reach zero at a higher pressure when there is collateral flow than when there is none (Fig. 2). Furthermore, there is more curvature at the lower end of the curve when collateral flow is absent than when it is present (Fig. 2). In these experiments, zero-flow pressure without collateral flow was low and almost the same as coronary sinus pressure. In support of this explanation is the finding in swine, which normally have almost no collateral flow, that zero-flow pressures were similar to right atrial pressures [16].

Even when the effects of capacitance and of collateral flow are avoided, virtually all studies have demonstrated a small (2–4 mmHg) pressure difference between zero-flow pressure and coronary sinus pressure. Whether this small difference is due to pressure gradients in the small veins, to surface tension effects, or to a vascular waterfall is unknown.

The relationship of coronary venous pressures to steady-state zero-flow pressures. Certain perturbations can raise the zero-flow pressures by affecting coronary venous pressures, even if right atrial pressures remain unchanged. Satoh et al. [17] showed in the arrested isolated canine heart with maximally dilated coronary vessels that zero-flow pressures were about 6 mmHg when pressures in both ventricles and around the heart were atmospheric. When pressure around the heart was raised to 22 and then 30 mmHg, the zero-flow pressure rose by equal amounts. When right ventricular pressure was raised to 22 and 30 mmHg, there was initially little change in zero-flow pressure, which then rose by an amount equal to the increment in right ventricular pressure. When left ventricular pressure was raised to 22 and then 30 mmHg, the zero-flow pressure rose about two-thirds of those amounts (Fig. 5). The effect of raising surrounding heart pressure was interpreted as being due to a direct effect on coronary veins which became classical vascular waterfalls. The effect of raising right ventricular pressure (little change in zero-flow pressure initially and then a parallel rise in the two pressures) was interpreted as being due to a waterfall effect which had previously been described in the coronary veins [18–20]; raising downstream (right ventricular) pressure had no effect on coronary venous pressure until some critical value was exceeded. The effect of raising left ventricular pressure was harder to explain, although their results were compatible with those obtained by Uhlig et al. [19] and Aldea et al. [21]. Both of these studies showed that a vascular waterfall existed in the coronary sinus and great cardiac vien, and also observed that when left ventricular diastolic pressure was elevated, the coronary sinus pressure rose by about two-thirds of the pressure rise in the ventricle. The probable explanation is that when the left ventricle is dilated acutely by high diastolic pressures, the epicardial veins may be squeezed between the connective tissue of the epicardium and the subjacent muscle, and transformed into vascular waterfalls with an external pressure that is less than the ventricular pressure because of pressure dissipation across the wall.

When coronary vessels are maximally dilated, zero-flow pressures are shifted

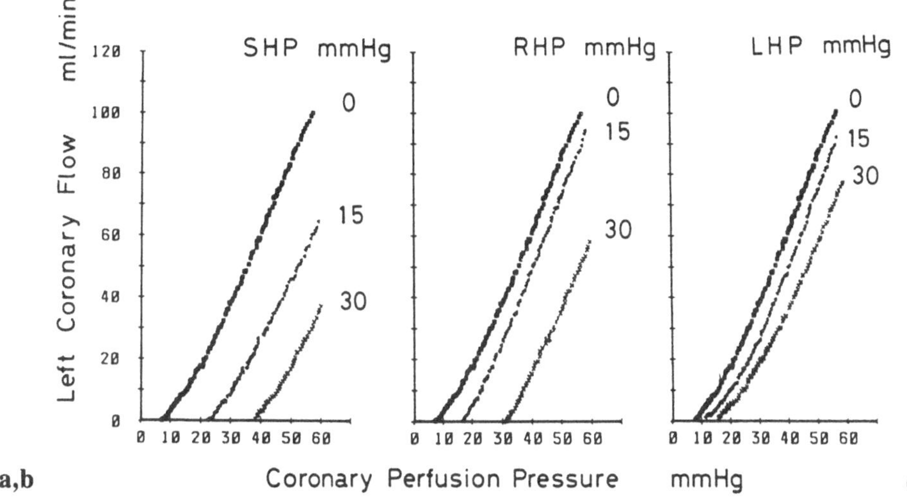

Fig. 5a–c. Diastolic pressure-flow relations in an isolated canine heart during cardiac arrest and maximal coronary vasodilatation. In each panel the control pressure-flow relation is denoted by *0*. Subsequent pressure-flow relations are obtained with pressures of 15 and 30 mmHg **a** around the heart (*SHP*); **b** in the right ventricle (*RHP*); and **c** in the left ventricle (*LHP*). Note the differences in the effects of the elevated pressures on the zero-flow pressures (see text). From [17], with permission

to the right if the coronary sinus pressures are raised [2, 19, 20, 22], and usually there is no accompanying change in the slope of the pressure-flow curve. On the face of it, these observations suggest that no change has occurred in coronary vascular resistance, but this is unlikely to be correct, because raising coronary venous pressures has been shown to decrease steady-state coronary vascular resistance, probably by distending the vessels [23]. The contradiction might be explained by the fact that in at least three of these studies the coronary sinus pressures were raised for only 5–15 s, a time that might have been inadequate to produce a steady-state change in the intramyocardial capacitance. Consequently, the unchanged pressure-flow relations at the input might not have reflected what was occurring at the microcirculatory level. The results of the steady-state study [23] also imply that any effect that distending the coronary venous system might have on increasing intramyocardial tissue pressures (which would increase resistance) was outweighed by the effect of distension in lowering resistance.

The extramural venous waterfall is not merely passive, because coronary veins are affected by vasoconstrictors and vasodilators [24]. The importance of this has been demonstrated experimentally [25]: in steady-state measurements made in an arrested isolated heart, the great cardiac vein had a waterfall pressure of 23 mmHg in the control state, from 24–36 mmHg after vasopressin infusion, and as low as 5–11 mmHg after an adenosine infusion. Thus, investigators should determine these venous pressures whenever interpreting coronary pressure-flow relations.

Diastolic Pressure-Flow Relations and Zero-Flow Pressures During Autoregulation

During autoregulation the diastolic pressure-flow relationship may or may not be linear, and zero-flow pressure can be very high when there is vasomotor tone. This high zero-flow pressure was interpreted originally as a vascular waterfall at arteriolar level [1], and part of the pressure required to open the arteriolar lumen would be related to overcoming smooth muscle tone. The consequence of this line of thought is that arterioles must be contracted to such an extent that the lumen would offer infinite resistance, even if only at one very localized point in the vessel.

An alternative hypothesis for the high zero-flow pressures during autoregulation is that put forward by Spaan and his colleagues [11, 26, 27], based on the intramyocardial capacitance. They pointed out that there is a large intramyocardial blood volume, and that the coronary vascular bed can be regarded as a series of compliant systems. The most proximal system consists of the extramural arteries, beyond which are intramyocardial vessels with arteriolar, capillary, and venular compartments. The intramyocardial system has a much larger compliance and a lower frequency response than has the extramural system. Therefore, even when steps are taken to eliminate the effects of compliance of the extramural arteries, the long time constant of the intramyocardial compartment leads to an overestimate of the zero-flow pressure. According to this view, the zero-flow pressure represents the pressure at the entrance to the intramyocardial compartment immediately after inflow has ceased in the extramural arteries. This pressure depends on many factors, including the time after cessation of the inflow that the readings are made, and the tone of the vascular bed. The higher the vascular tone, the higher the distal resistance of the arteriolar compartment, the longer the time constant for emptying all the intramyocardial compartments, and the higher the apparent zero-flow pressure.

It is thus apparent that many events once interpreted as being due to intravascular waterfalls can be equally well explained on the basis of intramyocardial compliance, and vice versa. Thus, Hoki et al. (personal communication) noted that sudden changes in pressure in the great cardiac vein did not alter the quasi-linear decay of pressure in the autoregulating left anterior descending coronary artery when the ventricles were arrested by vagal stimulation, and that the pressure-flow line reached zero pressure at about 40 mmHg. This finding is consistent with a vascular waterfall in the system, but can just as well be explained by a large intramyocardial capacitance and coronary resistance. Distinguishing between the two mechanisms requires further experiments.

An extra problem related to steady state in the autoregulated bed is the change in vasomotor tone that can occur during the determination of the diastolic pressure-flow line [5, 28]; linearity of the diastolic pressure-flow relation is not proof of a constant resistance. As an example, consider the linear pressure-flow relations that can be found during reactive hyperemia [1]. Assuming that these lines indicate a constant resistance implies that coronary vascular resistance changes only during the intervening systoles [29].

Shape of the Diastolic Pressure-Flow Relations

When the vessels are fully dilated, two other factors influence pressure-flow relations. One is that when the intravascular pressures decrease, transmural pressure across the vessel wall also decreases, and causes the elastic vessel to decrease its diameter. Consequently vascular resistance increases (Fig. 6), thereby accounting for a large part of the curvature of the pressure-flow relation observed at low perfusing pressures [23, 27]. The second determinant of pressure-flow relations is blood viscosity which in turn is largely a function of the hematocrit. An increase in hematocrit has been shown to increase coronary vascular resistance [30, 31]. At the shear rates prevailing in the microcirculation, an increase in hematocrit from 45% to 70% approximately doubles the effective blood viscosity and the vascular resistance, and therefore halves the flow at any given driving pressure (assuming that the distribution of resistance does not alter). The change in hematocrit has no [30] or little [7] effect on zero-flow pressures. In fact, changing from perfusion by blood to perfusion by Tyrode solution in an arrested, isolated cat heart with maximally dilated vessels does not lower zero-flow pressure, thus suggesting that the rheologic effects of red cells at low flow rates do not influence this measurement [32].

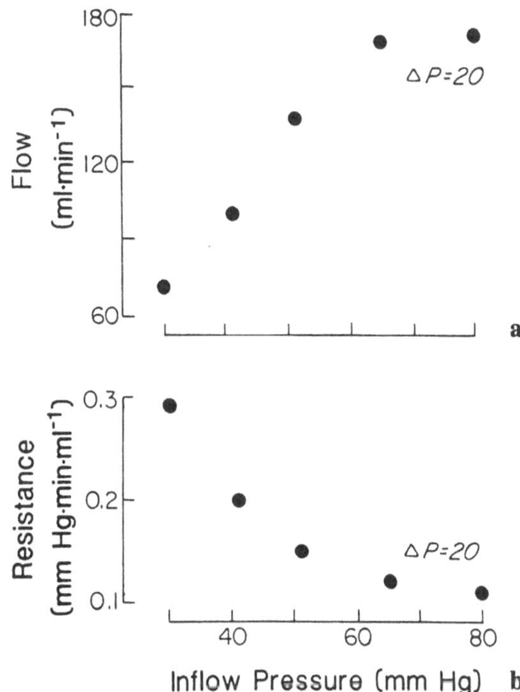

Fig. 6. Flow (**a**) and resistance (**b**) in the canine left main coronary artery during diastolic arrest and maximal vasodilatation. The pressure difference between inflow and venous outflow (ΔP) is kept constant at 20 mmHg as inflow pressure is changed. A decreased inflow pressure at a constant driving pressure is associated with an increased resistance. From [23], with permission

Differences in regional pressure-flow relations have also been invoked to explain the curvature at the lower end of the coronary pressure-flow relationship. Investigations of regional zero-flow pressures have indeed shown differences in different layers of the left ventricle, but only of the order of 2–4 mmHg at the most [33]. Furthermore, Hanley et al. showed that as driving pressures across the ventricular wall were lowered, flows decreased in all layers but did not go to zero in any layer, even at the lowest driving pressures [23]. Recent studies in our laboratory have shown that in the isolated heart with zero pressures in the ventricles and around the heart, zero-flow pressures do not differ substantially in layers across the left ventricular wall [21], although in any layer some individual small pieces of myocardium attained zero flow before others did. Therefore some of the curvature of the coronary pressure-flow relation in the normal heart may be due to cessation of flow at different pressures in different pieces of myocardium. The effect of cardiac contraction cannot be a major determinant of curvature because curvature of the pressure-flow relation is seen even when the heart is not beating [15]. The main cause of the curvature, however, is likely to be the increase in resistance that accompanies the decrease in vessel radius as transmural pressure falls.

Systolic Pressure-Flow Relations

The waterfall hypothesis has also been invoked to explain the decrease in coronary blood flow in systole. This was first done by Downey and Kirk [34] who postulated a gradient of systolic tissue pressure across the left ventricular wall such that at lower perfusing pressures, layers would drop out in sequence: first the layer with the highest tissue pressure, then the next one, and so on. When all the regional relationships were pooled, curvature concave towards the flow axis was observed at the lower end of the pressure-flow relationship. During cardiac arrest, resembling diastole, the pressure-flow relation that they observed was linear. Their model included a diode which would have prevented retrograde flow, and so cannot explain the many instances of systolic and diastolic [7] retrograde flow that have been observed.

There is another difficulty with a simple model that has a distribution of waterfall pressures across the wall in systole, because it cannot explain why there is little systolic perfusion from the extramural coronary arteries. If the waterfall hypothesis were true, then because tissue pressures in the subepicardium are substantially below the coronary arterial pressure, we would expect a large proportion of systolic flow from the extramural coronary arteries. In fact, forward systolic flow in the distal extramural coronary arteries is negligible or absent [35], even though there is continuous forward perfusion of the subepicardial vessels in systole [12, 36]. Recently, we have shown that about 30%–50% of subepicardial blood flow is derived from retrograde systolic flow coming from blood squeezed out of the subendocardial vessels [37], a finding incompatible with any of the current vascular waterfall hypotheses.

When pressure-flow relations have been examined in systole by taking instantaneous flows and pressures at the end of systole [2, 38], the line is shifted towards higher pressures than occur for the diastolic pressure-flow relation, as one might expect. In dogs with congenital infundibular pulmonic stenosis [38], the degree of shift of the right coronary arterial pressure-flow relation was greater for the dogs with higher right ventricular systolic pressures. The opposition to flow has usually been regarded as being related to intramyocardial pressure, which in turn is a function of cavity pressure in systole. There are, however, arguments against this hypothesis. Baird et al. [39] observed that intramyocardial pressures in the beating empty heart were similar to those occurring when the heart was ejecting. Krams et al. [40] found that the decrease in systolic flow was virtually unaffected by a change from isovolumic contraction which generated high pressures to an isobaric contraction which generated large volume changes. Based on these findings, they concluded that the coronary vessels had a variable elastance, similar to that postulated for the ventricular cavity. For a given value of elastance (defined as pressure divided by the difference between actual and rest volumes) there would be a given impediment to coronary systolic flow that would be independent of the left ventricular pressure or ejection fraction.

Choices Between Vascular Waterfall and Intramyocardial Compliance Hypotheses

Conclusive proof for the presence or absence of intramyocardial waterfalls is not yet available. It is possible that diastolic intramyocardial vascular waterfalls may be absent under normal physiologic circumstances yet present when, for example, left ventricular diastolic pressure is raised. It is also possible that even though zero-flow pressures are similar in all layers of the normal left ventricle, at higher pressures and flows there could be differences in back pressures in different layers. In other words, dynamic and static back pressures may be quite different. If indeed back pressures can vary throughout the cardiac cycle, then an added complication exists. In a study of pressure-flow relations in the lung, Ducas et al. [41] observed that the slope of the pressure-flow line was steeper if left atrial pressure was deliberately altered while pressure-flow relations were being determined, than if the measurements were made at constant left atrial pressure; as well, the intercept on the pressure axis was much lower than the actual back pressure. Most conclusions drawn from experiments on the coronary circulation implicitly assume a fixed back pressure, and will need reinterpretation if this is ever shown to be incorrect. Finally, the dense myocardial structure could lead to some complex interactions. For example, an increase in venous pressure might raise intramyocardial tissue pressures, perhaps unequally across the ventricular wall, and so induce vascular waterfalls that were absent at normal venous pressures [5, 11, 28].

It would be wrong to regard the vascular waterfall and intramyocardial compliance mechanisms as completely different, especially for static measurements; they are both, in fact, functions of the nonlinearity of vessel pressure-volume relations. Westerhof and his colleagues [42] analyzed the relationship between transmural pressure and cross-sectional area, and, with simplifying assumptions, transformed this into the relationship between pressure drop across the vessel and flow. As transmural pressure increased from zero, there was initially little change in cross-sectional area (or volume per unit vessel length), then area increased more rapidly, and at higher transmural pressures area remained almost constant because the vessel became stiff. The shapes of these pressure-area relations varied, depending on whether the vessel was dilated or constricted. When these findings were transformed into pressure-flow relations, a dilated vessel had an almost linear pressure-flow relation with a negligible intercept on the pressure axis, whereas a constricted vessel with tone showed a steep increase in pressure drop across the vessel at low flows, and an apparent intercept on the pressure axis. If the slope of the central portion of the transmural pressure-area curve was almost parallel to the area axis, then the increase in pressure drop was steeper at low flows, and the apparent intercept approached a real intercept on the pressure axis. A truly horizontal central portion in which transmural pressure was independent of cross-sectional area (that is, a vascular waterfall) had a positive intercept on the pressure axis.

Because the circulation through an organ consists of vessels with different pressure-volume relations and exposed to different transmural pressures, the pressure-flow relations become complex [43]. A very stiff vessel has little cross-sectional area or volume change as transmural pressure changes; it therefore has almost zero compliance, and flow through it is linearly related to perfusion pressure, with a zero intercept on the pressure axis. The extramural coronary arteries fit this pattern to some extent. A Starling resistor on the other hand acts as a pressure regulator because of its large variations of resistance, and perfusion pressure remains constant as flow increases, so that the device has infinite compliance. Some portions of the coronary microcirculation, perhaps arterioles or venules, may follow this behavior. If a stiff vessel and a Starling resistor are in series, then in this model the pressure-flow relation is a straight sloping line with an apparent positive intercept on the pressure axis; this intercept is set by the pressure in the compartment around the Starling resistor.

It is therefore possible to use basic concepts about nonlinear pressure-volume relations to model the coronary circulation, and it is clear that vascular waterfalls alone or in combination with nonlinear compliances of microvessels are capable of explaining the experimental results that have been described in the literature. It is equally clear that nonlinear compliances of the coronary microcirculation might explain the experimental findings without the need to invoke vascular waterfalls, providing the values of compliance, resistance, and volume are appropriate [27]. Whether indeed there are vascular waterfalls in the myocardium remains for future experiments to determine.

Acknowledgment. This work was supported in part by Program Project Grant HL-25847 from the National Institutes of Health. Portions of this Chapter appear in the review by Hoffman and Spaan (1990) Pressure-flow relations in the coronary circulation. Physiol Rev. 70: 331–389, by permission.

References

1. Bellamy RF (1978) Diastolic coronary pressure-flow relations in the dog. Circ Res 43: 92–101
2. Rouleau JR, White M (1985) Effects of coronary sinus pressure elevation on coronary blood flow distribution in dogs with normal preload. Can J Physiol Pharmacol 63: 787–797
3. Satoh S, Maruyama Y, Watanabe J, Keitoku M, Hangai K, Takishima T (1987) Comparison of zero-flow pressure (Pf = 0) in diastolic coronary pressure-flow relationship with diastolic pressures in outer (IMPo) and inner (IMPi) myocardial layers. Circulation 76 (Suppl IV): IV–489
4. Heineman FW, Grayson J (1985) Transmural distribution of intramyocardial pressure measured by micropipette technique. Am J Physiol 249 (Heart Circ Physiol 18): H1216–H1223
5. Eng C, Jentzer JH, Kirk ES (1982) The effects of the coronary capacitance on the interpretation of diastolic pressure-flow relationships. Circ Res 50: 334–341
6. Kirkeeide R, Puschmann S, Schaper W (1981) Diastolic coronary pressure-flow relationships investigated by induced long-wave pressure oscillations. Basic Res Cardiol 76: 564–569
7. Kajiya F, Tsujioka K, Ogasawara Y, Wada Y, Hiramatsu O, Goto M, Nakai M, Tadaoka S, Matsuoka S, Sha Y (1988) Effect of packed cell volume on diastolic coronary artery pressure-flow relations in the dog. Cardiovasc Res 22: 545–554
8. Klocke FJ, Mates RE, Canty JM Jr, Ellis AK (1985) Coronary pressure-flow relationships. Controversial issues and probable implications. Circ Res 56: 309–323
9. Canty JM Jr, Klocke FJ, Mates RE (1987) Characterization of capacitance-free pressure-flow relations during single diastoles in dogs using an RC model with pressure-dependent parameters. Circ Res 60: 273–282
10. Chilian WM, Marcus ML (1984) Coronary venous outflow persists after cessation of coronary arterial inflow. Am J Physiol 247 (Heart Circ Physiol 16): H984–H990
11. Spaan JAE (1985) Coronary diastolic pressure-flow relations and zero flow pressure explained on the basis of intramyocardial compliance. Circ Res 56: 293–309
12. Ashikawa K, Kanatsuka H, Suzuki T, Takishima T (1986) Phasic blood flow velocity pattern in epimyocardial microvessels in the beating canine left ventricle. Circ Res 59: 704–711
13. Chadwick RS, Tedgui A, Michel JB, Ohayon J, Levy B (1988) A theoretical model for myocardial blood flow. In: Brun P, Chadwick RS, Levy BI (eds) Cardiovascular dynamics and models. Les Editions INSERM, Paris, pp 77–90
14. Oddou C, Samaké G, Pelle G (1987) Coronary microvessels lumen deformation. Automedica 9: 54
15. Messina LM, Hanley FL, Uhlig PN, Baer RW, Grattan MT, Hoffman JIE (1985) Effects of pressure gradients between branches of the left coronary artery on the pressure axis intercept and the shape of steady state circumflex pressure-flow relations in dogs. Circ Res 56: 11–19
16. Pantely GA, Ladley HD, Bristow JD (1984) Low zero-flow pressure and minimal capacitance effect on diastolic coronary artery pressure-flow relations during maximal vasodilation in swine. Circulation 70: 485–494
17. Satoh S, Watanabe J, Keitoku M, Itoh N, Maruyama Y, Takishima T (1988) Influences of pressure surrounding the heart and intracardiac pressure on the diastolic coronary pressure-flow relation in the excised canine heart. Circ Res 63: 788–797
18. Scharf SM, Bromberger-Barnea B, Permutt S (1971) Distribution of coronary venous flow. J Appl Physiol 30: 657–662
19. Uhlig PN, Baer RW, Vlahakes GJ, Hanley FL, Messina LM, Hoffman JIE (1984) Arterial and venous coronary pressure-flow relations in anesthetized dogs. Evidence for a vascular waterfall in epicardial coronary veins. Circ Res 55: 238–248

20. Pantely GA, Bristow JD, Ladley HD, Anselone CG (1988) Effect of coronary sinus occlusion on flow, resistance, and zero flow pressure during maximum vasodilatation in swine. Cardiovasc Res 22: 79–86
21. Aldea G, Hoffman JIE, Husseini W, Mori H (1987) Determinants of regional myocardial blood flow. In: Brun P, Chadwick RS, Levy BI (eds) Cardiovascular dynamics and models; Proceedings of NIH-INSERM Workshops. Les Editions INSERM, Paris, pp 44–47
22. Bellamy RF, Lowensohn HS, Ehrlich W, Baer RW (1980) Effect of coronary sinus occlusion on coronary pressure-flow relations. Am J Physiol 239 (Heart Circ Physiol 8): H57–H64
23. Hanley FL, Messina LM, Grattan MT, Hoffman JIE (1984) The effect of coronary inflow pressure on coronary vascular resistance in the isolated dog heart. Circ Res 54: 760–772
24. Klassen GA, Armour JA (1983) Canine coronary venous pressures: responses to positive inotropism and vasodilation. Can J Physiol Pharmacol 61: 213–221
25. Klocke FJ, Mates RE, Canty JM Jr, Sekovski B, Gunawardane C, Baello EB, Hajduczok ZD (1986) Tone-dependent vascular waterfall behavior during forward coronary flow. Circulation 74 (Suppl II): 87
26. Spaan, JAE, Breuls NPW, Laird, JD (1981) Diastolic-systolic flow differences are caused by intramyocardial pump action in the anesthetized dog. Circ Res 49: 584–593
27. Bruinsma P, Arts T, Dankelman J, Spaan JAE (1988) Model of the coronary circulation based on pressure dependence of coronary resistance and compliance. Basic Res Cardiol 83: 510–524
28. Klocke FJ, Weinstein IR, Klocke JF, Ellis AK, Kraus DR, Mates RE, Canty JM, Anbar RD, Romanowski RR, Wallmeyer KW, Eclit MP (1981) Zero-flow pressures and pressure-flow relationships during single long diastoles in the canine coronary bed before and during maximal vasodilation: Limited influence of capacitive effects. J Clin Invest 68: 970–980
29. Spaan JAE (1979) Does coronary resistance change only during systole: Circ Res 45: 838–839
30. Jan KM, Chien S (1977) Effect of hematocrit variations on coronary hemodynamics and oxygen utilization. Am J Physiol 233 (Heart Circ Physiol 2): H106–H113
31. Baer RW, Vlahakes GJ, Uhlig PN, Hoffman JIE (1987) Maximal myocardial oxygen transport during anemia and polycythemia in dogs. Am J Physiol 252: H1086–H1095
32. Van Dijk LC, Krams R, Sipkema P, Westerhof N (1988) Changes in coronary pressure-flow relation after transition from blood to Tyrode perfusion. Am J Physiol 255 (Heart Circ Physiol 24): H476–H482
33. Ellis A, Klocke FJ (1980) Effects of preload on the transmural distribution of perfusion and pressure-flow relationships in the canine coronary vascular bed. Circ Res 46: 68–77
34. Downey JM, Kirk ES (1975) Inhibition of coronary blood flow by a vascular waterfall mechanism. Circ Res 36: 753–760
35. Chilian WM, Marcus ML (1982) Phasic coronary flow velocity in intramural and epicardial coronary arteries. Cire Res 50: 775–781
36. Tillmanns H, Ikeda S, Hansen H, Sarma JSM, Fauvel J, Bing RJ (1974) Microcirculation in the ventricle of the dog and turtle. Circ Res 34: 561–569
37. Flynn AE, Coggins DL, Aldea GS, Austin RE, Goto M, Husseini W, Hoffman JIE (1989) Ventricular contraction increases subepicardial blood flow: evidence for a deep myocardial pump. FASEB J 3: A1305
38. Bellamy RF, Lowensohn HS (1980) Effect of systole on coronary pressure-flow relations in the right ventricle of the dog. Am J Physiol 238 (Heart Circ Physiol 7): H481–H486
39. Baird RJ, Goldbach MM, de la Rocha A (1972) Intramyocardial pressure: the persistence of its transmural gradient in the empty heart and its relationship to myocardial oxygen consumption. J Thorac Cardiovasc Surg 64: 635–646

40. Krams R, Sipkema P, Westerhof N (1989) Varying elastance concept may explain coronary systolic flow impediment. Am J Physiol 257: H1471–H1479
41. Ducas J, Schick U, Girling L, Prewitt RM (1988) Effects of left atrial pressure on pulmonary vascular pressure-flow relationships. Am J Physiol 255 (Heart Circ Physiol 24): H19–H25
42. Sipkema P, Westerhof N (1989) Mechanics of a thin walled collapsible microtube. Ann Biomed Eng 17: 203–217
43. Westerhof N, Braakman R, Sipkema P (1988) Small vessel compliance and peripheral pressure-flow relations. Proceedings of the 9th international conference and satellite symposium, The Cardiovascular System Dynamics Society 27–30, Halifax, Nova Scotia

Effect of Ventricular and Extraventricular Pressure on the Coronary Artery Pressure-Flow Relationship

YUKIO MARUYAMA and TAMOTSU TAKISHIMA[1]

Summary. In this article, we firstly discuss effects of diastolic pressure on diastolic right and left coronary pressure-flow relationships. The effects of left heart pressure on left coronary flows were evident, especially through $P_{f=0}$ changes. An increase in right heart filling pressure beyond a certain value also impeded left coronary flow through an increased $P_{f=0}$. This suggests that a classic waterfall mechanism is conceivable in the coronary circulation system. If the magnitude of the raised filling pressure is the same in the right and left heart, the effects were larger on elevation of RHP. The effect of LHP on left coronary $P_{f=0}$ was enhanced in the presence of the pericardium, suggesting that a Starling resistor, if any, exists in the epicardial layer. On elevation of heart surrounding pressure, $P_{f=0}$ was definitely increased, and the possibility that the Starling resistor may reside in the epicardial site was supported. Right coronary artery slope changes in the right coronary pressure-flow relationship on elevation of diastolic filling pressure may be intimately related to increased strain of coronary vascular beds; when simultaneous increases in intra- and extra-cardiac pressures occur, no significant changes in the slope were observed. This fact is important for understanding the mechanism of coronary vascular resistance changes.

Secondly, effects of systolic compression on coronary pressure-flow relations were discussed. Although interesting new results have been provided by the excised, perfused heart preparations, many questions, i.e., the location of the vascular waterfall, propriety of the intramyocardial compliance hypotheses, and so on, remain to be clarified.

Key words: Preload—Coronary circulation—Intramyocardial pressure—Pericardium

Introduction

In 1984, Aversano et al. [1] reported the influence of left ventricular diastolic pressure on diastolic coronary pressure-flow relationships independently of effects of capacitive flow in the canine heart. According to their results, the capacitance-free coronary pressure flow relationship was curvilinear, and the degree of curvilinearity was accentuated and coronary conductance was decreased; whereas mean zero-flow pressure ($P_{f=0}$) increased from 14 to 23 mmHg as left ventricular diastolic pressure was raised from 6–10 to 31–35

[1] First Department of Internal Medicine, Fukushima Medical College, Fukushima 960-12, and Tohoku University School of Medicine, 1-1 Seiryo-machi, Sendai, 980 Japan

mmHg. Moreover, when left ventricular diastolic pressure was elevated, coronary sinus pressure rose by about two-thirds of the pressure rise in the ventricle [2]. Thus, similar mechanisms described in coronary sinus pressure elevation [3] may also work in the elevation of left ventricular diastolic pressure. However, there are several unresolved problems in previous reports with respect to preload elevation on coronary circulation.

First, it remains to be clarified why preload elevation accentuates the curvilinearity of the relationship and, if any, increases the coronary resistance. Second, since the pericardium was usually opened in previous experiments, effects of the pericardium on coronary circulation could not be evaluated. As previously reported [4], the pericardium protects against heart size enlargement; as a result, an increase in wall stress would be much smaller than that without the pericardium, indicating a small increase in extravascular compressive force. On the other hand, if the determinant of $P_{f=0}$ resides in the epicardial vein [2], effects of the pericardium on coronary circulation will be accentuated following preload elevation. Thus, it is unclear how two different preload elevations with and without the extraventricular pressure increases (which will be accompanied by ventricular volume enlargement or not, respectively) affect coronary circulation. Third, since venous blood flow from the left ventricle returns not only to the coronary sinus but also to the right ventricle directly, the $P_{f=0}$ value of the left coronary pressure-flow relationship obtained during outflow pressure elevation through raised coronary sinus pressure may differ from that in the in vivo condition. Thus, considering these problems and taking our recent work into account, the focus of attention in this article will be placed on the effects of physical factors, especially extracardiac and intracardiac loading, on coronary circulation from the standpoint of the pressure-flow relation in diastole.

Methods

An isolated canine heart preparation perfused with arterial blood from a support dog was developed in our experimental series because of the ease with which diastolic filling pressure of the right and left ventricles can be controlled [5–7] (Fig. 1). The methods of the isolated canine heart preparation have been previously described. Briefly, left and right coronary arteries of the canine heart were separately cannulated and perfused with the same perfusion pressure through a common compressed-air chamber. Right and left filling (i.e., atrial and ventricular) pressures (RHP, LHP) were independently changed by an individual variable-height reservoir connected to each atrial and ventricular chamber. Moreover, to investigate the effects of the pressure surrounding the heart on the coronary circulation, the excised, perfused heart was placed into an airtight plastic box, to which air was infused to attain a given pressure. In this experimental series, intracardiac and extracardiac pressures were elevated individually or simultaneously from 0 mmHg to 15 mmHg and then to 30 mmHg. After stopping the right ventricular pacing, the capacitance-free coronary pressure-flow relationship was obtained during long diastoles, decreasing perfusion pressure slowly (2–3 mmHg/s) from approximately 60 mmHg. Coronary

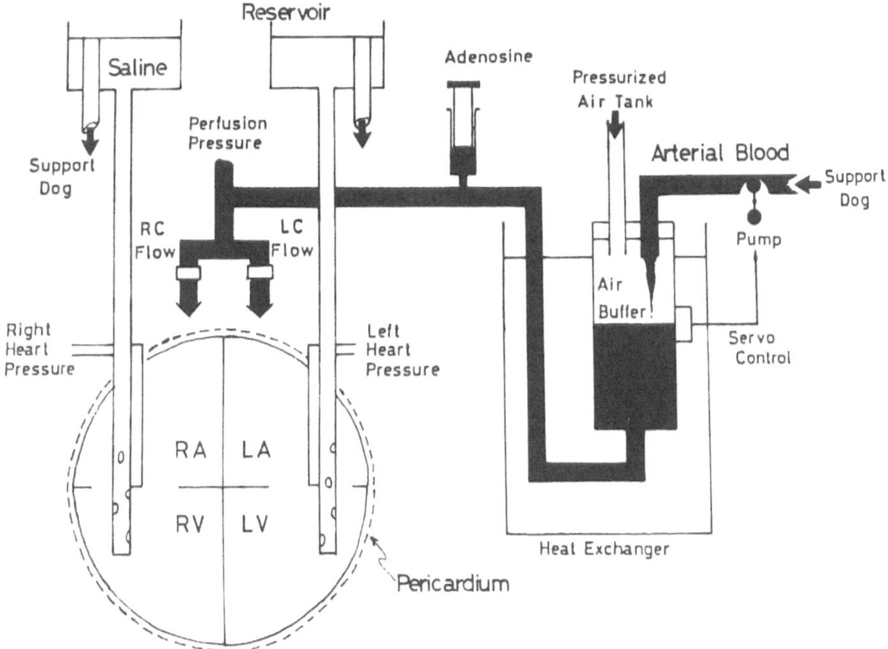

Fig. 1. Schematic drawing of the experimental arrangement. The right (*RC*) and left (*LC*) coronary arteries in an excised heart were perfused with the arterial blood of a support dog. *RA*, right atrium; *LA*, left atrium; *RV*, right ventricle; *LV*, left ventricle

perfusion pressure and coronary blood flow were digitized every 100 ms by a microcomputer (Fujitsu MB 25020) and recorded on an $x-y$ plotter (Graphtec DA 6000). In eight cases, intramyocardial pressure in outer and inner myocardial layers was measured with a needle-tip pressure transducer, placing a pair of the transducers 3.3 ± 1.2 and 8.0 ± 2.6 mm from the epicardium.

The zero-flow pressure ($P_{f=0}$) was determined from the actual intercept of the coronary pressure-flow relationship. Coronary pressure-flow relationship appeared to be linear, although we will discuss this point in more detail later. Thus, linear regression was adopted to determine the slope value of the relationship. Coronary resistance (mmHg/ml/min. 100 g heart weight) was calculated as slope^{-1} of the coronary pressure-flow relationship. All results are expressed as mean ± SEM. Statistical significance was determined using analysis of variance. A value of $P < 0.05$ was accepted as significant.

Results

Figure 2 shows typical tracings of left coronary pressure, flow, and ECG during beating and after the cessation of pacing. Cardiac contraction affects not only systolic coronary but also diastolic flow. That is, systolic compression significantly impeded coronary flow in systole, leading to a 20% – 30% mean flow reduc-

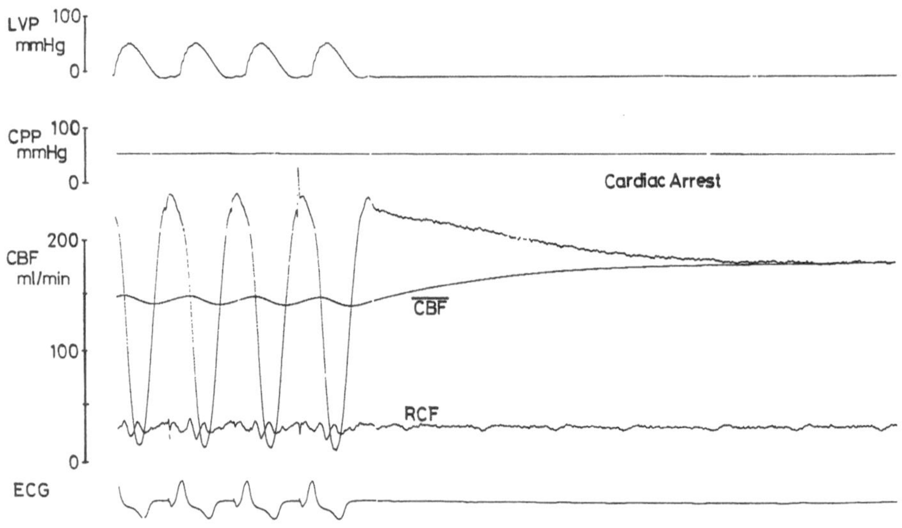

Fig. 2. Phasic and mean coronary flow (CBF) in the beating state and cardiac arrest, maintaining coronary perfusion pressure (*CPP*) constant. *LVP*, left ventricular pressure; *RCF*, right coronary flow

tion from the arrested state. It is interesting to note that diastolic flow in the beating condition is definitely larger than that in the arrested state, implying that heart contraction induces a pulsatile flow pattern during one cardiac cycle. This finding confirms recently published data [8]. Accordingly, the coronary pressure-flow relationship was obtained after reaching the plateau level of coronary flow.

As shown in Fig. 3, the shape of the left coronary pressure-flow relationship appears linear, although showing curvature in low perfusion pressure ranges, when left and right side intracardiac chamber pressures (LHP, RHP) were 0 mmHg. Preload elevation accentuated this curvilinearity [9]. Also, it is of interest that each $P_{f=0}$ value of the right and left coronary arteries was similar. Table 1 shows zero-flow pressure ($P_{f=0}$) and slope (i.e., coronary resistance) of the left coronary pressure-flow relationship at control LHP and RHP 0 mmHg and following an increased diastolic filling pressure of 15 and 30 mmHg in the right and left heart chambers. Simultaneous increases of LHP and RHP led to a small but significant further increase in $P_{f=0}$ value, although there was no significant change in left coronary resistance (Table 1).

Following an elevation of heart surrounding pressure, $P_{f=0}$ increased by approximately 6 mmHg more than the heart surrounding pressure without any significant change in shape, i.e., at heart surrounding pressures of 15 and 30 mmHg, $P_{f=0}$ values were 20.9 ± 2.4 and 35.6 ± 3.1 mmHg respectively [6]. Thus, this is in contrast to the results (Table 1) obtained following the same magnitude of increase in LHP. Those $P_{f=0}$ values also differed from the mean $P_{f=0}$ value of 16.0 ± 1.5 and 29.3 ± 1.5 mmHg with the same magnitude of RHP elevation. Moreover, $P_{f=0}$ values obtained with simultaneous increases of LHP and heart surrounding pressure at 15 and 30 mmHg (22.0 ± 2.9 and

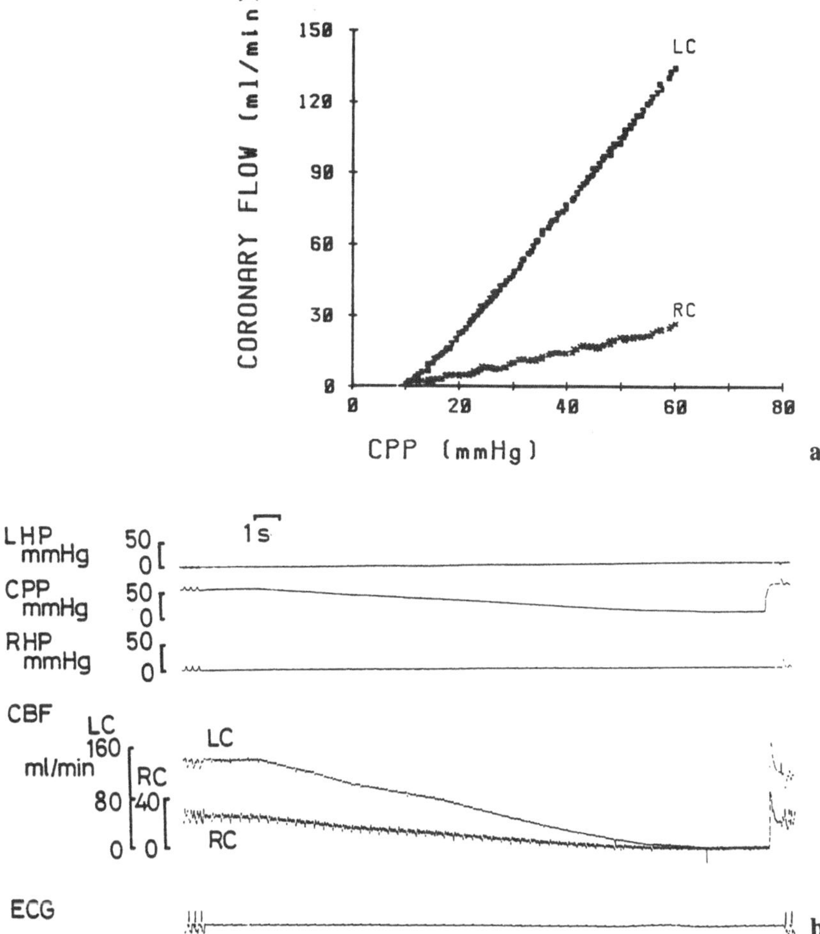

Fig. 3. a Diastolic left and right coronary (*LC*, *RC*) pressure-flow relationship. **b** Diastolic coronary flow changes are shown by decreasing coronary perfusion pressure (*CPP*) slowly from about 60 mmHg to zero-flow pressure ($P_{f=0}$). During this measurement, atrial and ventricular pressure in the left and right heart sides (*LHP, RHP*) were maintained at 0 mmHg. *CBF*, coronary blood flow

35.5 ± 3.2 mmHg, respectively) did not differ at all from corresponding values in the elevation of the latter alone, as described above. As for the slope of the coronary pressure-flow relation, curvilinearity in the lower ranges of perfusion pressure on elevation of left ventricular diastolic pressure also disappeared following a simultaneous increase in heart surrounding pressure [9–12]. This finding is similar to that shown in the presence of the pericardium [5].

To test how effects of ventricular stress itself on the shape of the right coronary pressure-flow relationship differ from those of ventricular strain increase, enhancement of right ventricular intramyocardial pressure without any dimensional change was also induced by simultaneously raising heart surrounding

Table 1. $P_{f=0}$ and coronary resistance changes in the left coronary artery following right and left heart pressure elevation

	Control (LHP = RHP = 0 mmHg)	LHP (mmHg) (RHP = 0 mmHg)		RHP (mmHg) (LHP = 0 mmHg)		LHP, RHP (mmHg) (LHP = RHP)	
		15	30	15	30	15	30
$P_{f=0}$ (mmHg)	7.0 ± 0.9	14.0 ± 1.4	16.9 ± 1.6	17.0 ± 1.1	30.4 ± 1.6	20.0 ± 1.1*	33.2 ± 1.1**
R (= 1/slope) (mmHg/ml/min · 100 g)	0.332 ± 0.037	0.361 ± 0.055	0.420 ± 0.061	0.366 ± 0.055	0.410 ± 0.076	0.390 ± 0.062	0.384 ± 0.054

Zero-pressure intercept ($P_{f=0}$) ($n = 9$) significantly increased, when left heart pressure (LHP) or right heart pressure (RHP) elevated to 15 and 30 mmHg. Simultaneous increases of LHP and RHP further elevated those values, but the magnitude was not additive of each value. On the contrary, there was no significant change among coronary resistance (R) values calculated by slope^{-1} of coronary pressure-flow relationship.
*$P < 0.01$ vs LHP 15 and 30 **$P < 0.05$ vs RHP 15 and 30

Table 2. Different effects of volume loading due to RHP elevation and loading without ventricular distension due to simultaneous increases (i.e., TP) of RHP, LHP, and heart surrounding pressure on diastolic right coronary pressure-flow relationship

	Control (TP = 0)	RHP (mmHg)		TP (mmHg)	
		15	30	15	30
Slope^{-1} (mmHg/ml mm · 100g)	0.620 ± 0.068	*1.135 ± 0.190	*1.004 ± 0.163	0.695 ± 0.123	0.748 ± 0.080
$P_f = 0$ (mmHg)	11.4 ± 2.1	*25.3 ± 2.5	*38.7 ± 2.8	*26.3 ± 1.5	*41.0 ± 1.7

RHP, right heart pressure; LHP, left heart pressure
*$P < 0.01$ vs control

pressure, RHP and LHP from 0 to 15 and to 30 mmHg, by putting the heart into the airtight chamber. $P_{f=0}$ values in the right coronary artery were greatly shifted to the right following these simultaneous pressure increases; however, the slope change was not significant (Table 2).

Discussion

Curvilinearity of Diastolic Left Coronary Pressure-Flow Relationship

It has been thought that a decrease in transmural pressure across the vascular wall induces a reduction of vessel diameter, leading to an increased coronary resistance in the lower perfusion pressure range. However, this explanation of the curvature of the pressure-flow relationship alone may not be sufficient in the pathological state of preload elevation because, when left ventricular filling pressure was elevated in the presence of the pericardium, the pressure-flow relation shifted greatly and the curvature of the relation lessened in the lower perfusion pressure range [5].

As for the mechanism(s) responsible for the lesser curvature, explanations are certainly conceivable. After pericardiectomy, the heart size increases and augments intramyocardial stress, in accordance with Laplace's law. Thus, it is likely that increased vascular resistance through the elongation and/or diameter decrease of resistance vessels with elevated left ventricular end-diastolic pressure is accentuated in the lower perfusion pressure range. Another possibility is that a ventricular transmural pressure gradient exists, and tissue pressure of the subendocardial layer is higher than that of the subepicardial layer [10]. Therefore, coronary blood flow in the latter area stops at a lower perfusion pressure than in the former [11]. The coronary pressure-flow relationship as a whole is the summation of each layer's pressure-flow relationship. Accordingly, when the ventricular transmural pressure gradient becomes larger following increased preload, and the difference between driving pressure and each layer's intravascular pressure disappears from the inner layer after the perfusion pressure becomes lower, the curvilinearity may be accentuated. In fact, the difference in intramyocardial pressure between outer and inner myocardial layers became larger when the left ventricular filling pressure increased, i.e., outer vs inner layer pressures were 9.1 ± 3.4 vs 7.4 ± 2.4 mmHg (NS) at LHP 0 mmHg, 13.4 ± 2.8 vs 18.9 ± 8.6 mmHg $P<0.05$) at LHP 15 mmHg, and 20.8 ± 5.4 vs 32.9 ± 13.0 mmHg ($P<0.01$) at LHP 30 mmHg [12]. Thus, when perfusion pressure decreases, coronary flow seems to stop at the inner layer first and, finally, at the outer, superficial layer. This hypothesis appears to be supported by the following evidence from the present study, i.e., the curvilinearity in lower range of perfusion pressure with elevation of LHP disappeared or decreased with the pericardium intact or following a simultaneous increase of heart surrounding pressure of the same magnitude with LHP to avoid the intramyocardial pressure difference [9, 12].

Accordingly, the phenomenon of the curvilinearity found in the lower perfusion pressure range may be related to the intramyocardial pressure difference.

By this characteristic feature of the pressure-flow relationship, the determination of coronary resistance from the slope^{-1} by linear regression analysis should be regarded as an approximation of the coronary resistance value, especially at higher preload levels. Also, even if the extrapolation of the pressure-flow relationship to obtain $P_{f=0}$ covers only a few mmHg, the curvature of the relationship at lower values of perfusion pressure will lead to a significant overestimation of the $P_{f=0}$ value.

$P_{f=0}$ and Slope^{-1} Changes Following an Elevation of Intra- and Extracardiac Pressures

The $P_{f=0}$ value at 0 mmHg of LHP, RHP, and heart surrounding pressure was smaller than previously reported values [3, 13–16]. There are several possible reasons why values in the present study were lower. First, we obtained them under zero diastolic pressure in four cardiac chambers, whereas in previous studies LHP as well as RHP, either of which affects the coronary pressure-flow relation [2, 5, 6, 11], showed physiological values. Second, if the pressure-flow relation is obtained using one coronary arterial branch, its $P_{f=0}$ value may be affected by the degree of collateral flow [17]. Third, we must consider the effect of the capacitance of coronary arteries [1, 5, 14, 15]. Due to this capacitive effect, coronary inflow in the major artery stops irrespective of the fact that at the microvessel level, flow is still continuing downstream and, therefore, the $P_{f=0}$ value is overestimated. Thus, $P_{f=0}$ of the coronary pressure-flow relation shifts to higher values when the perfusion pressure was decreased at higher rates [5]. Fourth, since the coronary pressure-flow relation is concave toward the flow axis at lower perfusion pressures, $P_{f=0}$ obtained by linearly extrapolating the pressure-flow line to the pressure axis may differ from that actually measured, such as ours. Accordingly, if the above-mentioned factors are taken into account, $P_{f=0}$ may be close to right atrial pressure in the physiological condition [18].

Previous studies suggest that $P_{f=0}$ is not the same in each myocardial layer, and the subepicardial muscle layer has the lowest $P_{f=0}$. Namely, it has been believed that $P_{f=0}$ determined by the coronary pressure-flow relation is the pressure of the last part of the myocardial layer to be perfused. When ventricular filling pressure increases, intramyocardial tissue pressure increases, especially in the inner myocardial layer as described above. As a result, this augments perivascular pressure which leads to an elevation of $P_{f=0}$ [3, 13]. Another possibility is an increase in coronary sinus pressure related to the left ventricular pressure rise, which raises effective back pressure. This is because a vascular waterfall probably exists in the coronary sinus and great cardiac vein [2]. Also, we must consider that the epicardial veins may be squeezed between the connective tissues of the epicardium and the subjacent muscle.

Then, when simultaneous diastolic pressure increases of four heart chambers occur, as found in heart failure, how are the shape and $P_{f=0}$ values of the left coronary pressure-flow relation modified from those obtained with right or left side heart pressure increase alone? The results were as follows, i.e., even if different sources of effective back pressure are contributory to raise $P_{f=0}$ independently, the amount of increase in the $P_{f=0}$ value was not additional, but governed by the more effective determinant of right heart filling pressure (Table 1).

It is worthy to note that $P_{f=0}$ definitely increases in the presence of the pericardium, compared to that without the pericardium (Table 1), when left ventricular filling pressure rises to the same levels, i.e., from 7 ± 1 at control of left heart pressure (0 mmHg) to 16 ± 1 and 28 ± 2 mmHg at left heart pressure 15 and 30 mmHg, respectively [5]. It remains to be clarified why the pericardium affects $P_{f=0}$ value, especially in the higher LHP range, However, two possibilities are conceivable. First, before pericardiectomy, LHP transmits across the ventricular wall and causes a uniform intramyocardial tissue pressure distribution. As the result of an increased compression of the subepicardial layer, $P_{f=0}$ showing the lowest value in the ventricular wall is believed to increase. Second, an increased intrapericardial and/or pericardial surface pressure compresses the epicardial vein, in which a vascular waterfall mechanism is assumed to exist [2], and this may lead to an evelation of $P_{f=0}$. Thus, we tried to test which possibility is more acceptable as the mechanism of raised $P_{f=0}$, by placing the excised, perfused heart into an airtight chamber and controlling heart surrounding pressure.

Measurements of intramyocardial pressure during elevation of heart surrounding pressure from 0 to 15 and 30 mmHg showed values higher by 4–6 mmHg in the outer layer of the myocardium, compared to the inner one. If $P_{f=0}$ is determined by the lower compressive forces, it must be close to the lower intramyocardial pressure of the inner layer rather than to the higher outer layer values. In contrast to this expectation, however, $P_{f=0}$ was determined by the higher compressive values of the outer myocardial layer. Thus, when we take all these results into account with no contradiction, $P_{f=0}$ is not simply determined by the smallest intramyocardial pressure value, but by the value at the epicardial site. This fact indicates that the location of classic Starling resistors may reside in the outer layer of the myocardium, implying that the second possibility mentioned above is the more likely.

In contrast to an increase in $P_{f=0}$ following preload elevation, the shape of the left coronary pressure-flow relationship as a whole was not influenced by RHP or LHP increase alone or also by simultaneous pressure increases in the four heart chambers. Again, it should be noted that this conclusion may be partly attributable to the method used to calculate the coronary resistance, in which the curvilinearity following preload elevation was neglected.

As shown in Fig. 3, the $P_{f=0}$ value of the left coronary artery was similar to that of the right coronary artery in the control state of RHP and LHP 0 mmHg. However, an elevation of $P_{f=0}$ following an increased filling pressure of the right atrial and ventricular chambers was greater in the right coronary artery. Also, the coronary vascular resistance was increased in the right coronary artery, although not significantly different in the left coronary artery (Table 2). The mechanism(s) governing why the shape of the relationship is modified on elevation of RHP, especially without the pericardium, in contrast to that of the left coronary artery, remains to be clarified.

Although the intramyocardial pressure difference was described as a possible

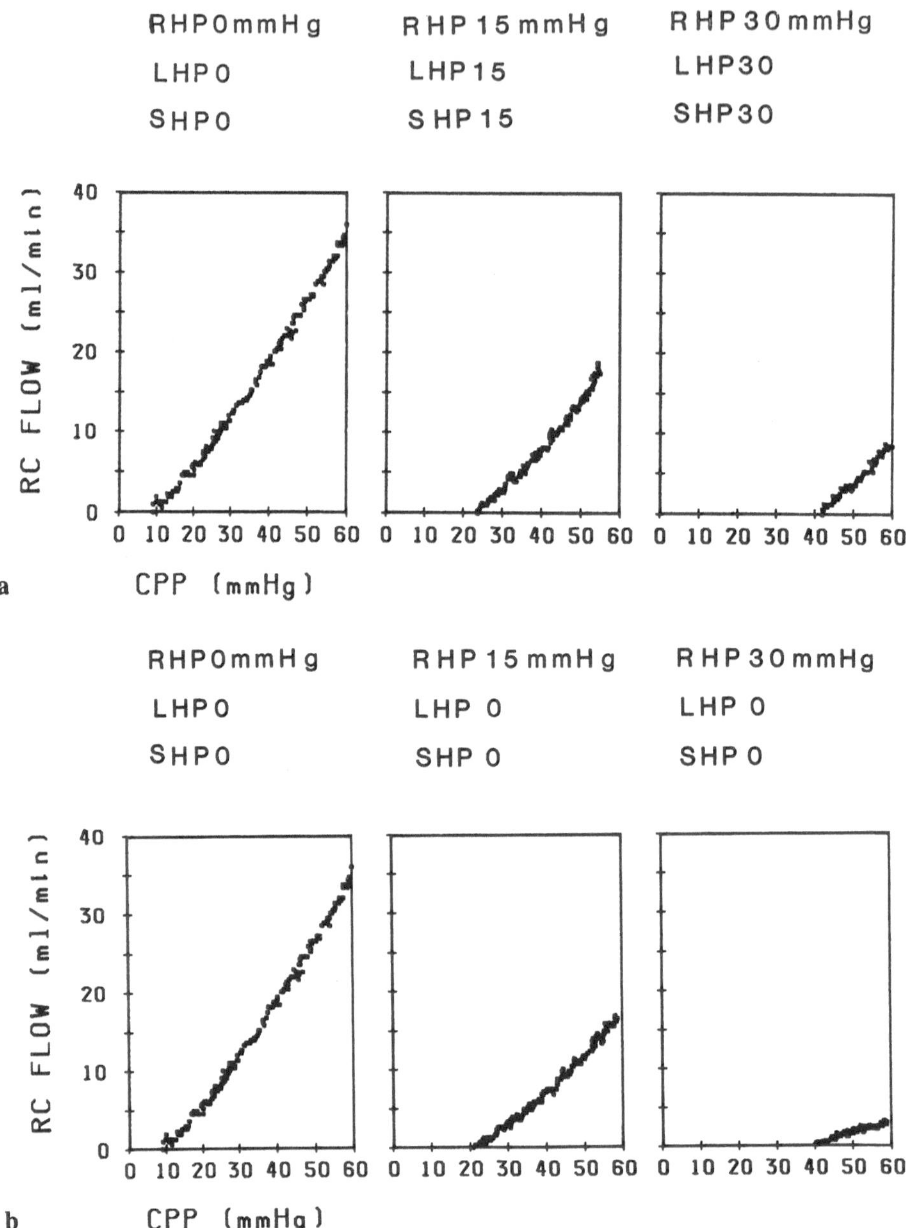

Fig. 4. a Diastolic right coronary (*RC*) pressure-flow relationships with right (*RHP*) and left (*LHP*) heart pressure and the pressure surrounding the heart (*SHP*) at 0 mmHg, 15 mmHg, and 30 mmHg. Following a simultaneous increase of intracardiac and extracardiac pressures, an elevation of $P_{f=0}$ with no significant slope change was induced. **b** Following RHP elevation alone, not only $P_{f=0}$ but also slope changes occurred. *CPP*, coronary perfusion pressure

cause of the curvilinearity of the left coronary pressure-flow relationship in left ventricular preload elevation, it is unlikely as a cause for the slope change of the right coronary artery since the intramyocardial pressure difference, if any, may be small in the thin right ventricle. Rather, the following possibility is available: Since the right ventricular wall is thin, coronary vascular beds are easily distended through stretching of the right ventricle with an elevation of diastolic filling pressure. If this factor plays an important role in determining the shape change, cancellation of muscle stretching due to simultaneous pressure increases from the inside and the outside of the right heart chamber suggests that the change in shape of the relation is no longer expected. This hypothesis was supported from results shown in Table 2 and Fig. 4. Namely, the percent changes in right coronary vascular resistance following an elevation of intra- and extraventricular pressures to 15 and 30 mmHg were 12% and 20%, respectively. These increases were not significant, whereas with an increase of RHP alone, these increases were 83% and 62%, respectively.

Left Coronary Pressure-Flow Relationship in Systole

Interaction between diastole and systole has been known in coronary circulation [19–21], and dynamic effective back pressure may differ from that of the steady state in long diastoles. That is, as shown in Fig. 2, maximally dilated steady flow in the arrested state is different from that of diastole in the beating state. As one might expect, systole impedes coronary flow by increasing intramyocardial pressure and at the same time systolic compressive force can squeeze blood out from the heart. Accordingly, it is necessary to observe the systolic coronary pressure-flow relationship to understand coronary flow dynamics. Linear relations were found between aortic pressure and right coronary flow during systole when peak ventricular pressure was less than systemic pressure [22]. According to a previous report [22], systole shifted the diastolic relation to a higher $P_{f=0}$ and did not alter the slope of the relation. Thus, in systole, a vascular waterfall mechanism that exists in diastole has been considered to also work in an augmented fashion. In contrast to this concept, our experiment, in which coronary pressure-flow relationships were investigated through one cardiac cycle, showed that systolic contraction seemed to change not only $P_{f=0}$, but also the slope of the relation. Furthermore, Westerhof et al. [23] observed that intermittent occlusion of coronary arteries during systole induces curvilinearity of the pressure-flow relation.

References

1. Aversano T, Klocke FJ, Mates RE, Canty JM Jr (1984) Preload induced alterations in capacitance-free diastolic pressure-flow relationships. Am J Physiol 246: H410
2. Uhlig PN, Baer RW, Vlahakes GJ, Hanley FL, Messian LM, Hoffman JIE (1984) Arterial and venous coronary pressure-flow relations in anesthetized dogs. Evidence for a vascular waterfall in epicardial coronary veins. Circ Res 55: 238–248
3. Bellamy RF, Lowensohn HS, Ehrlich W, Baer RW (1980) Effect of coronary sinus occlusion on coronary pressure-flow relations. Am J Physiol 239: H57–H64

4. Maruyama Y, Ashikawa K, Isoyama S, Kanatsuka H, Ino-Oka E, Takishima T (1982) Mechanical interactions between four heart chambers with and without the pericardium in canine hearts. Circ Res 50: 86–100
5. Watanabe J, Maruyama Y, Satoh S, Keitoku M, Takishima T (1987) Effects of the pericardium on the diastolic left coronary pressure-flow relationship in the isolated dog heart. Circulation 75: 670–675
6. Satoh S, Watanabe J, Keitoku M, Itoh N, Maruyama Y, Takishima T (1988) Influences of pressure surrounding the heart and intracardiac pressure on the diastolic coronary pressure-flow relationship in excised canine heart. Circ Res 63: 788–797
7. Maruyama Y, Nunokawa T, Koiwa Y, Isoyama S, Ikeda K, Ino-Oka E, Takishima T (1983) Mechanical interaction between the ventricles. Basic Res Cardiol 78: 544–559
8. Katz SA, Feigl EO (1988) Systole has little effect on diastolic coronary artery blood flow. Circ Res 62: 443–451
9. Watanabe J, Maruyama Y, Satoh S, Keitoku M, Itoh N, Hangai K, Takishima T (1987) Effects of the ventricular transmural pressure difference on the shape of the diastolic left coronary pressure-flow relationship. Automedica 9: 88
10. Archie JP Jr (1978) Transmural distrubution of intrinsic and transmitted left ventricular diastolic intramyocardial pressure in dogs. Cardiovasc Rec 12: 255–262
11. Ellis AK, Klocke FJ (1979) Effects of preload on the transmural distribution of perfusion and pressure-flow relationships in the canine coronary vascular bed. Circ Res 46: 68–77
12. Satoh S, Watanabe J, Maruyama Y, Keitoku M, Hangai K, Takishima T (1987) Close correlation of zero-flow pressure in diastolic coronary pressure-flow relationship with intramyocardial pressure. Automedica 9: 92
13. Bellamy RF (1987) Diastolic coronary pressure-flow relations in the dog. Circ Res 43: 92–101
14. Eng C, Jentzer JH, Kirk ES (1982) The effects of the coronary capacitance on the interpretation of diastolic pressure-flow relationships. Circ Res 50: 334–341
15. Dole WP, Bishop VS (1982) Influence of autoregulation and capacitance on diastolic coronary artery pressure-flow relationship in the dogs. Circ Res 51: 261–271
16. Klocke FJ, Weinstein IR, Klocke JF, Ellis AK, Kraus DR, Mates RE, Canty JM, Anbar RD, Romanowski RR, Wallmeyer KW, Eclit MP (1981) Zero-flow pressures and pressure-flow relationship during single long diastoles in the canine coronary bed before and during maximal vasodilation. Limited influence of capacitive effects. J Clin Invest 68: 970–980
17. Messina LM, Hanley FL, Uhlig PN, Baer RW, Grattan MT, Hoffman JIE (1985) Effects of pressure gradients between branches of the left coronary artery on the pressure axis intercept and the shape of steady state circumflex pressure-flow relations in dogs. Circ Res 56: 11–19
18. Pantely GA, Ladley HD, Bristow JD (1984) Low zero-flow pressure and minimal capacitance effect on diastolic coronary artery pressure-flow relations during maximal vasodilation in swine. Circulation 70: 485–494
19. Vergroesen I, Noble MIM, Spaan JAE (1987) Intramyocardial blood volume change in first moments of cardiac arrest in anesthetized goats. Am J Physiol 253: H307–H316
20. Spaan JAE, Breuls NPW, Laird JD (1981) Forward coronary flow normally seen in systole is the result of both forward and concealed back flow. Basic Res Cardiol 76: 582–586
21. Kajiya F, Tsujioka K, Goto M, Wada Y, Chen X-L, Nakai M, Tadaoka S, Hiramatsu O, Ogasawara Y, Mito K, Tomonaga G (1986) Functional characteristics of intramyocardial capacitance vessels in the dog. Circ Res 58: 476–485
22. Bellamy RF, Lowensohn HS (1980) Effect of systole on coronary pressure-flow relations in the right ventricle of the dog. Am J Physiol 238: H481–H486, 1980
23. Westerhof N, Sipkema P, van Huis GA (1983) Coronary pressure-flow relations and the vascular waterfall. Cardiovasc Res 17: 162–169

4. Transmural Myocardial Perfusion

Transmural Myocardial Perfusion

JULIEN I. E. HOFFMAN[1]

Summary. Left ventricular subendocardial ischemia occurs in many forms of heart disease, congenital and acquired, acute and chronic. In experimental studies, the ischemia is always related to decreases in perfusion, even though regional differences in myocardial oxygen demand may add to the ischemia. The current explanation for these changes relates to interaction between systole and diastole. At the onset of systole, there is a greater increase in intramyocardial pressure in subendocardial than in subepicardial muscle, so that intravascular pressures are increased more in the subendocardial muscle where they exceed pressures in the extramural coronary arteries. Although intravascular pressures increase also in the subepicardial vessels, these pressures remain lower than in the extramural arteries. The resulting pressure gradients permit blood to flow retrogradely from the subendocardial to the subepicardial vessels, but to flow anterogradely in the subepicardial vessels. By the end of systole, the subendocardial arteries are much narrower than they were at the end of diastole. Consequently, at the onset of diastole, blood entering the myocardium from the extramural arteries flows first to the most superficial myocardial vessels (which have the lowest resistance), then to those in the midwall, and last to those in the deep muscle which have the highest resistance. Any excessive reduction in diastolic perfusion time or pressure will thus result in decreased flow to the subendocardium and, if the flow reduction is marked, will cause subendocardial ischemia.

In addition to these regional inhomogeneities of flow, there is considerable inhomogeneity of flow and flow reserve in small (125 mg) adjacent regions, even within a layer. Therefore, reduced perfusion pressures beyond a stenosis may cause ischemia in some but not all regions in that territory. Furthermore, a reduced flow that seems adequate for an average piece of muscle may be inadequate for certain pieces of muscle that have a low reserve.

Key words: Subendocardial blood flow—Subepicardial blood flow—Inner: outer flow ratio—Systolic-diastolic interaction—Intramyocardial pressures—Regional zero-flow pressures

Introduction

It has long been known that the part of the left ventricle at greatest risk of ischemia is the subendocardium, that is, the deepest one-quarter or one-third of the ventricular wall. Documentation of this and reviews of the problem have

[1]Department of Pediatrics and Cardiovascular Research Institute, 1430 HSE, University of California, San Francisco, CA 94143, USA

appeared [1, 2]. More recently, emphasis has also been placed on differences between the pathologic findings in acute and chronic ischemia. When ischemia is acute and extensive, then the necrosis that occurs in one or more layers of the left ventricle is usually confluent. On the other hand, with chronic ischemia, as in angina pectoris or aortic stenosis, there is patchy necrosis or scarring, no matter what layers are involved.

Evidence for Subendocardial Ischemia

The pathology in animals and humans suggests that the lesions result from ischemia, implying an inadequate supply of oxygen for the needs of the myocardium and an inadequate flow of blood to carry away metabolites. Therefore regional subendocardial ischemia can be due to regional increases of myocardial oxygen consumption that are not matched by appropriate increases in regional blood flow, or by regional decreases in blood flow despite an unchanged or even increased regional demand for oxygen.

Oxygen consumption per unit mass in the subendocardium is about 20% greater than that in the subepicardium, perhaps because of greater work done by subendocardial muscle fibers which shorten more in systole than do those in the subepicardium. There are also differences between subendocardial and subepicardial layers for many metabolites and enzymes. These findings are reviewed in detail by Feigl [3]. Although these metabolic differences undoubtedly influence the risk of ischemia, most of the ischemic changes are associated with regional decreases of subendocardial blood flow. An inadequate regional blood flow is the primary determinant of myocardial ischemia, even though regional metabolic differences play a part in intensifying the resulting imbalance between demand and supply.

Although many of the reported studies observed reduced subendocardial flows, relatively few of them documented the presence of subendocardial ischemia. Nevertheless, subendocardial ischemia may be inferred from the dysfunction that occurs if the flow in the subendocardium falls slightly below its expected level. Strain gauge arches inserted in subepicardial and subendocardial layers of the canine left ventricle [4] demonstrated that a small reduction of coronary flow below the resting level reduced subendocardial contractility. Gallagher et al. [5] found that a reduction of 20%–25% in subendocardial flow decreased systolic wall thickening, a finding related to regional ischemia. Similarly, a 10%–20% reduction of subendocardial blood flow has decreased systolic shortening of the subendocardial fibers [6, 7] (Fig. 1).

In order to assess what subendocardial flow should be, the ratio of subendocardial to subepicardial flow per gram is used; this ratio is often termed the endo : epi or inner : outer flow ratio. The rationale for its use is that subepicardial flow is seldom inadequate so that it usually reflects the myocardial demand for blood flow. Thus as long as the inner : outer flow ratio remains constant, it is reasonable to infer that all layers are receiving an adequate blood flow. When the ratio falls it is likely that subendocardial flow is decreasing below a required level. This inference, however, cannot be made when the vessels are maximally dilated by pharmacologic agents, for then the inner : outer flow ratio depends on physical factors rather than on metabolic needs. When coronary vessels are maximally dilated, the ratio of flow per gram in subendocardial and subepicardial muscle varies with the coronary perfusing pressure. At perfusing pressures above normal, the inner : outer flow ratio usually is above 1.0, even in anesthetized animals [8], but at normal or low pressures the inner : outer ratio is usually under 1.0, and the lower the pressure the lower the flow ratio [8, 9]. Because the inner : outer flow ratio changes linearly with perfusion pressure when the vessels are maximally dilated, there is a pressure at which the flow ratios are equal

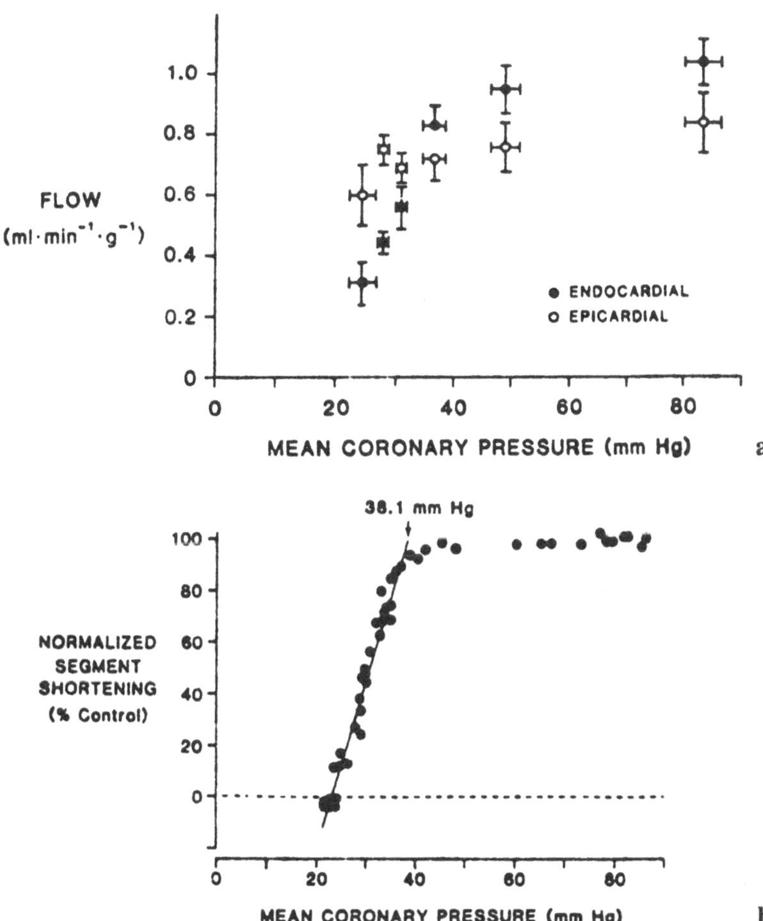

Fig. 1. **a** In conscious dogs, subendocardial flow decreased linearly when mean coronary pressure was reduced below 37 mmHg, whereas subepicardial flow was unchanged until mean coronary pressure was reduced below 25 mmHg. **b** When mean coronary pressure was lowered below the critical pressure at which subendocardial blood flow could be maintained, there was a linear decrease in function of the left ventricular free wall as shown by the decrease in normalized shortening of the subendocardial segment. $y = 6.70x - 155$; $r = 0.99$. From [7] by permission of the author and the publishers

whether the vessels are dilated or not; for unknown reasons, this pressure is usually near the normal operating pressure. Maximal flow at normal perfusing pressures is higher in subepicardial than in subendocardial muscle, so that the flow reserve is less in deep than in superficial layers [10].

Effects of Reduced Perfusing Pressures

If perfusing pressure is reduced in steps and regional flows are measured in the steady state at each pressure, important regional differences are revealed. At first, autoregulation occurs, and flows remain nearly normal in all layers. Then, as pressures are lowered still further, flow begins to decrease more rapidly, and this decrease is most marked in the subendocardium; as a result, a fall in perfusing pressure reduces flow reserve more in subendocardial muscle than in more superficial muscle [11–14] (Fig. 2).

The effects of gradual lowering of coronary perfusing pressure on regional flows are similar in all species, but the effects of sudden occlusion are species dependent. In dogs, transmural flow is much reduced in the ischemic region, with a greater decrease in subendocardial than in subepicardial regions [2, 15]. In swine [15, 16] and in baboons [17], almost no flow is seen in any layer immediately after acute occlusion or over the next few hours. The differences are probably due to the substantial collateral flow that is available in the dog heart at the time of occlusion [15, 18].

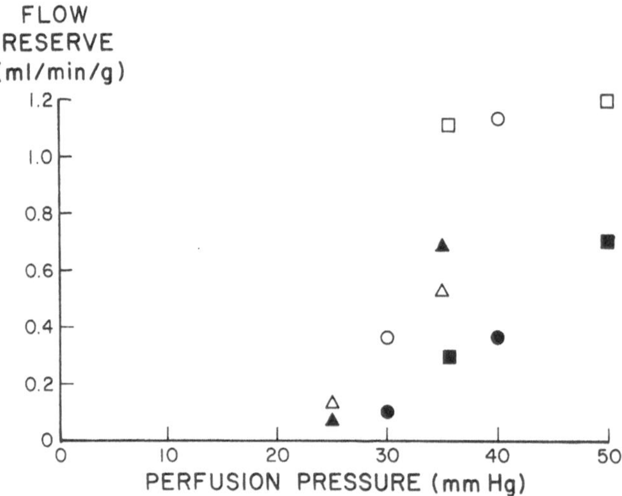

Fig. 2. The relationship between mean coronary perfusion pressure and the coronary flow reserve in the subendocardium (*closed symbols*) and the subepicardium (*open symbols*) taken from several studies reported in the literature. There was reserve in both layers at all pressures, and with one exception reserve was lower in the subendocardium. From [14] by permission of the authors and the publishers

Mechanisms of Subendocardial Ischemia

Three major mechanisms have been suggested to explain subendocardial ischemia. They are:

1. Differential systolic intramyocardial pressures causing differential systolic flows across the wall
2. Differential diastolic intramyocardial pressures and back pressures
3. Interactions between systole and diastole

Differential Systolic Intramyocardial Pressures

The initial explanation for subendocardial ischemia was based on the distribution of pressures across the left ventricular wall. In 1939 Johnson and DiPalma [19] attempted to measure pressures in different layers of the left ventricular wall by drawing an arterial segment tangentially through the wall at different depths from surface. Then they pressurized the artery to make it stiff, and recorded the changes in pressure in it during systole and diastole, or else measured the pressure generated in an external circuit when intramyocardial pressure was allowed to collapse the artery and displace its contents. They found that in systole the intramyocardial pressures were higher in the deeper than in the more superficial muscle, and that in the deep muscle the pressures exceeded systolic pressure in the left ventricular cavity. Since then many other attempts have been made to measure intramyocardial pressure [20]. Some have used variations on the original methods, some have used small solid state pressure transducers [21, 22], and at least two groups used vein segments pulled through the wall and measured the pressure at which liquid could be made to flow through them. All these studies agree that, in systole, intramyocardial pressures decrease from endocardium to epicardium, and that beneath the endocardium the pressure equals or exceeds that in the cavity. However, the absolute pressures reported vary widely from study to study. The likely reason for the variation is that inserting a large structure into dense myocardium distorts the tissue and artifactually raises pressure. Recently, intramyocardial pressures have been measured by micropipettes with tips under 15 μm in diameter [23]. This study agreed with those by Hamlin et al. [22] and others in finding a linear decrease in pressure from endocardium to epicardium, with systolic pressures almost equalling cavity pressures near the endocardium and decreasing to low pressures near the epicardium (Fig. 3). Pressures immediately beneath the endocardium and epicardium were not measured. Given the small size of the pipettes and the minimal tissue distortion caused by their insertion into the muscle, it is likely that the distribution and values of intramyocardial pressures found are correct. The observed distribution of intramyocardial pressures across the left ventricular wall is compatible with the radial stress distribution. This, however, does not solve completely the problem of what intramyocardial pressures are at different depths in the wall, because there are circumferential and longitudinal stresses that generate intramyocardial pressure; in the empty beating heart, intramyocardial pressures were found to be only slightly below those measured when the heart was generating a normal

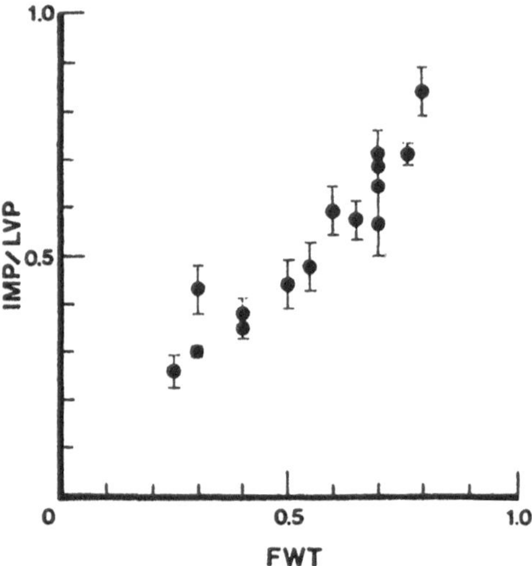

Fig. 3. The relationship between depth in the myocardium and the ratio of systolic intramyocardial pressure (*IMP*) to left ventricular cavity pressure (*LVP*). Mean ± SD. *FWT*, fractional wall thickness. From [23] by permission of the authors and the publishers

pressure [24]. Furthermore, because of the complex microstructure of the myocardium, the actual pressures exerted on microvessels of different sizes are unknown.

If we accept the intramyocardial pressure measurements at their face value, then it appears that in systole, subepicardial muscle could receive a lot of flow whereas subendocardial muscle would receive little flow. Therefore, subendocardial muscle has to get almost all its flow in diastole, and reduction of diastolic perfusion time or pressure will reduce flow more in the deep than in the superficial muscle. In this way, subendocardial vulnerability to ischemia is readily explained [1]. Unfortunately for this hypothesis, there is strong evidence against substantial systolic flow in any part of the left ventricle. The extramural branches of the left coronary artery have a large capacitance, and so would be expected to store some or all of the blood that enters the coronary artery orifice during sytole. This was shown directly by a study of flow in the small terminal extramural branches of the left coronary artery, which are distal to the main capacitance, or of the septal artery which is almost all intramural and has little capacitance [25]. These studies, made with small Doppler flow transducers, show that systolic flow after allowing for capacitance is only about 7% of the total flow during the cycle. Finally, direct examination of flow in subepicardial vessels by trans- or epi-illumination has demonstrated forward flow in arterioles in the outermost 300 μm of the left ventricular muscle; below that level, flow either ceases in systole or else becomes retrograde [26, 27]. From these studies it seems that at most only a thin rim of left ventricle is perfused in systole (about 4% of a 100 g left ventricle), so that differential systolic perfusion can no longer be used to explain subendocardial vulnerability to ischemia.

Differential Diastolic Intramyocardial Pressures and Back Pressures

The next explanation for subendocardial ischemia was based on the discovery by Bellamy [28] that if coronary flows and pressures were measured every 0.1 s during a long diastole, the pressure-flow points formed a straight line that intercepted the pressure axis at about 45 mmHg. During peak reactive hyperemia the pressure-flow points formed a straight line that, when extrapolated, intercepted the pressure axis at about 20 mmHg. These zero-flow pressures could well represent cessation of the last drop of flow through the last piece of muscle to be perfused, and flow might have stopped at higher pressures in other regions of the heart. Thus when pressures are lowered deliberately, the curvature that is seen at the lower end of the pressure-flow curve could represent serial dropping out (derecruitment) of different regions of the heart. If zero-flow pressures were indeed much higher in subendocardial than subepicardial muscle, not only would this explain curvature of the pressure-flow line (because of the increasing resistance as less and less muscle is perfused) but it might also explain the vulnerability of the subendocardium to ischemia. The driving pressure for the subendocardium would be less than that for the subepicardium, and if coronary perfusing pressure were low there would be more reduction of flow to the deep than to the superficial muscle. Although this theory is rational, available evidence does not support it. In recent studies [14, 29, 30], reduction of coronary pressure reduced flow in all layers of the left ventricle but flow did not cease prematurely in any layer. In other studies [31], pressure-flow curves were obtained in different layers of the left ventricle in dogs with normal or hypertrophied left ventricles. The zero-flow pressure was only about 2–3 mmHg higher in the subendocardium than in the subepicardium; so small a difference is unlikely to explain subendocardial vulnerability to ischemia.

Interaction Between Systole and Diastole

Because events in systole or diastole alone cannot explain subendocardial ischemia, the most likely hypothesis involves the interaction of the two periods [32–34]. There is no doubt that during systole there is an increase in intramyocardial pressure that is highest in the subendocardium and decreases linearly towards the epicardium (Fig. 4: end diastole and early systole). As soon as pressure rises outside a collapsible tube like a blood vessel, the pressure inside the vessel will rise by the same amount, because for a fraction of time the volume of the vessel will not have changed and so the transmural pressure will be constant (Fig. 4: early systole). Thus throughout most of the ventricular wall the intravascular pressure will exceed the coronary perfusing pressure, causing the blood flow to decelerate and even stop. Because the coronary vessels have no valves, the system is unstable, and blood will flow out of the vessels. Some blood will pass backwards (Fig. 4: mid-systole) where it may be detected as retrograde flow by a flowmeter; however, retrograde flow need not be detected by a flowmeter at the origin of the coronary artery, because the capacitance of the extramural arteries may prevent the passage of blood all the way back to the origin of the artery [35]. Some of the blood will move forward and appear as the increased systolic flow in

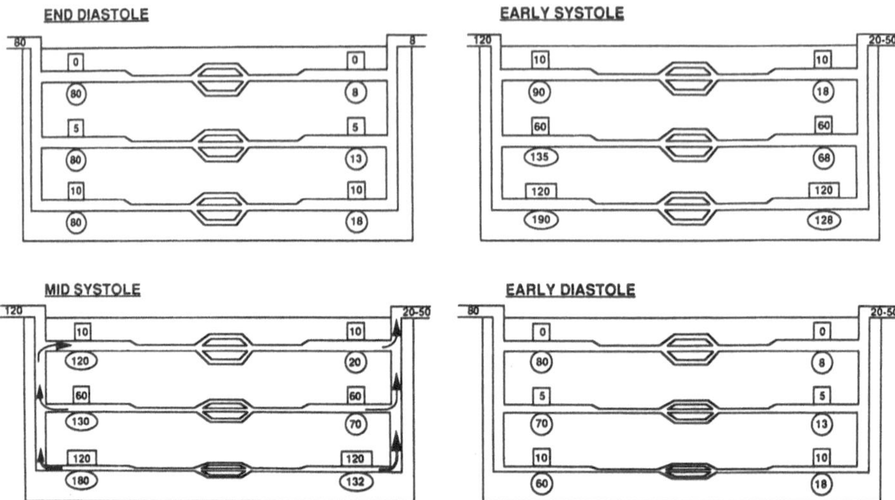

Fig. 4. Diagram to show interaction between the cardiac cycle, intramyocardial pressures, intravascular pressures, and regional anterograde and retrograde flows. The ventricular wall is shown in three layers, and in each there are arterial, capillary, and venous vessels. Assumed intramyocardial pressures are placed in *square* or *rectangular boxes above* the vessels, and the assumed intravascular pressures are placed in *round* or *oval boxes beneath* the respective vessels. Careful inspection of these pressures reveals some inconsistencies, related in part to ignoring the time course of intramyocardial pressures during systole and in part to the absence of certain critical measurements in the system

the coronary sinus [36]. The dividing line between anterograde and retrograde flow appears to be near the end of the arterioles [26]. At arteriolar level, the resistance of the more distal small vessels is high, so that it is easier for blood to pass backwards; on the other hand, in the capillaries and venules the resistance and pressure gradient are lower in front than behind, and this imbalance of pressures and resistances favors forward flow. As a result of movement of blood out of the vessels, they become narrower and their transmural pressure decreases. By the end of systole it is likely that the deepest vessels, especially the arterioles and venules, are the most narrowed (Fig. 4: early diastole). Consequently, at the onset of the next diastole, flow will be directed first to vessels with the lowest resistance, namely those in the superficial muscle; next the middle layer will receive flow, and only later will the subendocardial layer be perfused. Given enough time and perfusion pressure, all the muscle layers will receive an adequate flow, but should the duration of diastole or the diastolic perfusion pressure be reduced, then the subendocardial muscle will receive the least flow. Note that this theory implies that if the coronary perfusion pressure is very low, flow may be decreased in all layers, but will be lowest in the deep muscle.

Until it becomes possible to measure phasic pressures and flows in microvessels in each layer of the left ventricle, support for this hypothesis must necessarily be indirect. One prediction would be that in systole more blood will be squeezed out of the subendocardial than subepicardial vessels, and evidence supports this prediction. In one study, the volumes of blood in subendocardial and subepicardial vessels were measured in rabbit hearts arrested in systole or diastole [37]; there was a greater systolic decrease in vascular volume in the subendocardium than in the subepicardium. In a second study, comparisons of regional blood flows in the maximally dilated coronary vessels of the canine heart were made during cardiac arrest and at different heart rates while the heart was beating [38]. As heart rate increased from zero, the absolute subendocardial flow per gram decreased progressively, but subepicardial flow per gram increased as soon as the heart began to beat, and then remained constant at higher heart rates. There was no forward systolic flow into the myocardium from extramural arteries, and the data allowed the conclusion that about 30%–50% of the subepicardial flow came from blood squeezed retrogradely from the deeper vessels during systole (Fig. 4: mid-systole). An added prediction is that the amount of blood squeezed retrogradely would be increased if contractility increased, and conversely would decrease if the contractility were to decrease. This prediction is supported by the finding that there is less retrograde flow in the left coronary artery when contractility is depressed [1], and more retrograde flow when contractility is increased [35]. Furthermore, subendocardial blood flow is increased when contractility is depressed, and is decreased by increases in contractility]39].

Inhomogeneity of Regional Ischemia

The above discussion, like most similar discussions in the literature, deals with regional variation within the left ventricle as if flows in any one layer were homogeneous. There is, however, recent evidence that this is far from true. If the left ventricular muscle is cut into pieces that weigh about 100–150 mg, there are marked variations of resting flow, maximal flow, and flow reserve, from piece to piece within a layer [30, 40, 41]. As coronary perfusing pressure is decreased, the flow reserve decreases in all layers and all pieces. At low pressures (for example, 40 mmHg), some small pieces of muscle have no flow reserve left, whereas pieces nearby can have substantial reserve. The proportion of pieces with no reserve is greatest in the subendocardium, but some pieces in each layer do retain reserve. These findings may explain the apparent inability of ischemia to dilate myocardial vessels maximally. When 5 or 10 of the small pieces of muscle are combined into one larger piece with a weight of about 1 g (a weight commonly used in these investigations), then when adenosine is given at low perfusing pressures, there will be an increase in myocardial blood flow in the larger piece that has been interpreted previously as evidence for incomplete vasodilatation with ischemia [11–14]. In fact, when the smaller pieces are examined individually, some do not increase their flow after adenosine and so are fully vasodilated by ischemia, whereas other pieces that still have reserve are not ischemic and therefore are capable of vasodilating in response to adenosine [41]. It is possible that these local differences of reserve are responsible for the patchy necrosis that is so often seen when there is chronic ischemia; whatever the cause of the decreased flow or increased myocardial oxygen consumption, only certain pieces of tissue that have lost their reserve will become ischemic and necrotic. As mentioned before, there will be more of these ischemic pieces of tissue in the subendocardium than in more superficial muscle, thereby explaining the common observation that it is the subendocardium that becomes ischemic first.

Acknowledgment. This work was supported in part by Program Project Grant HL-25847 from the National Institutes of Health.

References

1. Hoffman JIE, Buckberg GD (1976) Transmural variations in myocardial perfusion. In: Yu P, Goodwin JF (eds) Progress in Cardiology. Lea and Febiger, Philadelphia, pp 37–89
2. Hoffman JIE (1987) Transmural myocardial perfusion. Prog Cardiovasc Dis 29: 429–464
3. Feigl EO (1983) Coronary physiology. Physiol Rev 63: 1–205
4. Downey JM (1976) Myocardial contractile force as a function of coronary blood flow. Am J Physiol 230: 1–6
5. Gallagher KP, Matsuzaki M, Osakada G, Kemper WS, Ross JJ (1983) Effect of exercise on the relationship between myocardial blood flow and systolic wall thickening in dogs with acute coronary stenosis. Circ Res 52: 716–729
6. Vatner SF (1980) Correlation between acute reductions in myocardial blood flow and function in conscious dogs. Circ Res 47: 201–207
7. Canty JM Jr (1988) Coronary pressure-function and steady-state pressure-flow relations during autoregulation in the unanesthetized dog. Circ Res 63: 821–836
8. Rouleau J, Boerboom LE, Surjadhana A, Hoffman JIE (1979) The role of autoregulation and tissue diastolic pressures in the transmural distribution of left ventricular blood flow in anesthetized dogs. Circ Res 45: 804–815
9. Bache RJ, Schwartz JS (1982) Effect of perfusion pressure distal to a coronary stenosis on transmural myocardial blood flow. Circulation 65: 928–935
10. Hoffman JIE (1987) A critical view of coronary reserve.Circulation 75 (Suppl I): I-6–11
11. Aversano T, Becker LC (1985) Persistence of coronary vasodilator reserve despite functionally significant flow reduction. Am J Physiol 248: H403–H411
12. Canty JM Jr, Klocke FJ (1985) Reduced regional myocardial perfusion in the presence of pharmacologic vasodilator reserve. Circlation 71: 370–377
13. Pantely GA, Bristow JD, Swenson LJ, Ladley HD, Johnson WB, Anselone CG (1985) Incomplete coronary vasodilation during myocardial ischemia in swine. Am J Physiol 249 (Heart Circ Physiol 18): H638–H647
14. Grattan MT, Hanley FL, Stevens MB, Hoffman JIE (1986) Transmural coronary flow reserve patterns in dogs. Am J Physiol 250 (Heart Circ Physiol 19): H276–H283
15. Sjöquist PO, Duker G, Almgren O (1984) Distribution of collateral blood flow at the lateral border of the ischemic myocardium after acute coronary occlusion in the pig and the dog. Basic Res Cardiol 79: 164–175
16. Fugiwara H, Ashraf M, Sato S, Millard RW (1982) Transmural cellular damage and blood flow distribution in early ischemia in pig hearts. Circ Res 51: 683–693
17. Lavallee M, Vatner SF (1984) Regional myocardial blood flow and necrosis in primates following coronary occlusion. Am J Physiol 246 (Heart Circ Physiol 15): H635–H639
18. Schaper W, Flameng W, Winkler B, Wüsten B, Turschmann W, Neugebauer G, Carl M, Pasyk S (1976) Quantification of collateral resistance in acute and chronic experimental coronary occlusion in the dog. Circulation 39: 371–377

19. Johnson JR, DiPalma JR (1939) Intramyocardial pressure and its relation to aortic blood pressure. Am J Physiol 125: 234–243
20. Van der Meer JJ (1972) Myocardial ischemia and epicardiectomy. An experimental study. Thesis, University of Groningen, The Netherlands
21. Stein PD, Marzilli M, Sabbah HN, Lee T (1980) Systolic and diastolic pressure gradients within the left ventricular wall. Am J Physiol 238 (Heart Circ Physiol 7): H625–H630
22. Hamlin RL, Levesque MJ, Kittelson MD (1982) Intramyocardial pressure and distribution of coronary blood flow during systole and diastole in the horse. Cardiovasc Res 16: 256–262
23. Heineman FW, Grayson J (1985) Transmural distribution of intramyocardial pressure measured by micropipette technique. Am J Physiol 249 (Heart Circ Physiol 18): H1216–H1223
24. Baird RJ, Goldbach MM, de la Rocha A (1972) Intramyocardial pressure: the persistence of its transmural gradient in the empty heart and its relationship to myocardial oxygen consumption. J Thorac Cardiovasc Surg 64: 635–646
25. Chilian WM, Marcus ML (1985) Effects of coronary and extravascular pressure on intramyocardial and epicardial blood velocity. Am J Physiol 248 (Heart Circ Physiol 17): H170–H178
26. Tillmanns H, Ikeda S, Hansen H, Sarma JSM, Fauvel J-M, Bing RJ (1974) Microcirculation in the ventricle of the dog and turtle. Circ Res 34: 561–569
27. Ashikawa K, Kanatsuka H, Suzuki T, Takishima T (1986) Phasic blood flow velocity pattern in epimyocardial microvessels in the beating canine left ventricle. Circ Res 59: 704–711
28. Bellamy RF (1978) Diastolic coronary pressure-flow relations in the dog. Circ Res 43: 92–101
29. Hanley FL, Messina LM, Grattan MT, Hoffman JIE (1984) The effect of coronary inflow pressure on coronary vascular resistance in the isolated dog heart. Circ Res 54: 760–772
30. Aldea G, Hoffman JIE, Husseini W, Mori H (1987) Determinants of regional myocardial blood flow. In: Brun P, Chadwick RS, Levy BI (eds) Cardiovascular dynamics and models; Proceedings of NIH-INSERM workshops. Les Editions INSERM, Paris, pp 44–47
31. Ellis A, Klocke FJ (1980) Effects of preload on the transmural distribution of perfusion and pressure-flow relationships in the canine coronary vascular bed. Circ Res 46: 68–77
32. Hoffman JIE, Baer RW, Hanley FL, Messina LM, Grattan MT (1985) Regulation of transmural myocardial blood flow. J Biomech Eng 107: 2–9
33. Arts T, Reneman RS (1985) Interaction between intramyocardial pressure (IMP) and myocardial circulation. J Biomech Eng 107: 51–56
34. Levy BI, Tedgui A, Michel JB (1985) A mechanical model of the dynamics of the coronary circulation in dog. J Theor Biol 116: 225–242
35. Kajiya F, Tomonaga G, Tsujioka K, Ogasawara Y, Nishihara H (1985) Evaluation of local blood flow velocity in proximal and distal coronary arteries by laser Doppler method. J Biomech Eng 107: 10–15
36. Kajiya F, Tsujioka K, Goto M, Wada Y, Tadaoka S, Nakai M, Hiramatsu O, Ogasawara Y, Mito K, Hoki N, Tomonaga G (1985) Evaluation of phasic blood flow velocity in the great cardiac vein by a laser Doppler method. Heart Vessels 1: 16–23
37. Levy BI, Samuel JL, Tedgui A, Kotelianski V, Marotte F, Poitevin P, Chadwick RS (1988) Intramyocardial blood volume in the left ventricle of rat arrested hearts. In: Brun P, Chadwick RS, Levy BI (eds) Cardiovascular dynamics and models. Les Editions INSERM, Paris, pp 65–71
38. Flynn AE, Coggins DL, Aldea GS, Austin RE, Goto M, Husseini W, Hoffman JIE (1989) Ventricular contraction increases subepicardial blood flow: evidence for a deep myocardial pump. FASEB J 3: A1305

39. Marzilli M, Goldstein S, Sabbah HN, Lee T, Stein PD (1979) Modulating effect of regional myocardial perfomance on local myocardial perfusion in the dog. Circ Res 45: 634–640
40. Austin RE Jr, Aldea GS, Coggins DL, Flynn AE, Hoffman JIE (1990) Profound spatial heterogeneity of coronary reserve: discordance between patterns of resting and maximal myocardial blood flow. Circ Res (in press)
41. Coggins DL, Flynn AE, Austin RE Jr, Aldea GS, Muehrcke D, Goro M, Hoffman JIE (1990) Non-uniform loss of regional flow reserve during myocardial ischemia in dogs. Circ Res (in press)

5. Coronary Microcirculation

Characteristics of Velocity Waveform and Radius in Epimyocardial Microvessels in Beating Left Ventricle

Kouichi Ashikawa[1], Hiroshi Kanatsuka[2], Toshimi Suzuki[2], and Tamotsu Takishima[2]

Summary. The phasic red cell velocity pattern and the phasic change in diameters in the epimyocardial microvessels were investigated in the beating canine left ventricle. We also studied the effects of nitroglycerin and dilazep, an adenosine potentiator, on the red cell velocity curves in epimyocardial microvessels and on the diameters of arterioles. Using a floating objective system and high-speed cinematography, we measured red cell velocities in arterioles, capillaries, and venules, and also microvascular diameters in the epimyocardium of left ventricle in open-chest anesthetized dogs. In arterioles and capillaries, peak red cell velocity occurred in mid-systole, followed by a decrease during diastole. In venules, red cell velocity reached its peak in late systole, followed by a gradual decline during diastole. An abrupt decline in red cell velocity and a momentary cessation or reverse flow were observed in these microvessels during the pre-ejection period. The internal diameter of small venules was significantly increased in late systole, while there was no significant change in the diameter of arterioles throughout the entire cardiac cycle. There was no significant change in the red cell velocity pattern in epimyocardial microvessels after administration of nitroglycerin. Nitroglycerin also failed to dilate the epimyocardial small arterioles. After administration of dilazep, red cell velocity curves in the epimyocardial microvessels were shifted upward throughout the entire cardiac cycle, with a significant increase in the total area under the velocity curves accompanied by significant dilation of small arterioles. These results suggest that the characteristics of velocity waveform in coronary microvessels in epimyocardium are quite different to those of endomyocardium in the beating left ventricle. Furthermore, using an intravital microscope system, we directly demonstrated that nitroglycerin has little effect on small arterioles and capillary flow, and that dilazep has a potent dilative effect on small arterioles and markedly increases capillary flow in the epimyocardium.

Key words: Phasic blood flow velocity—Coronary microvessels—Floating objective—In situ beating left ventricle—Vasoactive drugs

Introduction

It is well established that under normal conditions, coronary inflow in the left coronary artery occurs mainly during diastole and remains at a low level during systole [1]. This phasic change in coronary inflow is considered to be caused by cardiac contraction. Cardiac contraction has been reported to impede the coron-

Department of Cardiology, National Sendai Hospital[1], and First Department of Internal Medicine[2], Tohoku University School of Medicine, 1-1 Seiryo-machi, Sendai, 980 Japan

ary blood flow in the beating left ventricle [2]. Thus, coronary blood flow is understood to be greatest during diastole when the extravascular compression is minimal, and least during systole when the extravascular compression is greatest. However, it has been suggested that the phasic blood flow pattern in the epicardial coronary artery does not correctly reflect the intramyocardial flow pattern [2–4]. There is no doubt that direct and continuous observation of coronary microcirculation provides information useful in explaining this discrepancy. Thus, we have developed a new microscope system which allows direct and continuous observation of coronary microcirculation in the beating heart in situ [5]. Using this system, we determined the phasic patterns of red cell velocity in the epimyocardial microvessels and the phasic change in diameters of microvessels in the beating left ventricle.

Methods

Experimental Procedure

Young mongrel dogs of both sexes, weighing 4–10kg, were anesthetized with an intravenous injection of urethane (500 mg/kg) and chloralose (60 mg/kg) and ventilated with room air using a Harvard respirator. A positive end-expiratory pressure of 5 cm H_2O was introduced to prevent atelectasis of the lung. Arterial blood gases and pH were maintained within their normal range by adjusting ventilation volume and/or rate. Metabolic acidosis during the experiment was prevented by an infusion of sodium bicarbonate. A lead *II* EGG was monitored. A polyvinyl catheter was placed in the external jugular vein for intravenous injection. Aortic pressure was measured in the aortic root with a catheter passed through the right carotid artery. After a left thoracotomy through the 5th intercostal space, the heart was suspended in a pericardial cradle. A Tygon catheter was placed in the left atrium via the left atrial appendage. Left ventricular pressure was measured with a 16-gauge Teflon tube passed into the left ventricle through the appendage. Pressures were measured with a Statham strain gauge (Model P 23). The time delay of the fluid-filled systems for measurement of pressure was less than 2 ms compared with a solid-state transducer (Millar Instrument, Inc., Houston, TX, USA). Heart rate was kept constant at 140 beats per min by means of left atrial pacing after sinoatrial block, which was produced by injecting formaldehyde into the region of the sinus node. A plastic wrapping was used to separate the lung from the anterior aspect of the heart. The preparation was kept moist during the experiment by continuously dripping warm Krebs-Ringer solution (37°C, pH 7.400) on the cardiac surface. Rectal temperature was maintained at about 37°C by a heat blanket. The EGG and aortic and ventricular pressures were recorded on a Rectigraph at a paper speed of 100 mm/s as needed. Two 24-gauge steel needles were horizontally inserted through the mid-myocardium of the left ventricle, beneath the area of interest, to reduce excessive movement of the heart and to hold the area of interest in the microscopic field of view. Each end of the needle was fixed to a needle holder which was held with coil springs [5].

Microscope System

There are a number of methodological difficulties involved in the microscopic observation of coronary microcirculation in the beating heart in situ. The greatest difficulty is the alteration of the focal distance between the microscopic objective and the heart introduced by the cardiac contraction. This makes the direct and continuous observation of the coronary microcirculation almost impossible. To overcome this difficulty, we have reported a new microscope system using a floating objective. The theoretical considerations and design of the microscope system have been previously reported [5]. In brief, the floating objective consists of a pair of convex lenses which transmit the real image to a standard microscope without any change in magnification. The object at the front focus of the floating lens is transmitted to the back focus of the opposite lens. This real transmitted image is then observed with the objective of a standard microscope. The distance between the floating lens and the heart was adjusted to the focal distance of this lens. The floating objective is supported by low-resistance ball bearings and a weight-adjusting coil spring. The minimal weight of the floating objective (approximately 16 g in total weight) permits the lens to follow the cardiac motion, because the inertial force in vertical movement is minimal. The epimyocardium of the left ventricle was transilluminated by a light-conducting glass fiber inserted in a 20-gauge steel needle, and a xenon arc lamp. The needle holder was designed to move up and down in unison with the cardiac motion. Thus, in this floating objective system, the illuminating fiber, the floating objective, and the myocardium all move as a unit. Moreover, in order to prevent compression of the tissue, the floating lens was lifted by an arm of the objective lifter connected to the needle holder, and was lightly positioned just above the surface of the heart under microscopic observation. The beam splitter view of the microscopic image was monitored by means of a rotary shutter camera (shutter speed of 1/1800 s) and a video motion analyzer which was useful for confirming good focus on the film. After focusing, motion picture were taken at 500 frames per second with a 16-mm high-speed motion picture camera (Fig. 1). High-speed Ektachrome films were used with 100-μs exposure times. To verify the film speed and to correlate with hemodynamic data, timing flashes at 10-ms intervals and signals that were synchronized with the R wave of the ECG were simultaneously recorded on each edge of the film.

Measurement of Red Cell Velocity

Red cell velocities in coronary microvessels were calculated by frame-to-frame analyses from the distance of cell progressions on a projection screen and the numbers of frames needed. Red cell velocities could be calculated only in terminal arterioles (mean 12.8 ± 4.1 μm in diameter), capillaries, and collecting venules (mean 16.5 ± 6.5 μm in diameter). In this size of microvessels, it was possible to track cell-to-cell progression on the high-speed film. The distance of cell progression in serial frames was calculated as the difference in distances from a special marker on the vessel, usually the bifurcation, since the vessel was moving on the projection screen as in the heart. Arterioles could be easily differ-

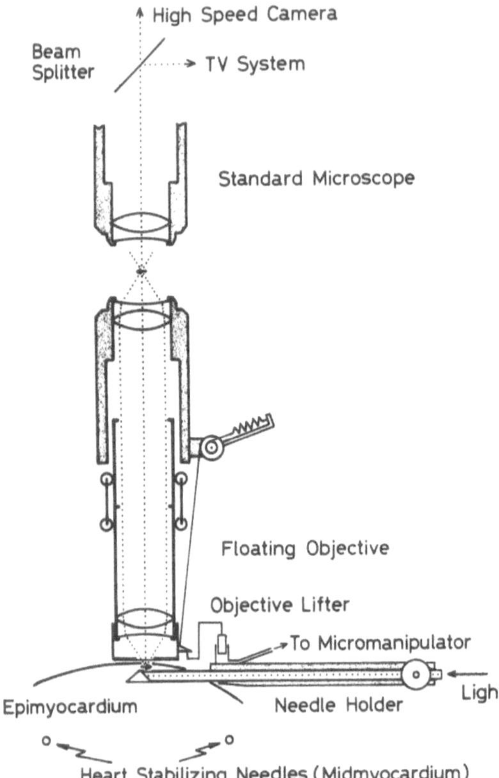

Fig. 1. Schematic illustration of the intravital microscope system for the coronary microcirculation in the beating left ventricle in situ. From [6]. Reproduced by permission

entiated from venules by the direction of flow at the bifurcation. The optical magnification on the projection screen and the film speed were confirmed by a reference scale and timing flashes, respectively. The red cell velocity curve was obtained from a plot of red cell velocity that was calculated every 20–40 ms during the entire cardiac cycle. The internal diameters of arterioles and venules were also measured during a cardiac cycle. Phasic changes in diameter were estimated in small arterioles (mean 24.9 ± 11.0 μm in diameter) and in small venules (mean 27.7 ± 11.6 μm in diameter). The velocity curve and the diameter were then correlated to ECG, aortic and left-ventricular pressure curves. The area under the red cell velocity curve was planimetrically measured during the different segments of a cardiac cycle: isovolumic systole, ejection phase, and diastole. The different segments of the cardiac cycle were determined from the aortic and left ventricular pressure recording. The areas under the red cell velocity curve during different segments of a cardiac cycle were expressed as percent of total area under the red cell velocity curve.

Fluorescence Microscope System

To verify the resolution of the measurements of the internal diameter of microvessels, we employed an incident light fluorescence method [6]. An incident

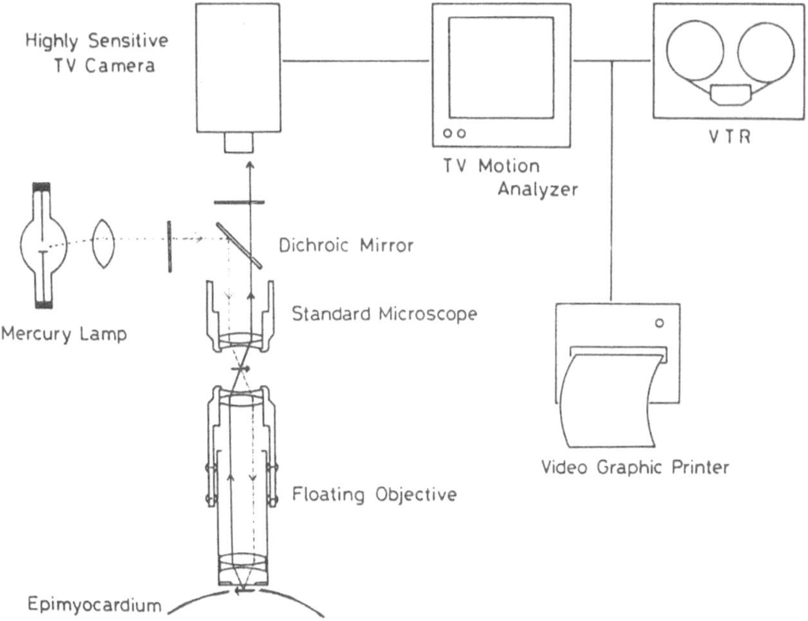

Fig. 2. Schematic diagram of the incident light fluorescence microscope system with the floating objective. *VTR*, videotape recorder

light fluorescence microscope was attached to our floating objective system (Fig. 2). The surface of the left ventricle was illuminated by incident light from a mercury lamp. Fluorescein isothiocyanate dextran (FITC dextran; molecular weight 40, 500) was injected into the left atrium to enhance contrast of the epimyocardial microvessels. By means of a highly sensitive television camera, the enhanced image was recorded on a videocassette recorder and printed out on a video-graphic printer for the diameter measurement. In 5 arterioles, the internal diameter was estimated by both cineangiographic and fluorescence methods. There was no significant difference between these two measurements of internal diameter. In addition, we estimated the effects of nitroglycerin (20 μg/kg, i.v.) and dilazep (50 μg/kg, i.v.), an adenosine potentiator [7], on red cell velocity curves in epimyocardial microvessels and on diameters of arterioles.

Results

Hemodynamic and arterial blood gases were kept in the normal range throughout the experiment. Heart rate was maintained constant at 140 beats per minute by means of left atrial pacing. Figure 3 shows mean red cell velocity curves in small arterioles (mean diameter 12.8 ± 4.1 μm), capillaries, and small venules (16.5 ± 6.5 μm) in the epimyocardium of the beating canine left ventricle. In small arterioles and capillaries, peak red cell velocity occurred in mid-systole, followed by a decrease during diastole. An abrupt decline in red cell velocity and

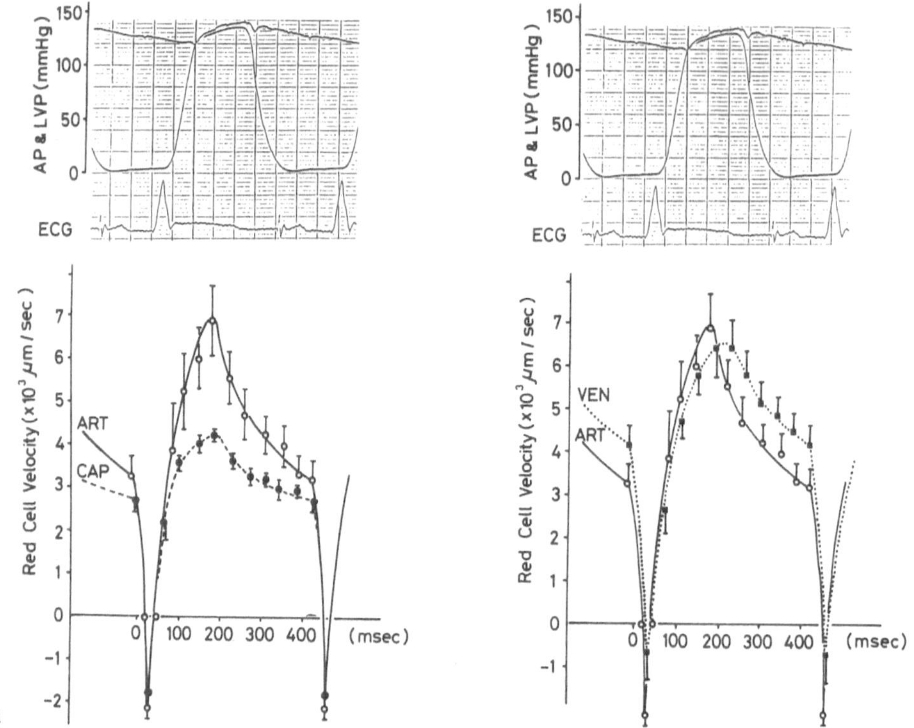

Fig. 3. a, b Phasic red cell velocity curves in the epimyocardial small arterioles (*ART*, 12.8 ± 4.1 μm in diameter) and capillaries (*CAP*) (**a**) and small venules (*VEN*, 16.5 ± 6.5 μm in diameter) (**b**) in the beating canine left ventricle. The *top graphs* illustrate simultaneously recorded aortic (*AP*) and left ventricular pressure (*LVP*) curves and *ECG*. The left atrium was electrically paced at 140 beats per minute. From [6]. Reproduced by permission

a momentary (20–30 ms duration) reverse flow were observed in these microvessels during the pre-ejection period (Fig. 3a). In small venules, red cell velocity reached its peak in late systole, followed by a gradual decline during diastole. Momentary cessation or reverse flow was also observed in small venules (Fig. 3b). Figure 4 illustrates the percentage of total area under the red cell velocity curve during different phases of the cardiac cycle in small arterioles, capillaries, and small venules. During isovolumic systole, the percent area under the velocity curve was minimum (mean ±SEM, 1.9% ± 1.9% in small arterioles, 4.4% ± 0.8% in capillaries, and 6.4% ± 1.0% in venules). During the ejection phase, a considerably higher percentage of the total area under the velocity curve occurred in the epimyocardial microvessels (51.4% ± 3.6% in small arterioles, 43.6% ± 2.0% in capillaries and 40.0% ± 1.8% in small venules). During diastole, approximately half of the total area under the velocity curve occurred in epimyocardial microvessels (mean ±SEM, 47.6% ± 5.5% in small arterioles, 52.5% ± 3.1% in capillaries and 55.2% ± 2.2% in small venules). There was no
small arterioles (mean 24.9 ± 11.0 μm in diameter at end-diastole). On the other

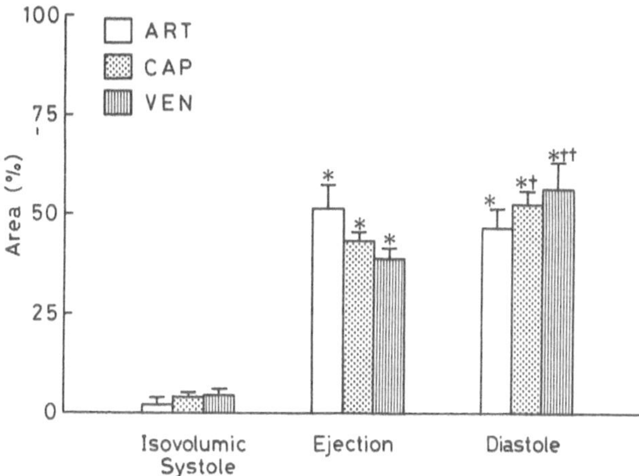

Fig. 4. Percentage of the area under the red cell velocity curves during different phases of the cardiace cycle in the epimyocardial small arterioles (*ART*), capillaries (*CAP*) and small venules (*VEN*). $^{*}P<0.005$ vs isovolumic systole; $^{++}P<0.005$ vs ejection; $^{+}P<0.005$ vs isovolumic systole; $^{++}P<0.005$ vs ejection; $^{+}P<0.05$ vs ejection. From [6]. Reproduced by permission

hand, small venules were significantly dilated in late systole when red cell velocity was maximum (27.3 ± 11.3 μm in diastole, 32.2 ± 14.1 μm in late systole; +14.2% ± 8.2%, $P<0.05$).

Administration of nitroglycerin and dilazep resulted in a temporary decline in aortic pressure. Systolic blood pressure returned to control level within 4 and 10 min, respectively. There was no significant change in diastolic aortic pressure following nitroglycerin injection. On the other hand, diastolic aortic pressure remained slightly but significantly lower following dilazep injection (−9.6%, $P<0.05$; Table 1). Heart rate was maintained at 140 beats/min throughout the experiment by means of the left atrial pacing. Measurements were performed after recovery of the systolic aortic pressure following administration of each drug.

After administration of nitroglycerin, there was no significant change in the red cell velocity pattern in the epimyocardial microvessels during a cardiac cycle. There was a tendency for the total area under the red cell velocity curve to increase following administration of nitroglycerin. However, the percent increase was scattered and was not significant in the epimyocardial microvessels (from −4.5% to +10.5%; Table 2). Administration of nitroglycerin also failed to dilate the epimyocardial small arterioles (26.5 ± 3.0 μm in control and 26.7 ± 3.3 μm following nitroglycerin) (Fig. 5a). After administration of dilazep, red cell velocity curves in the epimyocardial microvessels were shifted upward throughout the entire cardiac cycle, with a significant increase in the total area under the velocity curves (+37% vs control in arterioles, +30% in capillaries, and +28% in venules; Table 2). Figure 6 shows a typical response of a small arteriole after administration of dilazep. The internal diameters of small arterioles were significantly dilated following administration of dilazep (from 27.5 ± 3.6 μm to 35.8 ± 4.7 μm, +30.6%; $P<0.001$) (Fig. 5b).

Table 1. Change of aortic pressure after administration of nitroglycerin and dilazep

	Nitroglycerin (20 μg/kg, i.v.)		Dilazep (50 μg/kg, i.v.)	
	Before	After	Before	After
Systolic aortic pressure (mmHg)	128.5 ± 5.5	125.4 ± 5.8	138.0 ± 5.9	137.0 ± 4.9
Diastolic aortic pressure (mmHg)	85.4 ± 4.6	83.8 ± 6.8	94.4 ± 6.3	85.2 ± 5.3*

Data are means ± SEM.
* $P < 0.05$ vs before dilazep administration.

Table 2. Percent change of total area under the red cell velocity curve after administration of nitroglycerin and dilazep

	Nitroglycerin (20 μg/kg, i.v.)	Dilazep (50 μg/kg, i.v.)
Arteriole	+ 6.8 % ± 4.2 % (13.8 ± 2.1 μm)	+ 36.9 % ± 2.8 % * (14.0 ± 2.2μm)
Capillary	+ 6.7 % ± 5.1 %	+ 29.5 % ± 8.5 % *
Venule	+ 5.8 % ± 4.8 % (17.6 ± 2.3 μm)	+ 27.6 % ± 5.9 % * (18.5 ± 2.2 μm)

Data are means ± SEM. Parentheses show internal diameter.
* $P < 0.05$ vs before dilazep administration.

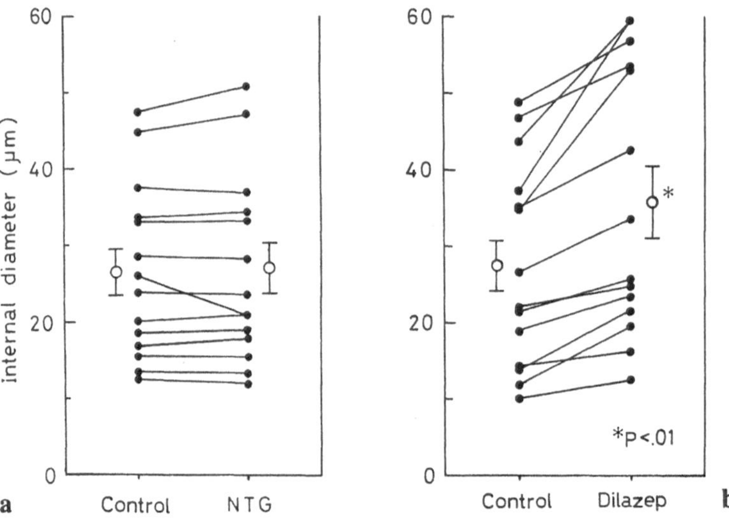

Fig. 5a, b. Effects of nitroglycerin (*NTG*, **a**) and dilazep (**b**) on the internal diameters of the epimyocardial small arterioles

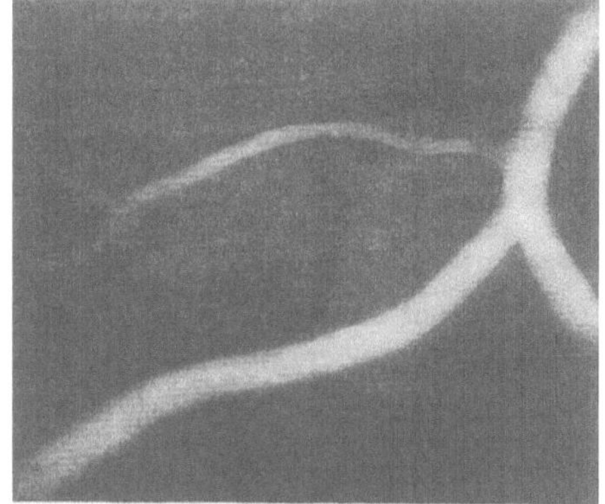

Fig. 6a, b. Vasodilator effect of dilazep on an epimyocardial small arteriole. The pictures were taken in a beating left ventricle before (**a**) and after (**b**) injection of dilazep (50 μg/kg, i.v.) using an incident fluorescence microscope attached to our floating objective system. Bar shows 100 μm

Discussion

The analysis of the blood flow velocity pattern in coronary microvessels was almost impossible mainly because of technical difficulties. By means of a floating objective system and high-speed cinematography, we analyzed the red cell velocity patterns in arterioles, capillaries, and venules of the epimyocardium in the beating canine left ventricle in situ.

The velocity waveform in the epimyocardial microvessels showed these characteristics: (1) the blood flood velocity was accelerated during systole, and reached a peak in mid-systole in small arterioles and capillaries and in late systole in small venules; (2) an abrupt decline of blood flow velocity and momentary cessation or reverse flow was observed in the epimyocardial microvessels. Furthermore, the effects of nitroglycerin and dilazep, an adenosine potentiator, on velocity patterns in these microvessels and on the internal diameter of small arterioles were directly estimated in the beating left ventricle in situ. As demonstrated in the present study, the phasic blood flow velocity patterns in the epimyocardial microvessels were apparently different from those in the epicardial coronary artery and the septal artery, which have been described by others [2, 3, 8, 9]. It is well known that under normal conditions, approximately 80% of coronary blood flow occurs during diastole. Using a pulsed Doppler flowmeter,

Marcus et al. [4, 10] reported that peak blood velocity was observed during diastole in the epicardial coronary artery and the septal artery in the beating canine and human heart. They also showed that the percentage of total coronary blood flow velocity occurring during diastole, per cardiac cycle, was significantly greater in the septal artery (92%) than in the left anterior descending artery (75%) in canine heart. More recently, using a laser Doppler velocimeter, Kajiya et al. [8, 9] reported that systolic forward flow was negligible in the septal artery. In the present study, we have shown that the peak blood flow velocity was observed in the ejection phase, and that a significantly higher percentage of the total area under the velocity curve occurred during the ejection phase in the epimyocadial microvessels of the beating left ventricle. These differences mainly result from the differences in measuring site, apparatus, and experimental preparation. Theoretical considerations and the limitations of our microscope system have been reported elsewhere [5]. In our experimental preparation, mechanical factors that disturb the coronary microcirculation are eliminated as much as possible. [5]. It has been suggested that the blood flow pattern in the epicardial artery does not necessarily represent the intramural flow, probably because of the capacitance of the epicardial arteries [2–4, 10] and an intramyocardial capacitance effect [11–16]. Reverse flow in the septal artery during isovolumic or early systole has been reported [3, 4, 9, 17]. It is well known that cardiac contraction inhibits coronary inflow [18] probably because of its extravascular compression effect. Chilian and Marcus [19] found that increasing extravascular forces augmented reverse flow in the septal artery, and suggested that the phasic nature of intramyocardial blood flow velocity is intimately related to extravascular pressure. Since systolic reverse flow has never been recognized in the epicardial coronary artery under normal perfusion, it has been supposed that negative systolic flow must be stored in the capacitance of the epicardial arteries. Moreover, it has been reported that the epicardial capacitance is large enough to store all of a normal systolic stroke flow [4, 20]. Systolic expansion of the epicardial coronary artery has also been detected by direct measurement of arterial diameter [21, 22]. Such effects of capacitance form a convenient concept for explaining why the epicardial coronary blood flow pattern does not reflect the intramyocardial flow pattern. On the other hand, the extravascular compression effect is a relatively minor determinant of myocardial blood flow in the outer layers of the ventricular wall [23, 24]. Using a direct observation method, we found no significant change in the internal diameters of epimyocardial small arterioles and capillaries during the cardiac cycle. This lack of significant change in diameter of the epimyocardial microvessels during the cardiac cycle seems to suggest that myocardial contraction has little effect on the microvascular diameter in the epimyocardium under normal conditions, probably because of lower intramyocardial pressure compared with a relatively high perfusion pressure throughout the entire cardiac cycle. This also suggests that the pressures in these microvessels are relatively constant during the course of the entire cardiac cycle and implies that there may be substantial capacitance-smoothing pressure changes in these small vessels, keeping pressures relatively constant. This may be compatible with the assumption of Spaan et al. [13] that coronary arterial resistance remains constant during the cardiac cycle under normal perfusion.

The lack of significant change in internal diameter indicates that red cell velocity patterns in small arterioles and capillaries reflect the patterns of volume flow in these microvessels. Accordingly, the present study demonstrates not only that blood flow velocity is accelerated during the ejection phase, but also that a considerable amount of myocardial perfusion occurs during systole in the epimyocardium (approximately 50% of a stroke flow in small arterioles). Thus, the present study indicates that characteristics of flow velocity waveform will differ considerably between the epimyocardium and the endocardium. Hamlin et al. [25] have reported that there is a linear reverse relationship between the intramyocardial pressure and the regional myocardial blood flow across the left ventricular wall during systole. Furthermore, we found that intravenous injection of nitroglycerin failed to dilate the small arterioles, or to increase the red cell velocities in microvessels in the epimyocardium. There was almost no effect on the red cell velocity profiles in the epimyocardial microvessels.

Nitroglycerin is well known as a potent dilator of the large coronary arteries, [26, 27] with little effect on the small resistance vessels [28, 29]. The present study has directly confirmed the indirect evidence, in the beating left ventricle. On the other hand, intravenous injection of dilazep, an adenosine potentiator, [7] markedly dilated the epimyocardial small arterioles. The red cell velocity curves in the epimyocardial microvessels were shifted upward with significant increase in the total area under the velocity curves and decrease in reverse flow during isovolumic systole, following administration of dilazep. The adenosine potentiator is known to dilate mainly small resistance vessels [26, 28]. By means of a direct observation method, we have confirmed that coronary small arterioles are markedly dilated for a long time (> 1h) following administration of dilazep, while pressure-rate products remained constant during the experiments. It has been reported that an increase of contractility, by administration of isoproterenol, caused some deformation of the blood flow velocity profiles [8]. In the present study, the characteristics of blood flow velocity profiles were not qualitatively different from control following administration of either nitroglycerin or dilazep.

The present study provides suggestive information that will be useful in considering the transmural distribution of phasic myocardial perfusion. First, reverse flow in the epimyocardial microvessels may suggest: (1) the compression of downstream veins during pre-ejection or isovolumic systole, or (2) the existence of an intramyocardial pumping action as suggested by Spaan et al. [13]. However, we were unable to observe the compression of downstream veins in our microscopic field of view during the cardiac cycle, although we did not observe more distal, large veins. As pointed out by Spaan et al. [13], a large change in volume may not be needed to explain this relatively small portion of reverse flow. This small amount of reverse flow in epimyocardial microvessels may reflect an abrupt increase of tension in the venous side, caused by the sudden increase of the intramyocardial compressive forces at the onset of isovolumic systole. Second, rapid acceleration of blood flow velocity following the reverse flow suggests that during early systole the epimyocardial layer of the left ventricle may be perfused by the reverse flow from the deeper myocardial layer. Thus, the reverse flow from the deeper myocardial layer will not only be stored in capacitance of the epicardial arteries, but also pour into the outer layer of the myocardium.

Third, after reaching its peak velocity, the blood flow velocity in the epimyocardium rapidly decreases, probably because blood flow pours into the deeper myocardial layer as soon as the extravascular compression is released in that region. As a result, a cyclic blood flow shift may occur between the superficial and the deeper myocardial layers in the left ventricular wall. Since the extent of extravascular compression is supposed to be different in each transmyocardial layer [25, 30], it is expected that the phasic blood flow pattern will be considerably different from the epimyocardium to the endomyocardium.

In summary, the present results suggest that it is difficult to predict the phasic blood flow pattern in each myocardial layer from measurements of the phasic blood flow pattern in the epicardial coronary arteries. Characteristics of the blood flow velocity profile in each myocardial layer will be determined by the complex of perfusion pressure, capacitance of epicardial arteries, intramyocardial capacitance, and intensity of intramyocardial pumping action in each myocardial layer.

References

1. Berne RM, Rubio R (1979) Coronary circulation. In: Berne RM, Sperelakis N (eds) Handbook of physiology, Cardiovascular system, vol I. American Physiological Society, Maryland, pp 873–952
2. Gregg DE, Green HD (1940) Registration and interpretation of normal phasic inflow into a left coronary artery by an improved differential manometric method. Am J Physiol 130: 114–125
3. Eckstein RW, Moir TW, Driscol TE (1963) Phasic and mean blood flow in the canine septal artery and an estimate of systolic resistance in deep myocardial vessels. Circ Res 12: 203–211
4. Chilian WM, Marcus ML (1982) Phasic coronary blood flow velocity in intramural and epicardial coronary arteries. Circ Res 50: 775–781
5. Ashikawa K, Kanatsuka H, Suzuki T, Takishima T (1984) A new microscope system for the continuous observation of the coronary microcirculation in the beating canine left ventricle. Microvasc Res 28: 387–394
6. Ashikawa K, Kanatsuka H, Suzuki T, Takishima T (1986) Phasic blood flow velocity pattern in epimyocardial microvessels in the canine left ventricle. Circ Res 59: 704–711
7. Sano N (1974) Enhancement of coronary vasodilation action of adenosine by dilazep and dipyridamole in the dog. Jpn J Pharmacol 24: 471–478
8. Kajiya F, Tomonaga G, Tsujioka K, Ogasawara Y, Nishihara H (1985) Evaluation of local blood flow velocity in proximal and distal coronary arteries by laser Doppler method. J Biomech Eng 107: 10–15
9. Mito K, Ogasawara Y, Hiramatsu O, Wada Y, Tsujioka K, Kajiya F (1988) Evaluation of blood flow velocity waveforms in intramyocardial artery and vein by laser Doppler velocimeter with an optical fiber. In: Manabe T, Zweifach BW, Messmer K (eds) Microcirculation in circulatory disorders. Springer-Verlag, Tokyo, pp 525–528
10. Marcus M, Wright C, Doty D, Eastham C, Laughlin D, Krumm P, Fastenow C, Brody M (1981) Measurements of coronary velocity and reactive hyperemia in the coronary circulation of humans. Circ Res 49: 877–891
11. Salisbury PF, Cross LE, Rieben PA (1961) Physiological factors influencing coronary blood volume in isolated dog hearts. Am J Physiol 200: 633–635
12. Scharf SM, Bromberger-Barnea B (1973) Influence of coronary flow and pressure on cardiac function and coronary vascular volume. Am J Physiol 224: 918–925

13. Spaan JAE, Breuls NPW, Laird JD (1981) Diastolic-systolic coronary flow differences are caused by intramyocardial pump action in the anesthetized dog. Circ Res 49: 584–593
14. Chilian WM, Marcus ML (1984) Coronary venous outflow persists after cessation of coronary arterial inflow. Am J Physiol 247: H984–H990
15. Spaan JAE (1985) Coronary diastolic pressure-flow relation and zero flow pressure explained on the basis of intramyocardial compliance. Circ Res 56: 293–309
16. Kajiya F, Tsujioka K, Goto M, Wada Y, Chen XL, Nakai M, Tadaoka G, Hiramatsu O, Ogasawara Y, Mito K, Tomonaga G (1986) Functional characteristics of intramyocardial capacitance vessels during diastole in the dog. Circ Res 58: 476–485
17. Carew TF, Covell JW (1976) Effect of intramyocardial pressure on the phasic flow in the interventricular septal artery. Cardiovasc Res 10: 56–63
18. Sabiston DC Jr. Gregg DE (1957) Effect of cardiac contraction on coronary blood flow. Circulation 15: 14–20
19. Chilian WM, Marcus ML (1985) Effects of coronary and extravascular pressure on intramyocardial and epicardial blood velocity. Am J Physiol 248: H170–H178
20. Douglas JE, Greenfield JC Jr (1970) Epicardial coronary artery compliance in the dog. Circ Res 27: 921–929
21. Vatner SF, Pagani M, Manders WT, Pasipoularides AD (1980) Alpha-adrenergic vasoconstriction and nitroglycerin vasodilation of large coronary arteries in the conscious dog. J Clin Invest 65: 5–14
22. Tomoike H, Ootsubo H, Sakai K, Kikuchi Y, Nakamura M (1981) Continuous measurement of coronary artery diameter in situ. Am J Physiol 240: H73–H79
23. Hess DS, Bache RJ (1976) Transmural distribution of myocardial blood flow during systole in the awake dog. Circ Res 38: 5–15
24. Russell RE, Chagrasulis RW, Downey JM (1977) Inhibitory effect of cardiac contraction on coronary collateral flow. Am J Physiol 233: H541–H546
25. Hamlin RL, Levesque MJ, Kittleson MD (1982) Intramyocardial pressure and distribution of coronary blood flow during systole and diastole in the horse. Cardiovasc Res 16: 255–262
26. Fam WM, McGregor M (1968) Effect of nitroglycerin and dipyridamole on regional coronary resistance. Circ Res 22: 649–659
27. Feldman RL, Pepine CJ, Conti CR (1981) Magnitude of dilatation of large and small coronary arteries by nitroglycerin. Circulation 64: 324–333
28. Winbury MM, Howe BB, Hefner MA (1969) Effect of nitrate and other coronary dilators on large and small coronary vessels. Hypothesis for the mechanism of action of nitrates. J Pharmacol Exp Ther 168: 70–95
29. Cohen MV, Kirk ES (1973) Differential response of large and small coronary arteries to nitroglycerin and angiotensin. Circ Res 33: 445–453
30. Stein PD, Marzill M, Sabbah HN, Lee T (1980) Systolic and diastolic pressure gradients within the left ventricular wall. Am J Physiol 238: H625–H630

Evaluation of the Velocity Waveform in Intramyocardial Small Vessels

OSAMU HIRAMATSU, KEIICHIRO MITO, and FUMIHIKO KAJIYA[1]

Summary. We measured the blood flow velocities in intramyocardial septal arteries and intramyocardial small veins by our laser Doppler velocimeter (LDV) with an optical fiber. The blood flow velocity pattern in the septal artery was almost exclusively diastolic and always accompanied by a reverse flow during isovolumic contraction. The blood flow velocity pattern of the intramyocardial vein increased rapidly during the isovolumic contraction phase and decelerated after the beginning of diastole. Following nitroglycerin administration, the forward flow in the intramyocardial small vein shows a reciprocal relation with the reverse flow in the intramyocardial artery, indicating that the phasic blood flow velocity patterns in the intramyocardial artery and vein are subject to the effect of myocardial contraction and relaxation on the intramyocardial capacitance vessels.

Key words: Laser Doppler velocimeter—Blood flow velocity—Septal artery—Intramyocardial small vein

Introduction

The epicardial coronary arteries function as capacitors for blood flow, in addition to their function as conduits. This capacitance effect of epicardial coronary vessels may lead to misinterpretation of the actual inflow pattern to the myocardial large coronary vessels [1–4]. Furthermore, flow in large coronary vessels does not provide direct information on the spatial distribution of intramyocardial flow, e.g., epicardial vs endocardial flow. Therefore, measurements of intramyocardial blood flow are needed for better understanding of coronary circulatory physiology. There have been several approaches to the measurement of intramyocardial blood flow, i.e., diffusible indicator techniques, radioactive microspheres, and direct visualization techniques. By applying the tracer methods, it was suggested that the myocardial flow has temporal heterogeneity which may be caused by twinkling of precapillary sphincters, and that there is also spatial heterogeneity [5–7]. Another matter of great interest, the direct measurement of intramyocardial blood velocity and pressure, has been ham-

[1]Department of Medical Engineering and Systems Cardiology, Kawasaki Medical School, 577, Matsushima, Kurashiki, 701-1 Japan

pered by excessive movement of microvessels due to the beating of the heart. However, several investigators have succeeded in measuring them by employing new techniques, i.e., the direct visualization technique and the laser Doppler method with an optical fiber probes [8–11]. We measured the blood velocity in the intramyocardial small veins and arteries [10, 11], and deep sites of the septal artery by Access Route 3 described in Kajiya et al., this volume, pp. 43–53.

Measurement of Blood Velocity Waveforms

Intramyocardial Small Artery and Vein

Four mongrel dogs (15–25 kg) were anesthetized with sodium pentobarbital and the chests were opened. The heart was cradled in the open pericadium and the proximal portion of the septal artery and left anterior descending artery (LAD) were dissected free. A small artery just before the site of penetration into the myocardium and a vein just after its appearance from myocardium were carefully investigated as to whether they were appropriate for velocity measurements. After the selection of a vessel to be measured, the fine optical fiber was inserted carefully into the intramyocardial vessels (the septal artery, a small coronary artery, and vein) through a prepunctured site on the epicardial vascular wall made by a puncher with a similar diameter to the fiber. The fiber was fixed at the epicardial surface with a drop of cyanoacrylate after confirmation of stable Doppler signals of blood flow velocity. The blood flow rate of the LAD was measured using the electromagnetic flowmeter. The left ventricular pressure, aortic pressure, and ECG were monitored. Interventricular and intramyocardial pressures were measured using a needle tip micromanometer (Millar PR230).

Intramyocardial Septal Artery

Figure 1 shows an example of the velocity pattern in the intramyocardial septal artery. The depth of the measured position in this particular case was 18 mm from the orifice of the septal artery. The blood velocity waveform was almost exclusively diastolic, accompanied by a reverse flow during the isovolumic contraction phase. During nitroglycerin administration, the reverse flow in the isovolumic contraction phase was enhanced, and a mid- and/or late- systolic reverse flow component was also present.

Intramyocardial Small Vein

Figure 2 shows the blood velocity pattern in the intramyocardial small vein (approximately 2 mm beneath the cardiac surface). The blood velocity waveform indicated a systolic-predominant pattern, which is compatible with the flows in the great cardiac vein and coronary sinus. However, the onsets of the flow acceleration and deceleration were earlier than those in the epicardial vein, i.e., the velocity increased with the ventricular isovolumic contraction phase and decreased rapidly with beginning of ventricular relaxation. A reverse flow was observed frequently during diastole, indicating a suction-effect of myocardial relaxiation. Administration of nitroglycerin increased the mean forward flow and the systolic flow wave was divided into two components, i.e., the first peak during isovolumic phase and the second peak in mid- and/or late-systole.

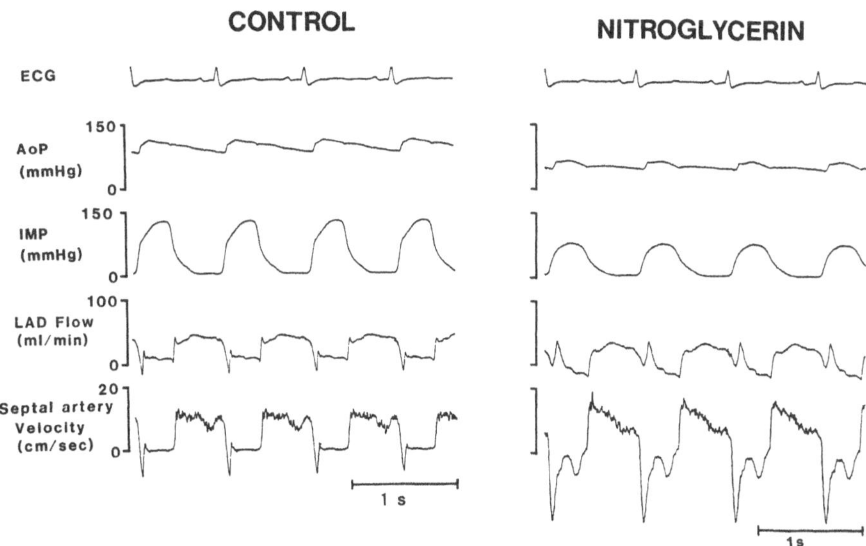

Fig. 1. The blood flow velocity waveform in the deep septal artery. Control vs nitroglycerin administration. *AoP*, aortic pressure; *IMP*, intramyocardial pressure; *LAD*, left anterior descending coronary artery

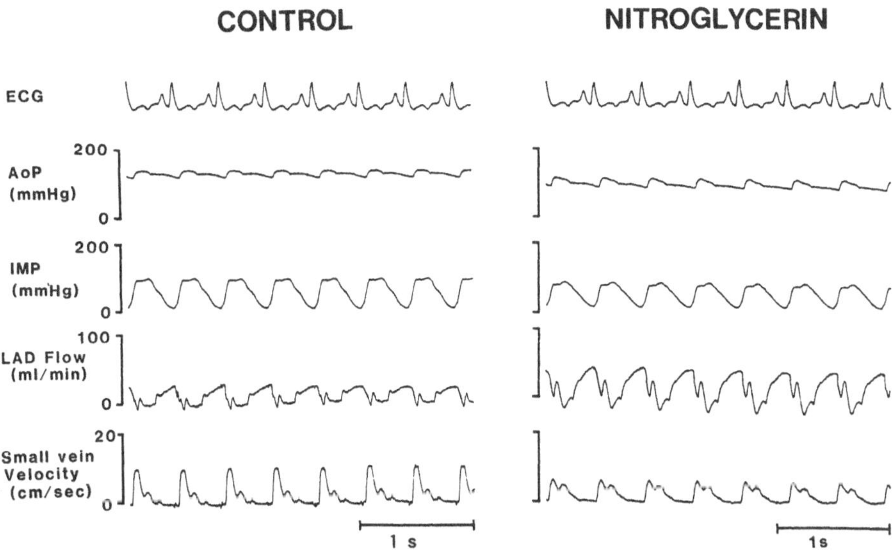

Fig. 2. The blood flow velocity waveform in the intramyocardial small vein. Control vs nitroglycerin administration. *AoP*, aortic pressure; *IMP*, intramyocardial pressure; *LAD*, left anterior descending coronary artery

Concluding Remarks

From our preliminary experimental results, the characteristics of the blood velocity in the intramyocardial small artery and vein were summarized as follows. The intramyocardial coronary artery flow was almost exclusively diastolic and always accompanied by reverse flow during early systole. On the other hand, the intramyocardial vein flow showed a systolic-predominant pattern with a small diastolic flow component and a reverse flow during early half diastole was observed frequently. Following nitroglycerin administration, the forward flow in the intramyocardial coronary vein showed a reciprocal relation with the reverse flow in the intramyocardial coronary artery. These observations may indicate the importance of the mechanical force acting on intramyocardial capacitance vessels. The blood in the deeper portion may be squeezed out by cardiac contraction and the blood may be translocated into the superficial layer, in which extravascular compressive force may be much smaller than that of the deeper layer, since the blood velocity waveforms in superficial arterioles, capillaries, and venules have a significant systolic forward flow component (see Ashikawa et al., this volume, pp. 155–167. During diastole, there may be suction of the blood in superficial vessels into the deeper portion.

References

1. Chilian WM, Marcus ML (1984) Effects of coronary and extravascular pressure on intramyocardial and epicardial blood velocity. Am J Physiol 248: H170–H178
2. Kajiya F, Hoki N, Tomonaga G, Nishihara H (1981) A laser-Doppler-velocimeter using an optical fiber and its application to local velocity measurement in the coronary artery. Experientia 37: 1171–1173
3. Kajiya F, Tomonaga G, Tsujioka K, Ogasawara Y, Nishihara H (1985) Evaluation of local blood flow velocity in proximal and distal coronary arteries by laser Doppler method. Trans ASME J Biomech Eng 107: 10–15
4. Kajiya F, Hiramatsu O, Mito K, Ogasawara Y, Tsujioka K (1987) An optical-fiber laser Doppler velocimeter and its application to measurements of coronary blood flow velocities. Med Prog Technol 12: 77–85
5. Falsetti HL, Carroll RJ, Marcus ML (1975) Temporal heterogeneity of myocardial blood flow in anesthetized dogs. Circulation 52: 848–853
6. Marcus ML, Kerber RE, Erhardt JC, Falsetti HL, Davis DM, Abboud FM (1977) Spatial and temporal heterogeneity of left ventricular perfusion in awake dogs. Am Heart J 94: 748–754
7. Sestier FJ, Mildenberger RR, Klassen GA (1978) Role of autoregulation in spatial and temporal perfusion heterogeneity of canine myocardium. Am J Physiol 235 (Heart Circ Physiol 4): H64–H71
8. Tillmanns H, Ikeda S, Hansen H, Sarma JSM, Fauvel J-M, Bing RJ (1974) Microcirculation in the ventricle of the dog and turtle. Circ Res 34: 561–569
9. Ashikawa K, Kanatsuka H, Suzuki T, Takishima T (1986) Phasic blood flow velocity pattern in epimyocardial microvessels in the beating canine left ventricle. Circ Res 59: 704–711
10. Kajiya F, Mito K, Ogasawara Y, Tsujioka K (1987) Evaluation of phasic blood flow in the coronary circulation by laser Doppler method. In: Tsuchiya M, Asano M, Mishima Y, Oda M (eds) Microcirculation. Excerpta Medica, Amsterdam, pp 352–355
11. Mito K, Ogasawara Y, Hiramatsu O, Wada Y, Tsujioka K, Kajiya F (1987) Evaluation of the velocity wave-forms in intramyocardial small arterial and vein by laser Doppler method. In: Tsuchiya M, Asano M, Mishima Y, Oda M (eds) Microcirculation. Excerpta Medica, Amsterdam, pp 189–190

Microvascular Pressure Profiles in the Left and Right Coronary Circulations

William M. Chilian[1], Susan M. Layne[1], and Stephen H. Nellis[2]

Summary. Most of our knowledge concerning regulation of coronary vascular resistance has been based on measurements obtained from epicardial coronary arteries, and/or measurements of myocardial perfusion. A primary limitation of measurements obtained in this manner is that they cannot distinguish control mechanisms at different levels in the coronary microcirculation. Recently, technological advancements have enabled direct measurements of coronary pressures throughout the microcirculation in the beating heart. These advances have enabled documentation of the microvascular pressure profile in both the left and right coronary circulations during a variety of physiological and pharmacological interventions. The pressure profile can provide an index of vascular resistance, because the steepness of the pressure gradient reflects directly the vascular resistance of a particular vascular segment. The pressure profiles of the coronary microcirculation exhibit several important characteristics. First, coronary arteries greater than 140–180 μm in diameter constitute approximately 20% of total vascular resistance. Second, vascular resistance in these relatively large coronary arterial vessels is due to active vasomotor tone, because it can be lessened substantially with the pharmacological vasodilator, papaverine. Third, small coronary arterioles less than 150 μm in diameter constitute the majority of coronary resistance. Fourth, resistance in these small arteriolar vessels is due primarily to active vasomotor tone, because the vasodilator dipyridamole will decrease resistance in this segment 12-fold. Fifth, the distribution of microvascular resistance in the left and right coronary circulations is similar. This paper discusses the techniques that have been developed to enable measurements of microvascular pressures in the beating heart, and implications of these measurements towards understanding regulatory mechanisms involved in the control of coronary vascular resistance.

Key words: Servo-null technique—Microcirculation—Arterioles—Coronary resistance —Microvascular pressure

Introduction

Historically, direct studies of the coronary microcirculation in the beating heart have been hampered by movement of microvessels during cardiac contraction. Most studies of the coronary microcirculation have involved indirect approaches

[1]Department of Medical Physiology, Microcirculation Research Institute, Texas A&M University College of Medicine, College Station, Texas 77843, USA
[2]Division of Cardiology, Department of Medicine, University of Wisconsin School of Medicine, Madison, WI 53706, USA

in which cardiac motion does not influence the measurement, e.g., indicator-dilution techniques [1–3], nuclide-labeled microspheres [4–5], and plasma-lymph solute concentrations [6]. These approaches do not require actual visualization of the coronary microcirculation, but are restricted by some limitations. First, most indirect measurements are based on principles that require steady state conditions. The time constant of these measurements, therefore, is too large to analyze transient microvascular dynamics. Second, the sample volume of indirect measurements is too excessive to document discrete microvascular phenomena, such as patterns of capillary red cell velocity. Third, indirect approaches do not enable the study of segmental control mechanisms at different levels within the coronary microcirculation. For instance, regulatory factors involved in vascular tone of arterioles versus small arteries cannot be detected. With the advent of new technological developments that have occurred in the last several years, it is now possible to obtain direct measurements of several variables in the coronary microcirculation of the beating or arrested heart, e.g., microvascular diameters [7–9], red cell velocity [10, 11], transcapillary flux of small molecules [12, 13], and microvascular pressures [7, 8, 14, 15].

The intent of this review is to focus primarily on the distribution of microvascular pressures in the left and right coronary circulations during basal conditions and during various experimental interventions that were employed to alter the pressure profile. Since the coronary microcirculation is composed of a series of interconnecting microvessels, the steepness of the pressure gradient in the direction of blood flow directly reflects the distribution of vascular resistance. Thus estimates of relative vascular resistance can be made from the apportionment of microvascular pressures. To provide a perspective from which the profile of coronary microvascular pressures and resistance can be discussed, the microvascular pressure profile in four different tissues will be discussed.

Distribution of Microvascular Pressures in the Brain, Cheek Pouch, and Skeletal Muscle

Figure 1 illustrates the microvascular pressure profiles in cheek pouch (cutaneous microcirculation) [16], pia mater (cerebral microcirculation) [17], and cremaster [18] and tenuissimus muscles [19] (skeletal muscle microcirculations).

The cheek pouch of the hamster is a preparation that has been used for many microvascular studies since the pioneering work of Fulton, et al. [20] approximately 40 years ago. Despite the plethora of studies which have described mechanisms of blood flow regulation [21], oxygen sensitivity, [22, 23], and pharmacological responses [24], only recently has the pressure profile throughout the entire cheek pouch circulation been described [16]. Figure 1 illustrates the pressures in the aorta and throughout the cheek pouch microcirculation. Fifty percent of the total pressure drop in the cheek pouch occurred in small arteries between 250 and 80 μm in diameter. Arterioles between 15 and 80 μm in diameter only constituted about 25%–30% of the total pressure drop, with the remaining pressure dissipation (10%–15%) due to venous resistance.

The cremaster preparation has been widely used for studies of skeletal muscle

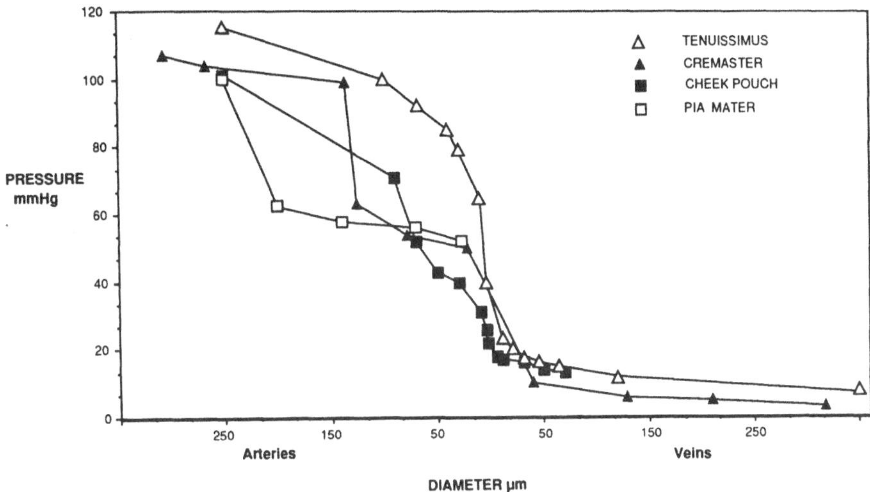

Fig. 1. The distribution of microvascular pressures in four different organ systems: cheek pouch [16], pial microcirculation [17], cremaster microcirculation [18], and tenuissimus [19]. Note the marked differences of the pressure profile among the various organ systems, with the pial microcirculation demonstrating the most significant proportion of resistance residing in relatively large vessels. In contrast, the steepness of the pressure profile on the tenuissimus muscle is greatest in the smallest arterioles, indicating that the majority of vascular resistance occurs at that site. The pressure profiles of the cheek pouch and cremaster lie between those represented by the pial and tenuissimus microcirculations

microcirculation [25]. The pressure profile of the cremasteric microcirculation is shown in Fig. 1. The pressure drop across small arteries, 125–175 μm in diameter, was 30–40 mmHg, implying that 40% of vascular resistance is between these large vessels. Much of the pressure drop, however, occurred across the arteriolar, capillary, and venular network.

Also shown in Fig. 1 are microvascular pressures in another skeletal muscle microvascular preparation, the cat tenuissimus muscle [16]. In constrast to the cremasteric microcirculation, most of the resistance in this system was found in vessels smaller than 80 μm in diameter.

Of the four tissues represented in Fig. 1, the pial circulation has the most significant resistance distributed in the largest arteries [17]. Approximately 50% of the total pressure drop occurred between the aorta and arterial vessels larger than 200 μm in diameter. Interestingly, there was almost no pressure drop along pial arteries and arterioles between 200 and 50 μm. The pial and the cremaster circulation have bimodal resistance distributions with a peak in resistance occurring in larger arteries and a second peak within the arteriolar, capillary, and venular network.

Collectively, the data presented in Fig. 1 show the marked variations of microvascular pressure distribution in different organ systems. The data also challenge the classical concept that small arterioles are the primary site of vascular resistance. In the brain, cremaster, and cheek pouch, about 40%–50% of the total pressure dissipation and vascular resistance was possessed by small arteries.

Distribution of Coronary Pressures

The pressure profile in the coronary circulation has previously been estimated by two experimental approaches. In the first approach, pressure at an epicardial coronary artery branch was used to divide the coronary circulation into proximal (aorta to branch) and distal (branch to vein) coronary vascular resistances [26–29]. This classical approach has been used to document the effects of several different pharmacological vasodilators and vasoconstrictors on the coronary circulation. For example, adrenergic activation in epicardial and distal coronary arteries with epinephrine or norepinephrine produced a decrease in distal artery resistance, but an increase in epicardial artery resistance [26]. On the other hand, Kelley and Feigl reported that α-adrenergic activation caused proportionate increases in epicardial artery and distal vessel coronary resistance [27]. Some pharmacological vasodilators such as nitroglycerin were found to reduce the pressure gardient across the epicardial coronary arteries, but others, such as dipyridamole, greatly augmented the pressure drop [28, 29]. This segmental approach divides the entire coronary circulation into only two vascular segments and thus cannot document discrete vascular responses within the microcirculation. Thus, a primary limitation is that the entire distal vasculature, *all* vessels less than about 500 μm in diameter, is viewed as a single vascular compartment. This approach, therefore, is too insensitive for the study of vasoactive responses at specific microvascular locations, e.g., 50 and 100 μm diameter arterioles.

The second, more recent approach for studying segmental vascular responses was been developed by Klassen and colleagues [30, 31]. This technique involves the insertion of small catheters into coronary arteries and veins, then feeding the catheters upstream and downstream to provide several pressure recording sites. Figure 2 shows examples of the pressure gradients between different measurement sites. These gradients are (1) aorta to branch of an epicardial artery, (2) branch of an epicardial artery to pressure in an obstructed coronary artery, (3) obstructed coronary artery to upstream-directed epicardial venous catheter, and (4) upstream venous catheter to epicardial vein. Interventions that influence the systolic (S) and diastolic (D) pressure gradients are also shown. It is worth noting that the major pressure drop occurred between the epicardial arteries and obstructed arteries and that the various interventions influenced the pressure gradients between the different vascular sites in a heterogenous manner. Although interesting information can be elucidated by this technique, there are some noteworthy considerations. It is problematic to determine which vascular location is represented by the pressure measurement, especially considering the changes in vascular tone which occur with the presence of the catheters and occlusion of arterioles. In the occluded artery, pressure undoubtedly represents an anatomical location somewhere in the microcirculation, but the exact location is uncertain and, more importantly, depends on the extent of the collateral circulation. These investigators succeed in dividing the complete coronary circulation into only four resistance segments and thus assume considerable spatial homogeneity.

In the aggregate, these two methods for examining segmental coronary vascular responses provides interesting and useful data, but do not provide insight into

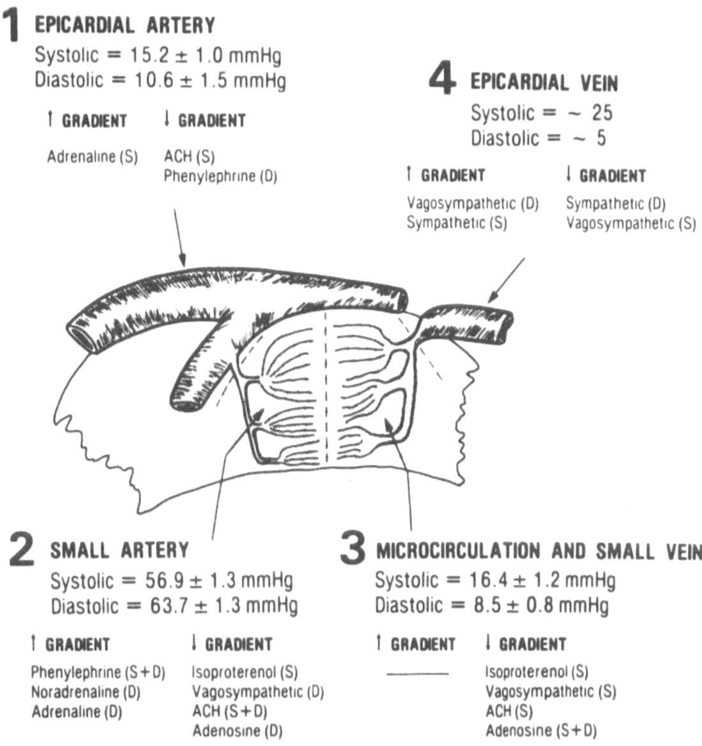

Fig. 2. An illustration of the four vascular sites in the coronary circulation represented by the measured pressure gradients. Each of the different vascular regions shows the gradients during systole (*S*) and diastole (*D*) under baseline conditions and the influence of various pharmacological interventions on the pressure gradients. *ACH*, acetylcholine. This figure was originally presented in [30] and is published with permission of Dr. Gerald A. Klassen and the Medical Research Council of Canada

discrete coronary microvascular phenomena, occurring at different vascular levels. The data developed by these techniques could be interpreted better if pressure profiles within the coronary microcirculation were known. With the development and application of micropuncture techniques in the coronary circulation by Tillmanns [32] and Nellis [8] and their respective colleagues, it is now possible to obtain direct measurements of coronary microvascular pressure in the beating heart. Thus, the distribution of coronary microvascular resistance can be assessed in the contracting myocardium.

Distribution of Coronary Microvascular Pressures

Prior to 1978 the distribution of microvascular pressures within the left and right coronary circulations was completely unknown. Since that time there have been two methodologies employed to acquire microvascular pressure measurements in the coronary circulation [8, 14, 15, 32]. One method involves nearly total

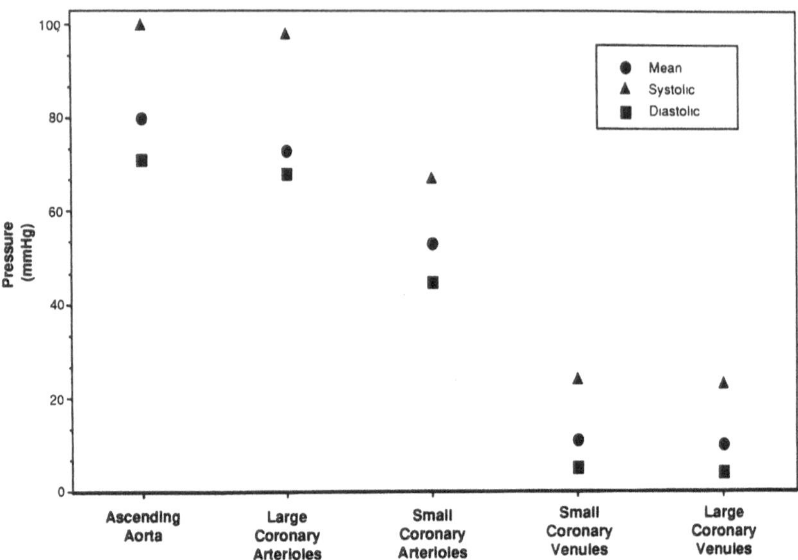

Fig. 3. An illustration of the distribution of systolic, diastolic, and mean microvascular pressures in coronary microvessels of the rat. The large arteriolar compartment is composed of vessels ranging from 150–300 μm in diameter, the small arteriolar compartment is composed of vessels between 25 and 100 μm in diameter, the small venular compartment is represented by veins between 25 and 100 μm in diameter, and the large venular compartment is composed of veins between 100 and 200 μm in diameter. Note the absence of a pressure gradient between the aorta and the large coronary arterioles. Pressures in small coronary arterioles, as well as coronary veins, were found to be significantly lower than that in the aorta. This figure is adapted from the results originally presented by Tillmanns et al. [32]

immobilization of a small portion of the left ventricle and the other enables measurement of microvascular pressures in the beating heart.

Tillmanns and colleagues [32] mechanically immobilized a portion of the left ventricle to minimize movement. These investigators have successfully measured microvascular pressures in coronary arterioles and venules within this region in both rat and cat hearts [32]. The largest pressure drop was downstream from small coronary arterioles, less than 100 μm in diameter. Thus a small fraction of total coronary resistance resided in arteries greater than 100 μm in diameter. Coronary venous pressures were relatively high, with peak venous pressure often in excess of 20 mmHg. Pressures in large coronary arterioles (150–300 μm) averaged only 4% lower than that in the aorta, and pressures in small coronary arterioles (25–100 μm) averaged 33% lower than aortic pressure (Fig. 3). The pressure drop between small arterioles and small venules was approximately 40 mmHg. In this preparation the smallest coronary vessels accounted for well over 50% of the coronary resistance. Interventions such as intravenous infusion of norepinephrine (5–50 μg/min) (Fig. 4) or nitroglycerin administration did not influence the pressure drop from the aorta to the site of micropressure measurement. Thus, neither drug caused a redistribution of resistance.

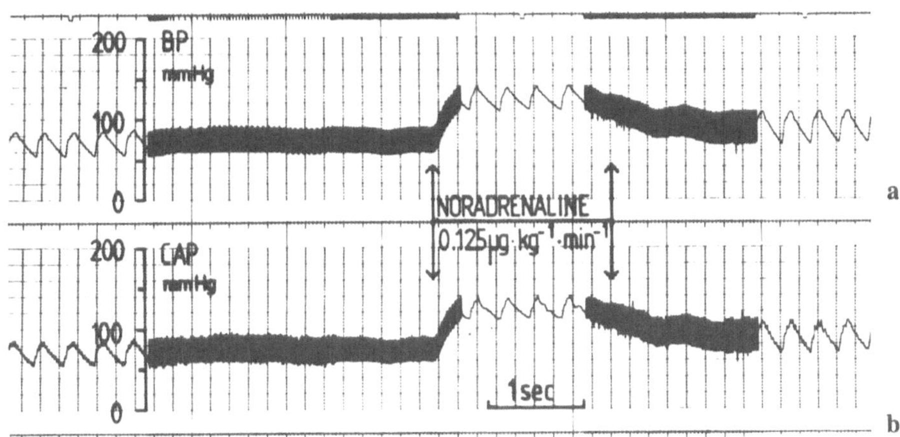

Fig. 4a, b. Phasic recordings of pressures in **a** the aorta (*BP*) and **b** in a large epicardial arteriole (150 µm in diameter, *CAP*) in the cat heart. Under baseline conditions no significant pressure difference was observed, and following application of norepinephrine a parallel rise in both the pressures was observed. These data were originally presented by Tillmanns et al. [32] and are published by permission of the American Heart Association, Inc.

Dipyridamole also was reported not to influence the pressure gradient between the aorta and large coronary arterioles, but did increase pressure in coronary veins. This latter result suggests that dipyridamole causes vasodilation of vessels upstream from the coronary venous system, presumably precapillary arterioles. Although it is difficult to unequivocally ascertain the effects of the immobilization procedure on the microcirculation within the restrained region, it is without question that the results of Tillmanns and colleagues [32] have added significantly to our knowledge concerning the distribution of pressures within the coronary microcirculation.

Advances in technology have permitted microvascular pressure measurements in the beating heart. This approach, developed by Nellis and colleagues, utilized a computer-controlled instrument to move a micropipette in synchrony with the beating right ventricle [8]. Stroboscopic transillumination synchronized to the cardiac cycle (one flash per heartbeat) allowed visualization of the heart. Chilian and colleagues have adapted this approach for micropuncture and epi-illumination of the left ventricle [14, 15].

Figure 5a illustrates the distribution of pressures of the beating rabbit right ventricle under baseline conditions. Pressures in the arteries greater than 140 µm in diameter were about 20 mmHg lower than aortic pressures. Coronary venular pressures averaged about 6 mmHg, with a 6 mmHg pulse pressure. Figure 5b illustrates mean microvascular pressure plotted as a percentage of mean aortic pressure. Pressures in the smaller arteries less than 140 µm in diameter averaged 40% lower than aortic pressure. Larger arteries (greater than 140 µm in diameter) constituted a significant proportion (15%–20%) of vascular resistance. Most of the coronary resistance (60%) was downstream from the

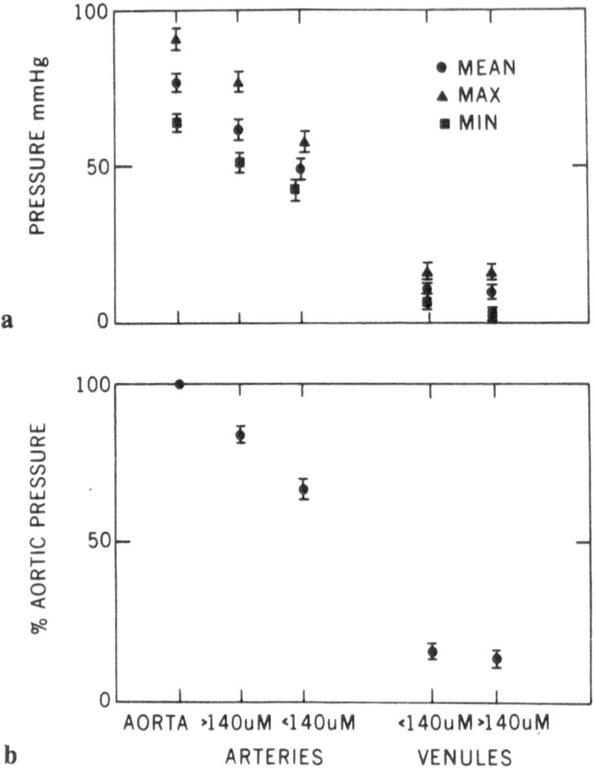

Fig. 5a, b. Systolic (*max*), diastolic (*min*), and mean pressures in rabbit right ventricular coronary arteries and venules. Data were divided according to microvascular diameter into two groups for arteries and veins. **a** The absolute pressure measurements are given along with the standard error of the mean, and **b** mean microvascular pressure is plotted as a percentage of mean aortic pressure. Note in the relatively large coronary arterial vessels (greater than 140 μm) pressures were substantially lower than that in the aorta. These data were originally presented by Nellis et al. [8] and are published by permission of the American Heart Association, Inc.

smaller class of arteries. Coronary veins and venules constituted a minor fraction of total coronary resistance (less than 10%).

Application of this technique to the left ventricle [14, 15] required partial restraint to minimize the "up and down" motion produced by left ventricular contraction. This procedure, however, did not produce changes in coronary hemodynamics because both resting and maximal blood flow in this region was comparable to that in unrestrained regions of the left ventricle [14]. Figure 6 illustrates the distribution of microvascular pressures in the coronary circulation of the cat under control conditions and during intense coronary vasodilation with dipyridamole, sufficient to increase blood flow sixfold. Under control conditions no appreciable drop in pressure was apparent in arteries greater than 270 μm in diameter. Pressures in arteries about 170 μm in diameter were about 20 mmHg lower than that in the aorta. Coronary venous pressures were very low (5–6 mmHg) under control conditions. This figure also illustrates an important fea-

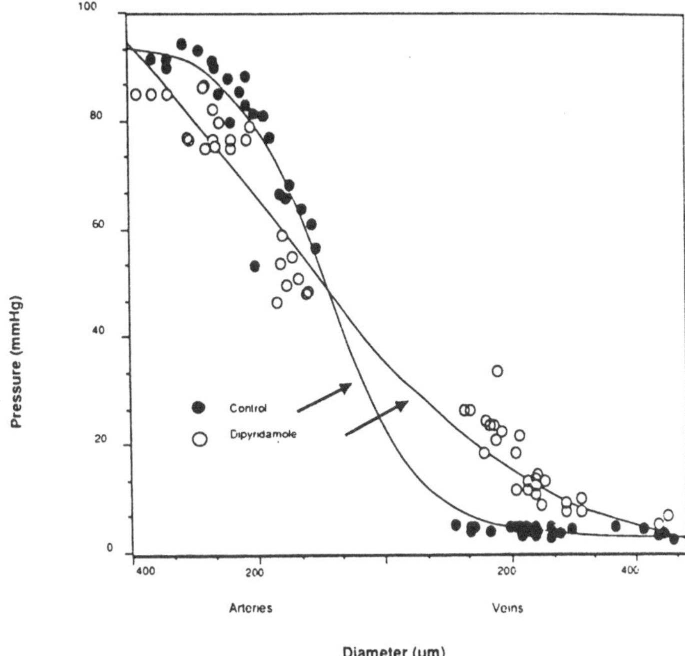

Fig. 6. Mean microvascular pressures in various sizes of coronary microvessels during control conditions with intact coronary vasomotor tone and during coronary vasodilation produced by dipyridamole (0.4 mg·kg^{-1}·min^{-1}). Sigmoidal relationships best describe the pressure distribution data using equations that are fit to the absolute range of diameter values (0–826 μm). Relationships between control conditions and during dipyridamole-induced coronary vasodilation were found to the significantly different ($P < 0.05$). Note that under control conditions the pressure difference between small coronary arterioles and coronary veins was approximately 50 mmHg, but was significantly reduced during dipyridamole administration. These data were originally presented by Chilian et al. [15] and are published with permission from the American Physiological Society

ture of the coronary microcirculation; namely, coronary microvascular pressure distribution is not fixed, rather it can be dramatically changed during alterations in coronary vasomotor tone. Dipyridamole significantly reduced pressures in arteries 170 μm in diameter and substantially augmented coronary venous pressures. The pressure gradient across the coronary microcirculation (between the smallest arteriole and venous vessels) decreased from about 60 mmHg under control conditions to 25 mmHg during coronary vasodilation. This dramatic decrement in resistance must have been due to active vasodilation of coronary microvessels and potential recruitment of parallel vascular channels.

Figure 7 illustrates the effects of papaverine on the fractional distribution of coronary microvascular pressures with mean microvascular pressure plotted as a percentage of mean aortic pressure. Papaverine was infused systemically and caused a threefold increase in coronary blood flow. In contrast to the effects of dipyridamole (Fig. 6), papaverine increased pressures in arteries 170–180 μm in diameter. Under control conditions pressures in these vessels were approxi-

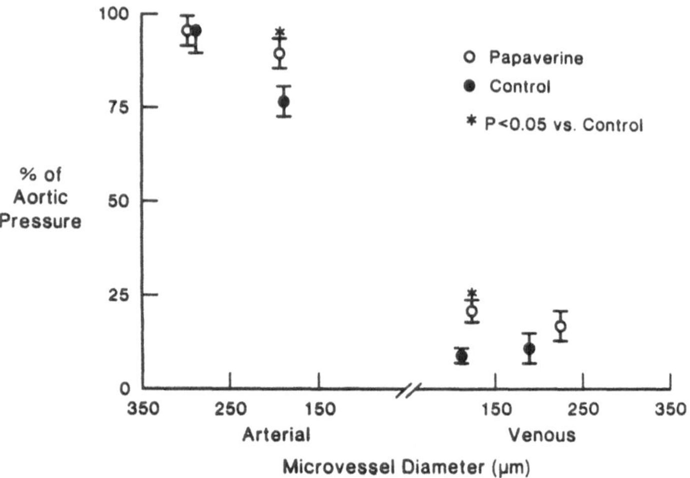

Fig. 7. The effect of papaverine on the fractional distribution of coronary microvascular pressures is shown. Mean microvascular pressure is plotted as a percentage of mean aortic pressure during control conditions and papaverine infusion in four classes of coronary microvessels. Note that papaverine significantly raised pressures in the smaller arterial microvessels and in the venous vessels. This figure was originally presented by Chilian et al. [14] and is published with permission from the Amercian Physiological Society

mately 25% lower than that in the aorta, whereas during papaverine-induced coronary vasodilation, pressures were only 10% lower than aortic pressure. Papaverine also caused venous pressures to increase, but not to the same extent as that during dipyridamole-induced coronary vasodilation. These results suggest strongly that coronary arteries between 170 and 200 μm in diameter have active vasomotor tone during control conditions.

Figure 8 illustrates the coronary vascular resistance divided into three vascular compartments: (1) large artery compartment (resistance of coronary arteries larger than 170 μm in diameter); (2) microvessel compartment (resistance between coronary arterioles less than 170 μm in diameter and coronary venules less than 150 μm in diameter); and (3) venous compartment (resistance of coronary veins larger than 150 μm in diameter). Under control conditions the large artery compartment constituted 25% of coronary resistance. Papaverine and dipyridamole had disparate effects on the contribution of vascular resistance from this segment. Papaverine decreased the relative resistance to 10%, whereas dipyridamole increased it to approximately 40%. Approximately 70% of the resistance in the coronary circulation occurred in the microvessel segment under control conditions. This is elevated slightly during papaverine-induced coronary vasodilation, but reduced significantly (to 25%) during dipyridamole-induced coronary vasodilation. The coronary veins represented about 8% of the normal pressure dissipation in the coronary vasculature under control conditions, but both dipyridamole and papaverine significantly augment the pressure drop across this segment of the coronary circulation. These data illustrate that different interventions may have markedly divergent effects on the distribution of coronary micro-

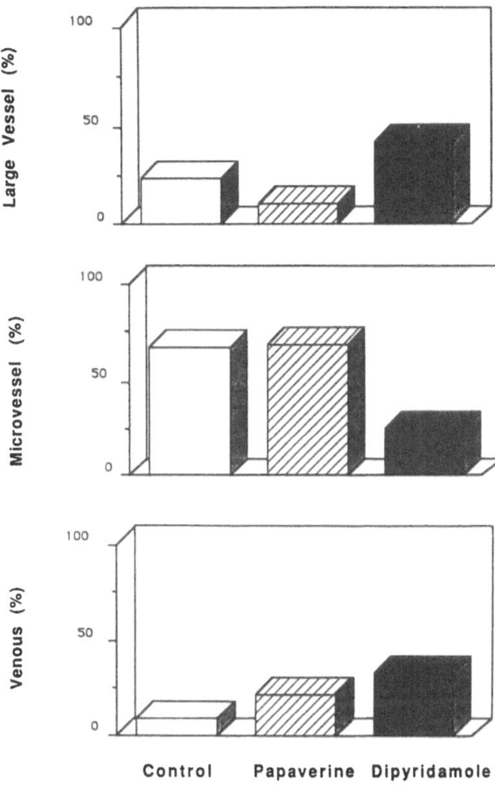

Fig. 8. The fractional dissipation of coronary microvascular pressure in three classes of vessels is shown (see text for definition of the different vessel classes). Note that dipyridamole and papaverine have contrasting effects on the dissipation of pressures in large vessels and microvessels

vascular resistance by influencing coronary vasomotor tone at different vascular locations. These results also suggest that there may be several independent regulatory mechanisms responsible for controlling coronary vasomotor tone.

Table 1 provides a perspective for the fractional distribution of coronary imcrovascular pressures in different animal species and preparations [8, 14, 15, 32]. The major difference among the various studies relates to the relative contribution of large and small arterioles to total coronary vascular resistance. Tillmanns et al. [32] found almost no pressure drop across coronary arteries and large arterioles. In contrast, Nellis et al, [8] and Chilian et al. [14, 15] found pressures in large arterioles significantly less than that in the aorta. Reasons which may account for the differences in results include immobilization of a small area of the superficial left ventricle and species differences. The coronary circulation of most mammalian species is characterized by a high resting resistance, which can decrease sufficiently to produce a 400%–600% increase in coronary blood flow [33]. The rat, however, has high resting blood flow, limited coronary vasodilator reserve, and can only increase blood flow by 200%–250% [34]. Thus, comparisons of the distribution of coronary microvascular pressures in species char-

Table 1. Mean coronary pressure as a percentage of mean aortic pressure under baseline experimental conditions

Laboratory	Arteries	Large arterioles	Small arterioles	Preparation
Tillmanns [32]	98% (200–300 μm)	96% (150–200 μm)	67% (25–100 μm)	Rat; immobilization of a portion of the left ventricle
Nellis [8]	—	83% (< 140 μm)	57% (< 140 μm)	Rabbit; freely beating right ventricle
Chilian [14]	93% (220–440 μm)	77% (210–160 μm)	56% (70–130 μm)	Cat; partially restrained beating left ventricle
Chilian [15]	98% (450–275 μm)	79% (220–170 μm)	—	Cat; partially restrained, beating left ventricle

acterized by different levels of baseline resistance and flow reserve should be exercised with caution. Also, it may be erroneous to compare vessels of similar diameters from animals of greatly different size. A 200 μm arteriole vessel in the cat is most likely at a far different branching order than a similarly sized vessel in the rat. In view of the differences among the preparations and animal species it is not surprising that there were variations reported for the respective coronary microvascular pressure profiles. However, it is worth emphasizing an important *similarity* of the experimental results and conclusions. All the investigators agree that the majority of coronary vascular resistance and the greatest pressure dissipation under baseline experimental conditions occurs in the smallest coronary arterioles [8, 14, 15, 32].

Conclusions

The pressure profiles of the coronary circulation exhibit several salient characteristics. *First*, coronary arteries greater than 140–180 μm in diameter constitute approximately 20% of total coronary vascular resistance. This concept, however, is somewhat controversial because in one animal species, the rat, large coronary arteriolar and arterial vessels do not appear to constitute a major fraction of resistance. *Second*, these relatively large coronary arterial vessels have significant vasomotor tone, because papaverine lessened the pressure drop across this segment. *Third*, small coronary arterioles less than 100–150 μm in diameter constitute the majority of coronary vascular resistance. *Fourth*, resistance in the small arteriolar vessels can be dramatically influenced following administration of the vasodilator dipyridamole. During a 600% increase in blood flow, the pressure drop across these vessels can be decreased from about 60 mmHg to 25 mmHg. *Fifth*, the distribution of microvascular resistance in the left and right coronary circulations in animal species of comparable sizes is markedly similar.

It appears that the distribution of microvascular pressures in the coronary circulation is not unique, rather it follows the distribution of microvascular pressures of some of the organ systems shown in Fig. 1, in which relatively large arterial vessels comprise a significant portion of vascular resistance. The importance of vascular resistance residing throughout the network, rather than largely in one location remains to be unequivocally elucidated. It is tempting to speculate that the resistance in large arterioles and small arteries communicates with the downstream small arterioles through flow-mediated changes [35] or myogenic mechanisms, which have been recently demonstrated in isolated coronary arterioles [36]. For instance, vasodilation of the smallest arterioles associated with increased myocardial oxygen demands would cause an increase in blood flow. This would not only increase the shear forces exerted on these relatively large coronary arteriolar vessels and cause flow-mediated vasodilation, but also decrease intravascular pressures and potentially cause myogenic vasodilation.

All pressure measurements of the coronary microcirculation, to date, have been confined to vessels in the epicardial surface. Whether or not the distribution of microvascular pressures in mid-myocardial or subsendocardial vessels is similar to that in the epicardium is unresolved. This represents an important future direction for our work, especially since pathological events following myocardial ischemia or coronary hypoperfusion occur initially in the subendocardium. It is important, therefore, to elucidate the distribution of subendocardial vascular resistance and hopefully delineate mechanisms for the deleterious events that occur following pathophysiological disturbances in the coronary microcirculation.

Acknowledgments. The authors' studies were partially supported by US Public Health Service Grants from the National Heart, Lung, and Blood Institute HL32788, HL01570, HL20827, HL29917, and the Council for Tobacco Research, Inc. Grant #1709.

References

1. Duran WM, Marsicano TH, Anderson RW (1977) Capillary reserve in isometrically contracting dog hearts. Am J Physiol 233: H276–H281
2. Bassingthwaighte JB, Goresky CA (1984) Modelling in the analysis of solute and water exchange in the microvasculature. In: Renkin EM, Ichel CC (eds) The cardiovascular system, vol 4, Microcirculation. Williams and Wilkins, Baltimore, pp 549–626
3. Ziegler WM, Goresky CA (1971) Transcapillary exchange in the working left ventricle of the dog. Circ Res 29: 181–207
4. Falsetti HL, Carroll RJ, Marcus ML (1975) Temporal heterogeneity of myocardial blood flow in anesthetized dogs. Circulation 52: 848–853
5. King RB, Bassingthwaighte JB. Hales JRS, Rowell LB (1985) Stability of heterogeneity of myocardial blood flow in normal awake baboons. Circ Res 57: 285–295
6. Laine GA, Granger HJ (1985) Microvascular, interstitial, and lymphatic interactions in normal heart. Am J Physiol 249 (Heart Circ Physiol 18): H834–H842
7. Chilian WM, Layne SM, Eastham CL, Marcus ML (1987) Effects of epinephrine on coronary microvascular diameters. Circ Res 61 (Suppl II): II-47–53
8. Nellis SH, Liedtke AJ, Whitesell L (1981) Small coronary vessel pressure and diameter in an intact beating rabbit heart using fixed-position and free-motion techniques. Circ Res 48: 342–353
9. Chilian WM, Layne SM., Eastham CL, Marcus ML (1989) Heterogeneous microvascular coronary alpha-adrenergic vasoconstriction. Circ Res 64: 376–388
10. Steinhausen M, Tillmanns H, Thederan H (1978) Microcirculation of the epimyocardial layer of the heart. I. A method for in vivo observation of the microcirculation of superficial ventricular myocardium of heart and capillary flow pattern under normal and hypoxic conditions. Pflügers Arch 378: 9–14
11. Ashikawa K, Kanatsuka H, Suzuki T, Takishima T (1986) Phasic blood flow velocity pattern in epimyocardial microvessels in beating canine left ventricle. Circ Res 59: 704–711
12. McDonagh PF (1983) Both protein and blood cells reduce coronary microvascular permeability to macromolecules. Am J Physiol 245: H698–H706
13. McDonagh PF, Roberts DJ (1986) Prevention of transcoronary macromolecular leakage after ischemia-reperfusion by the calcium entry blocker nisoldipine. Direct observations in isolated rat hearts. Circ Res 58: 127–136
14. Chilian WM, Eastham CL, Marcus ML (1986) Microvascular distribution of coronary vascular resistance in the beating left ventricle. Am J Physiol 251 (Heart Circ Physiol 20): H779–H788
15. Chilian WM, Layne SM, Klausner EC, Eastham CL, Marcus ML (1989) Redistribution of coronary microvascular resistance produced by dipyridamole. Am J Physiol 256 (Heart Circ Physiol 25: H383–H390)

16. Davis MJ, Ferrer PN, Gore RW (1986) Vascular anatomy and hydrostatic pressure profile in the hamster cheek pouch. Am J Physiol 250 (Heart Circ Physiol 19): H291–H303
17. Shapiro HM, Stromberg DD, Lee DR, Wiederhielm CA (1971) Dynamic pressures in the pial arterial circulation. Am J Physiol 221: 279–283
18. Meininger GA, Fehr KL, Yates MB (1987) Anatomic and hemodynamic characteristics of the blood vessels feeding the cremaster skeletal muscle in rat. Microvasc Res 33: 81–97
19. Fronek K, Zweifach BW (1975) Microvascular pressure distribution in skeletal muscle and effective vasodilation. Am J Physiol 228: 791–796
20. Fulton GP, Jackson RG, Lutz BR (1946) Cinephotomicroscopy of normal blood circulation in the cheek pouch of the hamster, *Cricetus auratus*. Anat Rec 96: 537
21. Lombard JH, Duling BR (1977) Relative contributions of passive and myogenic factors to diameter changes during single arteriole occlusion in the hamster cheek pouch. Circ Res 41: 365–373
22. Davis MJ, Joyner WL, Gilmore JP (1981) Microvascular pressure distribution and responses of pulmonary allografts and cheek pouch arterioles in the hamster to oxygen. Circ Res 49: 125–132
23. Duling BR (1972) Microvascular responses to alterations in oxygen tension. Circ Res 31: 481–489
24. Duling BR, Berne RM (1970) Propagated vasodilation in the microcirculation of the hamster cheek pouch. Circ Res 26: 163–170
25. Baez S (1973) An open cremaster muscle preparation for the study of blood vessels by in vivo microscopy. Microvasc Res 5: 384–394
26. Malindzak GS Jr, Kosinski EJ, Green HD, Yarborough GW (1978) The effects of adrenergic stimulation on conductive and resistive segments of the coronary vascular bed. J Pharmacol Exp Ther 206: 248–258
27. Kelley KO, Feigl EO (1978) Segmental alpha-receptor-mediated vasoconstriction in the canine coronary circulation. Circ Res 43: 908–917
28. Cohen MV, Kirk ES (1973) Differential response of large and small coronary arteries to nitroglycerin and angiotensin. Autoregulation and tachyphylaxis. Circ Res 33: 445–453
29. Fam WM, McGregor M (1968) Effects of nitroglycerin and dipyridamole on regional coronary resistance. Circ Res 22: 649–659
30. Klassen GA, Armour JA, Garner JB (1987) Coronary circulatory pressure gradients. Can J Physiol Pharmacol 65: 520–531
31. Sylven, JCH, Armour JA, Klassen GA (1984) Flow and pressure responses of coronary arteries and veins to vasodilating agents. Can J Physiol Pharmacol 62: 1365–1373
32. Tillmanns H, Steinhausen M, Leinberger H, Thederan H, Kübler W (1981) Pressure measurements in the terminal vascular bed of the epimyocardium of rats and cats. Circ Res 49: 1201–1211
33. Marcus ML (1983) The coronary circulation in health and disease. McGraw-Hill, New York
34. Chilian WM, Wangler RD, Peters KG, Tomanek RJ, Marcus ML (1985) Thyroxin-induced left ventricular hypertrophy in the rat. Anatomical and physiological evidence for angiogenesis. Circ Res 57: 591–598
35. Segal SS, Duling BR (1987) Propagation of vasodilation in resistance vessels of the hamster: Development and review of a working hypothesis. Circ Res 61 (Suppl II): II-20–25
36. Feigl EO (1983) Coronary physiology. Physiol Rev 63: 1–205
37. Kuo L, Davis MJ, Chilian WM (1988) Myogenic activity in isolated subepicardial and subendocardial coronary arterioles. Am J Physiol 255 (Heart Circ Physiol 24): H1558–H1562

6. Control of Coronary Circulation

Mechanisms of Control in the Coronary Circulation

MELVIN L. MARCUS[1]

Key words: Coronary circulation—Myocardial perfusion—Autoregulation—Coronary endothelium—Angiogenesis

Introduction

The purpose of this brief introductory review is to place into perspective the relative importance of several mechanisms that regulate myocardial perfusion. I will also discuss some older concepts that have been the subject of stringent criticism and new developments in research concerning regulation of myocardial perfusion.

Dominant Regulatory Mechanisms

In a normal heart there are three dominant mechanisms that regulate myocardial perfusion: (1) metabolic control, (2) autoregulation, and (3) extravascular compressive forces. In addition, there are three mechanisms of lesser importance: (1) neural, (2) humoral, and (3) myogenic control. Finally, in the past several years it has become clear that the anatomy and functional integrity of the coronary endothelium may play a critical role in coronary regulation particularly in pathological states. All of these factors interact in complex ways.

Although the relative importance of these control mechanisms is well established in normal hearts, in the presence of a pathologic state, a relatively minor control mechanism can become extremely important. An excellent example of this is alpha-adrenergic stimulation. In the normal heart, alpha-adrenergic stimulation produces a minor increase in coronary vascular resistance and has no apparent effect on the transmural distribution of myocardial perfusion. In the presence of severe coronary stenosis the effects of alpha-adrenergic stimulation are far more prominent [1, 2].

[1]Department of Internal Medicine and The Cardiovascular Center, College of Medicine, University of Iowa, E325-1, and The Veterans Administration Hospital, Iowa City, IA 52242, USA

Outdated Concepts

During the past decade at least two concepts have been severely criticized: the adenosine hypothesis and critical closing or stop-flow pressure.

The results of many studies with adenosine antagonists and adenosine deaminase in sufficient concentration to prevent the effects of pharmacologic doses of infused adenosine, suggest that exercise-induced coronary dilation [3], autoregulation [4], and reactive hyperemia [5] are largely unrelated to adenosine. Furthermore, the old classical adenosine experiments involving measurements of tissue adenosine or adenosine concerntration in coronary venous blood have been severely challenged on methodological grounds.

When Bellamy published his initial paper [6] on stop-flow pressure in the coronary circulation, he stimulated the interest of many established coronary physiologists in this area. As a result, several laboratories began to study the "back pressure" in coronary circulation. Eventually the high stop-flow pressures reported by Bellamy were shown to be related to coronary capacitance or due to methodological problems. Direct measurements of red cell velocity in coronary microvessels in the beating heart made by Kanatsuka and co-workers [7] have established that critical closure of conronary arterial microvessels does not occur and that stop-flow pressure is similar to right atrial pressure and probably identical to coronary microvenous pressure. Bellamy should be given credit for stimulating studies that brought to the forefront the potential importance of coronary capacitance in understanding certain aspects of coronary circulatory control.

New Developments

Studies in Humans

In 1989, it is now possible to perform in humans almost all of the animal studies originally done by Dr. Gregg and his co-workers. Besides measuring systemic hemodynamics, it is possible in humans to now measure coronary dimensions of conduit vessels, phasic coronary velocity, the velocity profile in normal and diseased coronary vessels, coronary reactive hyperemia, coronary reserve, and responses to neural and humoral agents. Studies in isolated coronary vascular rings from humans are now being done in many laboratories. In addition, cardiac chamber volume, regional ventricular function, and valvular function can now be measured with impressive accuracy. Furthermore, drugs and humoral agents can now be administered directly into coronary vessels. In the near future, it will be possible to measure coronary collateral perfusion accurately and regional perfusion of the left ventricle including the transmural distribution of flow. Furthermore, it will be possible to measure regional metabolism and the transmural distribution of high energy phosphates. Thus, a virtual explosion in technology will allow clinical investigators to test the validity of concepts about the coronary circulation developed in the animal laboratories in the clinical arena.

In my view this trend is welcome. It will always be difficult to extrapolate findings from animal experiments to patients. Furthermore, there are now many examples, especially in disease states, where studies in animals and humans have yielded widely disparate results mainly because the animal models of disease do not closely simulate clinical disease states.

Future studies of the coronary circulation in humans are likely to yield a vast amount of new insight about mechanisms that control the coronary circulation in patients in normal and diseased states.

Heterogeneity of Coronary Responses

The father figures of modern coronary physiology (Katz, Gregg, and Berne) considered the coronary circulation to be divided into two main components—large conduit coronary vessels and all other smaller coronary arteries and veins. Implicit in this concept is the notion that these two components function in a homogeneous manner. It has now been shown that there are many exceptions to this idea [8–12]. Studies of the in vivo coronary microcirculation suggest that numerous regulatory mechanisms that modulate coronary vascular resistance exert very different effects on coronary arterial microvessels smaller or larger than 100–150 μm in diameter. In addition, mature coronary collaterals and coronary veins have different regulatory mechanisms than other vessels in the coronary tree.

The concept that marked heterogeneity of regulatory mechanisms exists in various segments in the coronary vasculature to a large variety of stimuli is very important. This area deserves much attention from investigators.

Coronary Endothelium

In the 1980s there has been an enormous interest in the role of the endothelium in modulating coronary vasomotor tone. It is now established that many humoral substances—serotonin, acetylcholine, vasopressin, etc.—influence coronary tone in part by their effects on the coronary endothelium. In addition, many disease states—diabetes, hypertension, cardiac transplantation, and atherosclerosis—appear to modify endothelial function. It has been shown that the endothelium releases both potent dilator and constrictor factors that can influence the tone of both large conduit coronary vessels and coronary microvessels. This exciting area of research is likely to bring forth substantial new information on regulation of the coronary circulation both in normal and pathological states.

Coronary Vascular Growth and Remodeling

It is now clear that several disease states stimulate vascular growth in the coronary circulation in an effort to compensate for the adverse effects of the disease process. There is increasing evidence that vascular growth factors may influence coronary vascular resistance by stimulating growth of vessels in response to ischemia (coronary collaterals) and possibly cardiac hypertrophy [13, 14] In addition, there is evidence that enlargement or remodeling of coronary conduit vessels may be an important compensatory mechanism in obstructive coronary atherosclerosis [15]. This is a fertile area of new investigation that should be vigorously pursued.

Effects of Hypertrophy on the Coronary Circulation

Many studies in animals and patients have now established the concept that in the presence of a stimulus that produces cardiac hypertrophy, increases in muscle mass and the coronary vasculature do not occur at similar rates. This usually results in a net increase in minimal coronary vascular resistance in hypertrophied ventricles and a decrease in coronary reserve [16]. Furthermore, autoregulation is impaired in pressure-induced hypertrophy [17], and there may be alterations in vascular permeability. The complex interaction between coronary vascular and cardiac muscle in hypertrophied ventricles has important clinical implications because most cardiac disease processes are associated with cardiac hypertrophy and, consequently, major alterations in myocardial perfusion. Although many descriptive studies in this area have been reported, the basic mechanisms responsible for the complex interaction between cardiac muscle and coronary vascular growth remains to be elucidated. Furthermore, the effects of regression of hypertrophy on regulation of the coronary vasculature is a largely unexplored but important area.

Conclusion

Impressive technological improvements and exciting new concepts developed in the 1980s have given coronary physiologists vast new fields to explore. Given the powerful research techniques at hand, it is likely that in the 1990s there will be substantial progress in increasing our understanding of basic mechanisms that regulate myocardial perfusion in normal and diseased states.

References

1. Chilian WM, Ackell PH (1988) Transmural differences in sympathetic coronary constriction during exercise in the presence of coronary stenosis. Circ Res 62: 216–225
2. Heusch G, Deussen A (1983) The effects of cardiac sympathetic nerve stimulation on perfusion of stenotic coronary arteries in the dog. Circ Res 53: 8–15
3. Bache RJ, Dai X, Schwarts JS, Homans DC (1988) Role of Adenosine in coronary vasodilation during exercise, Circ Res 62: 846–853
4. Dole WP, Yamada N, Bishop VS, Olsson RA (1985) Role of adenosine in coronary blood flow regulation after reductions in perfusion pressure. Circ Res 56: 517–524
5. Saito D, Steinhard CR, Nixon DG, Olsson RA (1981) Intracoronary adenosine deaminase reduces canine myocardial reactive hyperemia. Circ Res 49: 1262–1267
6. Bellamy RF (1978) Diastolic coronary artery pressure-flow relations in the dog. Circ Res 43: 92–101
7. Kanatsuka H, Ashikawa K, Suzuki T, Komaru T, Takishima T (1990) Diameter change and pressure-red cell velocity relationships in coronary microvessels during long diastoles in the canine left ventricle. Circ Res 66: 503–510
8. Chilian WM, Eastham CL, Marcus ML (1986) Microvascular distribution of coronary vascular resistance in beating left ventricle. Am J Physiol (Heart Circ Physiol) 20: H779–H788

9. Chilian WM, Layne SM, Eastham CL, Marcus ML (1987) Effects of epinephrine on coronary microvascular diameters. Circ Res 61(5): II-47–II-53
10. Chilian WM, Eastham CL, Layne SM, Marcus ML (1986) Non-uniform α-adrenergic constriction in the coronary circulation. Circulation 74(4): II-242
11. Lamping KG, Layne SM, Eastham CL, Marcus ML, Chilian WM (1987) Segmental differences in the response of the coronary arteriolar microcirculation to vasopressin. Fed Proc 46(4): 1532
12. Kanatsuka H, Lamping KG, Eastham CL, Clothier JL, Marcus ML (1987) Metabolic stimulation produces heterogeneous dilation in different size coronary microvessels. Circulation 76: IV-147
13. D'Amore PA, Thompson RW (1987) Mechanisms of angiogenesis, Annu Rev Physiol 49: 453–464
14. Tomanek RJ, Schalk KA, Marcus ML, Harrison DG (1989) Coronary angiogenesis during long-term hypertension and left ventricular hypertrophy in dogs. Circ Res 65: 352–359
15. McPherson DD, Hiratzka LF, Meng R, Armstrong ML, Marcus ML, Kerber RE (1987) Coronary arterial remodeling precedes collateral formation in obstructive coronary arteriosclerosis: High frequency intraoperative epicardial echocardiographic studies. Circulation 76: IV-42
16. Marcus ML, Doty DB, Hiratzka LF, Wright CB, Eastham CL (1982) Decreased coronary reserve—A mechanism for angina pectoris in patients with aortic stenosis and normal coronary arteries. N Engl J Med 307: 1362–1367
17. Harrison DG, Cooper SM, Brooks LA, Marcus ML (1988) Effect of hypertension and left ventricular hypertrophy on the lower limit of coronary autoregulation. Circulation 77: 1108–1115

Sympathetic Control of the Coronary Circulation

Ishio Ninomiya, Tsuyoshi Akiyama, Toshihiro Honda, and Naoki Nishiura[1]

Summary. For better understanding of the neural coronary regulatory mechanism, the efferent pericoronary nerve activity (PCNA), cardiac sympathetic nerve activity (CSNA), and coronary vascular resistance were measured in the anesthetized animal. The presence of tonic sympathetic PCNA, which shows a temporal and spatial nonuniformity and induces a vasoconstrictor tone, was clarified. Baroreflex control of sympathetic PCNA, but not of parasympathetic PCNA, was observed. During systemic hypotension, a reflex increase in CSNA and sympathetic PCNA resulted in a biphasic change (increase and decrease) in coronary vascular resistance due to a combined activation of alpha- and beta-adrenergic receptors in the coronary vessel and myocardium.

Key words. Sympathetic control—Pericoronary nerve activity—Cardiac sympathetic nerve activity—Coronary vascular resistance—Hypotension—Arterial baroceptor control

Introduction

It is well known that the coronary vasculature is innervated by both the sympathetic and parasympathetic divisions of the autonomic nervous system [1–3]. The dual innervation indicates that the coronary circulation would be controlled directly by neural signals transmitting through the perivascular sympathetic, parasympathetic, or both nerve fibers. However, information on neural signals in the pericoronary nerve is still lacking because of the absence of an appropriate recording method.

For better understanding of the neural coronary regulatory mechanism, it is important to clarify the presence of discharges in the sympathetic and parasympathetic pericoronary nerve fibers under various physiological and pathophysiological states, and is necessary to determine the relative contribution of sympathetic and parasympathetic nerve activities to changes in coronary blood flow, regional flow distribution, coronary vascular resistance, or coronary arterial diameter.

Previously, the neural control of coronary vessels was mainly studied by

[1]Department of Cardiac Physiology, National Cardiovascular Center Research Institute, 5-7-1 Fujishirodai, Suita, 565 Japan

measurements of changes in coronary blood flow, coronary vascular resistance, or large coronary diameter, in response to denervation [4–7] or to stimulation of nerve fibers to the heart [8–14]. Reflex effects of natural stimuli, acting through receptors such as the baroreceptor [15–19], chemoreceptor [20], or cardiopulmonary receptor [20, 21], on the coronary circulation have been studied. It has been suggested that any direct neural effect on the coronary circulation is overshadowed by a parallel change in cardiac metabolism due to activation of the cardiac sympathetic nerve [18, 22]. The question of whether or not the reflex changes in neural signals to the coronary vasculature are in similar to those to the heart remains to be solved.

In this paper, in addition to a brief review, new information is presented concerning sympathetic control of coronary circulation. First, to clarify the presence of tonic sympathetic and/or parasympathetic activity involved in regulating the coronary vascular tone, we have measured efferent discharge from pericoronary nerve fibers (PCNA) in the left anterior descending coronary artery. Second, to understand the baroreflex control of the coronary circulation concomitantly with that of the cardiac function, we have measured simultaneously cardiac sympathetic nerve activity (CSNA) and coronary resistance during systemic hypotension, and determined a relation between those two variables. Finally, we have attempted to review the physiological role of the PCNA and CSNA on the coronary circulation.

Tonic Pericoronary Nerve Activity

As pointed out by Feigl [22], with measurements of changes in the coronary circulation in response to denervation, it is not possible to draw a definite conclusion concerning the presence of either sympathetic or parasympathetic vasomotor tone in the coronary vascular bed. To clarify directly the presence of both the sympathetic and parasympathetic pericoronary nerve discharges which determine vasomotor tone, the following experiment was conducted.

Recording Site and Measurements

Perivascular plexus, consisting of both sympathetic and parasympathetic nerve fibers, are situated in the adventitial layer along the larger coronary artery in the dog [2, 3, 23, 24]. After periarterial neurotomy, the total disappearance of the innervation of the conduit coronary segment as well as its branches was observed [23, 24]. This indicates that all neural signals to the coronary artery are transmitted through the pericoronary nerve. However, partial degeneration of the myocardial innervation within the area supplied from this artery was also observed. It is suggested that the pathways of the vasomotor nerve to the coronary vessel differ from that of the cardiomotor nerve to the myocardium, and that the vasomotor nerve to the large coronary artery is located deeper in the adventitial layer.

In this study, using a dissecting microscope in open-chest, pentobarbital-anesthetized dogs, we carefully separated a fine nerve strand (about 0.1 μm in diameter) located in the inner region of the adventitial layer of the large left anterior descending (LAD) coronary artery. However, the possibility that the pericoronary nerve strand contains both efferent and afferent fibers could not be neglected.

To record only efferent nerve discharges, we sectioned the pericoronary nerve strand and measured efferent pericoronary nerve activity (PCNA in impulses/s) from the central cut end of the nerve strand using our published method [25]. We assumed that the majority of PCNA reflected neural discharges controlling the vasomotor tone of the LAD coronary artery and its branched arterial tree, but may not reflect neural discharges controlling myocardial contractility.

Pattern of Tonic Pericoronary Nerve Activity

We measured simultaneously PCNA, electrocardiogram (ECG) and arterial blood pressure (AP) in the anesthetized, open-chest animal. Figure 1 shows a typical example of tonic PCNA, ECG, and AP under control conditions. Tonic PCNA was observed in all five dogs.

In all experiments, the tonic PCNA showed grouped discharges synchronized with the cardiac cycle and respiration. In many cardiac cycles, PCNA increased maximally during part of early systole, decreased during diastole, and then tended to increase during late diastole. The magnitude of cardiac-related PCNA varied among the different cardiac cycles with or without respiratory modulation. In Fig. 1, the cardiac-related PCNA increased to the maximum value of

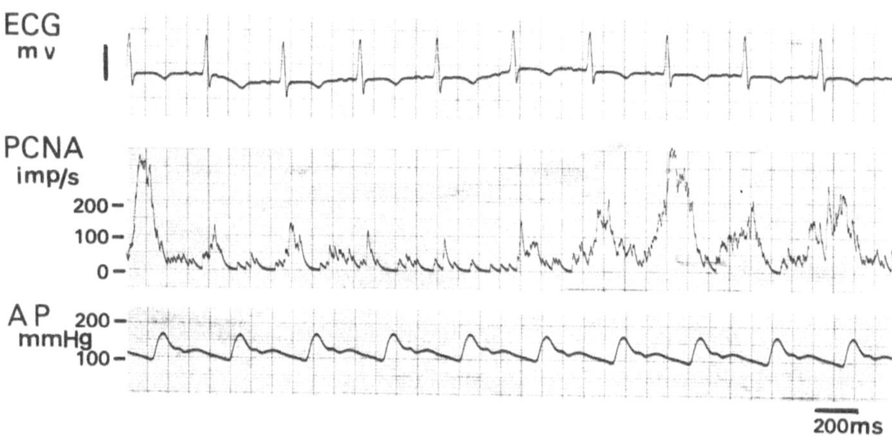

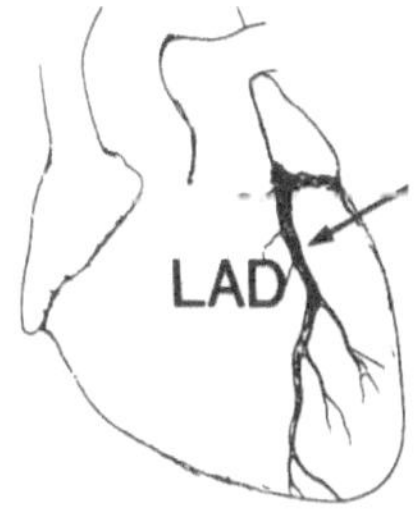

Fig. 1. Electrocardiogram (*ECG*), pericoronary nerve activity (*PCNA*) and arterial blood pressure (*AP*) were measured simultaneously in the open-chest dog anesthetized with sodium pentobarbital. PCNA was recorded from a fine nerve strand located in the inner site of adventitial layer of the large left anterior descending (*LAD*) coronary artery. Cardiac-related changes in PCNA are observed. *imp*, impulses

about 500 impulses/s at the 8th cardiac cycle, whereas it decreased to the minimal value of about 50 impulses/s at the 5th cardiac cycle (average tonic PCNA was 80 impulses/s). This suggested that the LAD coronary artery tree as a whole was regulated by a temporal nonuniform PCNA.

As compared with the frequency of discharge in a single fiber preparation, the larger discharge frquency seen in the multifiber preparation showed that the magnitude of PCNA reflected a number of active fibers. For example, in Fig. 1, the number of active fibers at the 8th cardiac cycle increased as compared with that at the 5th cardiac cycle. This suggested that tonic PCNA distributed with a spatial nouniformity in the LAD coronary vascular tree.

Origin of Tonic Pericoronary Nerve Activity

We consider the possibility that tonic PCNA contains both sympathetic and parasympathetic nerve activity. To determine the origin of tonic PCNA, first, we have blocked bilateral vago-aortic nerves at the middle of the neck by local application of 4% xylocaine or by surgical section. Upon blocking those nerves, the magnitude of the tonic PCNA did not decrease but rather increased. A significant cardiac-related grouped PCNA could still be observed. This finding indicated that the majority of tonic PCNA originated from efferent sympathetic nerve fibers. The rather-enhanced PCNA following vago-aortic denervation may be produced by a lack of inhibitory baroceptor inputs through vago-aortic afferents.

Second, after administration of a ganglionic blockade (hexamethonium bromide, 2 mg/kg, i.v.), the grouped discharges of PCNA disappeared completely, decreasing nearly to the noise level. This finding indicated that the majority of tonic PCNA originated from postganglionic sympathetic fibers but not from parasympathetic fibers.

Influence of Anesthesia on PCNA

In this study, we examined the dose effect of sodium pentobarbital on the magnitude of PCNA. The tonic PCNA usually increased during light anesthesia but decreased with an additional administration of sodium pentobarbital (deep anesthesia). At the end of an experiment, after the administration of large doses of sodium pentobarbital, PCNA disappeared completely, and this level was defined as the noise level.

Recently, different anesthetic effects on tonic and reflex renal sympathetic nerve activity were reported in the conscious cat [26]. It is suggested that the magnitude of sympathetic PCNA altered depend upon the degree and type of anesthesia.

Tonic Pericoronary Nerve Activity: Summary

Tonic PCNA in the LAD coronary artery was observed in the sodium pentobarbital anesthetized dog. The majority of PCNA originated from the postganglionic sympathetic fibers but not from the parasympathetic nerve fibers. Tonic sympathetic PCNA showed grouped discharges related to cardiac and respiratory cycles. Our results suggested that under the present experimental conditions, the neural signal regulating the vasomotor tone of the LAD coronary artery and its branched artery was the tonic sympathetic PCNA, which showed not only temporal but also spatial nonuniformity. The magnitude of tonic sympathetic PCNA altered with the degree of anesthesia.

Arterial Baroceptor Control of the Pericoronary Nerve Activity

Baroreflex Effect on Coronary Hemodynamics

As reviewed previously [1, 22, 27], baroreflex effects on the coronary circulation were studied by many investigators. Usually, carotid sinus hypotension resulted in no change [15] or decrease [18, 19, 22] in coronary vascular resistance. However, with administration of a beta-blocker, carotid sinus hypotension resulted in sympathetic coronary vasoconstriction that was independent of myocardial metabolism, aortic perfusion pressure, and autoregulation [15, 18, 19]. We assumed that during carotid sinus hypotension, the sympathetic PCNA was increased reflexly and in turn caused the alpha-adrenergic vasoconstriction.

On the other hand, carotid sinus hypertension resulted in coronary vasoconstriction [15] or coronary vasodilation [17]. It is suggested that the reflex coronary vasodilation is mediated primarily by activation of parasympathetic vasodilator fibers, but also by inhibition of sympathetic vasoconstrictor fibers [16].

Baroreflex Inhibition of Sympathetic PCNA

The baroreflex effect on sympathetic nerve activity to various organs is quantitatively nonuniform [28, 29]. The majority of sympathetic nerve activity to the kidney was inhibited by the baroreflex, whereas that to skin vessels was not inhibited [28].

To examine the reflex effect of the arterial baroceptor on sympathetic PCNA, arterial blood pressure (AP) was decreased by temporal ligation of the inferior vena cava (IVC), or was elevated with administration of norepinephrine (1 μg/kg, i.v.) or by temporary ligation of the descending aorta (DA) in anesthetized, vagotomized dogs. Figure 2A shows an example of the data obtained. Immediately after the IVC ligation, mean arterial blood pressure (MAP) decreased rapidly and reached the lowest level within 15–20 s. Concomitantly, mean pericoronary nerve activity (MPCNA) increased in inverse proportion to the MAP change, and reached the maximum level. Immediately after the release of the ligation, MAP increased to above the control level, whereas MPCNA decreased to below the control level. During falling MAP, the cardiac- and respiratory-related grouped discharges tended to decrease. On the other hand, when MAP was elevated by the DA ligation, MAP increased suddenly and reached its highest level within 5–8 s. Concomitantly, MPCNA decreased and reached a level near to noise level. Immediately after the release of the ligation, MAP decreased to below the control level, whereas MPCNA increased to above the control value. During rising MAP, the cardiac-related grouped discharges disappeared.

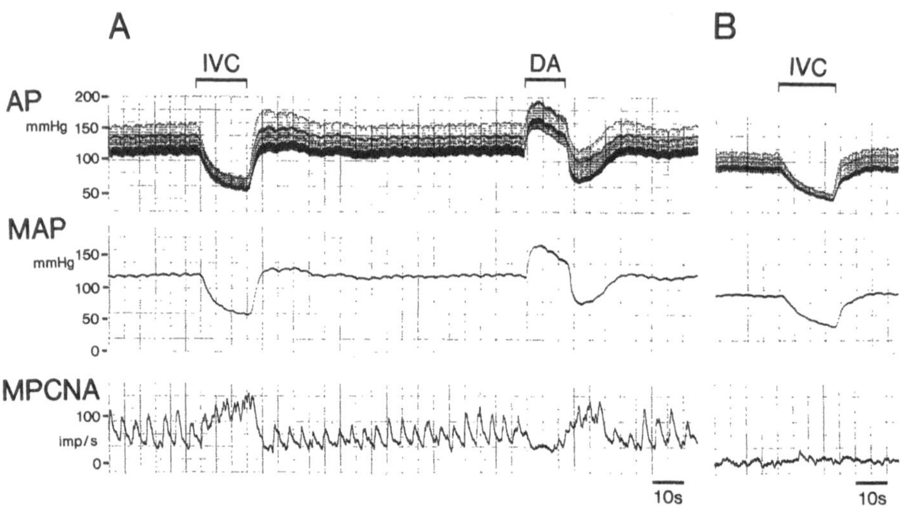

Fig. 2A, B. Arterial blood pressure (*AP*), mean AP (*MAP*) and mean pericoronary nerve activity (*MPCNA*) were measured simultaneously during ligation of the inferior vena cava (*IVC*) and of the descending aorta (*DA*) before (**A**) and after (**B**) administration of hexamethonium bromide. In **A**, tonic MPCNA altered with changes in MAP. In **B**, tonic MPCNA decreased to near the noise level and did not increase with a fall in MAP during the IVC ligation. *imp*, impulses

After pretreatment with ganglion blockade (hexamethonium bromide, 1 mg/kg, i.v.), the control level of MPCNA decreased significantly, to near the noise level (Fig. 2B). When MAP was reduced by the IVC ligation, MPCNA did not increase (Fig. 2B). This finding showed that the majority of increased PCNA, induced by withdrawal of baroceptor input during hypotension, was made up of the activity from the postganglionic sympathetic fibers.

An Inverse Relationship Betweem MAP and MPCNA

Figure 3 shows the relationship between MAP and MPCNA obtained during rising and falling MAP phases due to the IVC and DA ligations. During the rising MAP phase, MPCNA decreased in proportion with the increase in MAP. During the falling MAP phase, MPCNA increased in proportion with the decrease in MAP. From overall data during rising and falling MAP phases, the MAP-MPCNA relationship was analyzed by a least squares linear regression. The correlation coefficient between MAP and MPCNA was significant ($r = -0.93$, $P < 0.01$), and the regression line was $\text{MPCNA} = -1.05 \cdot \text{MAP} + 187$. The high negative correlation between MAP and MPCNA showed that the majority of sympathetic MPCNA was inhibited in proportion to the baroceptor input within a MAP range of 65–165 mmHg.

The inverse relationship between MAP and MPCNA was qualitatively similar to that between MAP and mean cardiac sympathetic nerve activity [25, 29], and between MAP and mean renal nerve activity [25, 26, 28, 29]. This suggested that during hypotension and hypertension, the majority of sympathetic PCNA was

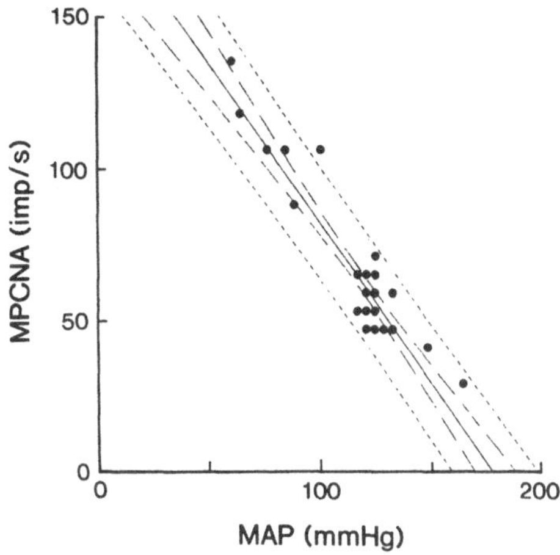

Fig. 3. Mean values of responses in pericoronary nerve activity (*MPCNA*) and arterial blood pressure (*MAP*) are used to analyze an interrelationship between MAP and MPCNA during systemic hypotension and hypertension. Percent changes in MPCNA are expressed as a function of MAP. The *solid line* is the linear regression line. The *inner* pair of *dashed curves* is the 95%-confidence zone of the regression line. The *outer* pair of *dotted curves* is the 95%-confidence zone of sample data. The correlation coefficient between MAP and MPCNA was significant ($r = -0.93$; $P < 0.01$). *imp*, impulses

controlled by the baroreflex, as with sympathetic nerve activity to the heart and kidney.

Baroreflex Excitation of Parasympathetic PCNA

Parasympathetic coronary vasodilation is reflexly activated by carotid baroceptor discharge during carotid sinus hypertension in α-chloralose-anesthetized dogs [13, 16, 21]. We considered that parasympathetic PCNA could be induced reflexly by an increase in baroceptor input when MAP was elevated above the control level, although we could find the tonic parasympathetic PCNA at the normal MAP level.

To examine the presence of parasympathetic PCNA during systemic hypertension, MAP was elevated by DA ligation in the sympathectomized condition with ganglion blockade. When MAP was elevated to 130–165 mmHg by DA ligation, a slight bradycardia occurred, but an increase in MPCNA could not be detected. This suggested that parasympathetic PCNA was not increased reflexly by elevation of MAP up to 130–165 mmHg in the anesthetized, sympathectomized state.

Baroceptor Control of PCNA: Summary

The majority of tonic sympathetic PCNA was inhibited by the elevation of MAP to 160–180 mmHg. Sympathetic PCNA changed with a strong inverse relation to MAP in a range from 65 to 165 mmHg. On the other hand, parasympathetic PCNA did not appear with the elevation of MAP (130–165 mmHg) in the sympathectomized state. Under the present experimental conditions, the baroreflex control of coronary vasomotor tone was mainly mediated by changes in sympathetic PCNA, but not by changes in parasympathetic PCNA.

Cardiac Sympathetic Nerve Activity and Coronary Vascular Resistance During Systemic Hypotension

Cardiomotor and vasomotor fibers in the cardiac sympathetic nerve innervate the myocardium and coronary vessels, respectively [23, 24]. Cardiac sympathetic nerve activity (CSNA) was increased by withdrawal of baroceptor input during systemic hypotension [29]. Sympathetic PCNA in the LAD pericoronary nerve, part of the vasomotor fibers of the cardiac sympathetic nerve, increased during hypotension. Therefore, activity in the cardiomotor fiber may also increase in a similar manner during systemic hypotension.

A parallel increase of CSNA to the myocardium and coronary vessels suggested that the direct effect of sympathetic PCNA on coronary vasomotor tone is altered by the effect of CSNA on myocardial metabolism. Indeed, during carotid sinus hypotension, coronary vascular resistance is decreased, secondary to metabolic vasodilation due to the increased myocardial contractility and heart rate, but not by vasodilation due to autoregulation [18, 22]. The question is whether or not the effect of the metabolic vasodilation always exceeds that of the direct sympathetic vasoconstriction.

To solve this question, it is necessary to analyze quantitatively the relationship between changes in CSNA and coronary vascular resistance during hypotension. The following experiment was carried out.

Measurements

To serve the left cardiac sympathetic nerve intact, a small branch of the cardiac sympathetic nerve running from the right stellate ganglion, crossing the vagal trunk, to the heart was prepared. Efferent cardiac sympathetic nerve activity (CSNA) was recorded from the central cut end of the cardiac sympathetic nerve in 11 anesthetized cats. Analysis of CSNA was made by the method published [25, 29].

In the same animals, mean coronary sinus outflow (MCSF) and mean aortic blood pressure (MAP) were measured. Coronary vascular resistance (CVR; $\text{mmHg} \cdot \text{ml}^{-1} \cdot \text{min}$) is calculated from the ratio of MAP to MCSF. In this study, the inferior vena cava (IVC) was ligated temporarily and partially to induce systemic hypotension (50–70 mmHg). The mean cardiac sympathetic nerve activity (MCSNA), MAP, MCSF, and mean coronary vascular resistance (MCVR) were obtained over a period of 30 s before and during the IVC ligation.

Changes in MCSNA, MAP, MCBF, and CVR due to the IVC Ligation

Before the IVC ligation, MAP was 116 mmHg. Tonic CSNA which showed cardiac- and respiratory-related grouped discharges was observed in all animals.

Immediately after the IVC ligation, MAP suddenly decreased, whereas MCSNA and heart rate increased. At 20–30 s, MAP reached the minimum level of 54 mmHg (46% of the control level), while MCBF significantly decreased, to 50% of that under the control conditions with the IVC ligation. MCSNA increased significantly, by 202%.

On the other hand, the changes in the CVR varied among animals. In 5 of 11 cats, CVR decreased; but in one other cat CVR was unchanged; and in the other 5 cats, CVR increased. On an average, the CVR decreased by 7%. However, the decreases were not statistically significant ($P > 0.05$).

Relationship Between Percentage Changes in MCSNA and MCVR

The percentage change in MCVR is plotted as a function of the percentage change in MCSNA (Fig. 4). In 4 of 5 cases, when MCSNA increased by 75%–178% above the control value, MCVR was increased by about 6%–19%. However, in 6 cases, when MCSNA increased further by 208%–409%, MCVR increased only in 1 case. In the other 4 cases, MCVR decreased significantly, by 10%–32%.

For a better understanding of this biphasic relationship between the changes in these two variables, in Fig. 4, instead of a least squares regression line, the range of CVR responses is arbitrarily shown by the *dotted area*. This suggested that the CSNA caused a biphasic CVR response, i.e., increase and decrease, depending on the magnitude of increased MCSNA.

In summary, during systemic hypotension, CBF decreased whereas CSNA increased reflexly with an inverse relation to the fall in MAP. On the other hand, CVR showed a biphasic change, i.e., increase and decrease, depending on the magnitude of increased MCSNA.

Sympathetic Coronary Regulatory Mechanism and Its Physiological Importance

Based on the present study, the relationship between CSNA and CVR is shown schematically by an operating curve (*heavy solid line*) in Fig. 5. Alpha-adrenergic action of CSNA on CVR, obtained after administration of a beta blocker, is shown by the *thin line*. Beta-adrenergic action (*dashed line*) is simply estimated by subtraction of the alpha-adrenergic line from the operating curve.

Sympathetic Vasomotor Tone, CSNA, and PCNA

When autonomic tone of coronary vasculature was examined indirectly by analysis of coronary vascular responses to denervation, many conflicting results (increase, no change, or decrease in vascular tone) were reported [4–7, 20, 22].

In Fig. 5, at the control CSNA (100%), CVR was about 110%. CVR reached the maximum of 115% at 150% CSNA. When CSNA altered over a range of 0%–275%, CVR increased over a corresponding range of 0%–15%. Our results are in good agreement with the data in which sympathetic coronary vasoconstric-

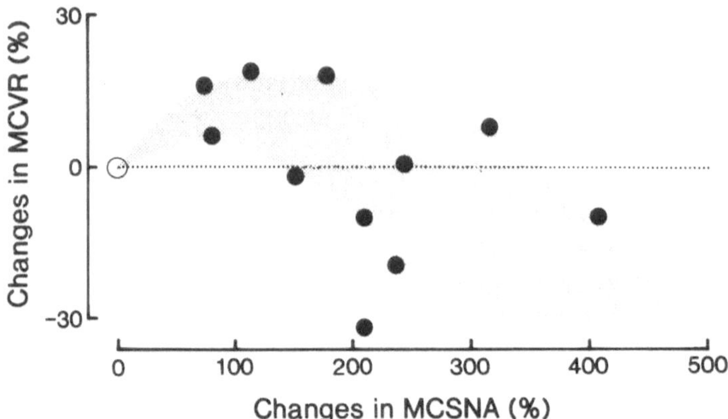

Fig. 4. Percent changes in mean cardiac sympathetic nerve activity (*MCSNA*) and percent changes in mean coronary vascular resistance (*MCVR*) were measured during systemic hypotension. Percent changes in MCVR are plotted as a function of percent changes in MCSNA. A nonlinear relationship between MCSNA (% change) and MCVR (% change) is shown by the *dotted area*. This suggested that a biphasic change, i.e., increase and decrease, of MCVR was produced by the increase in MCSNA

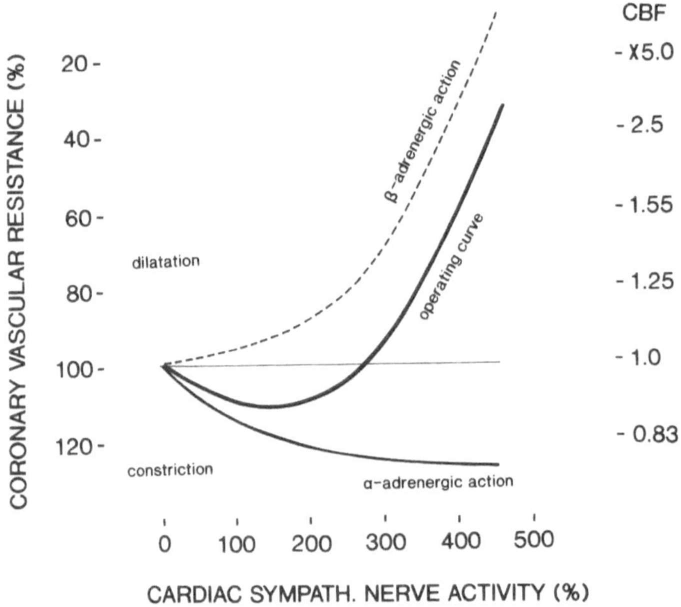

Fig. 5. Relationships between coronary vascular resistance (%) and cardiac sympathetic nerve activity (%) under three different states are schematically shown: *heavy solid curve*, operating curve; *thin solid curve*, alpha-adrenergic action; and *dashed curve*, beta-adrenergic action. In the operating curve, CVR altered over a range of 100%–115%, when CSNA increased over a range of 0%–275%. However, when CSNA increased further, up to 400%, CVR did not increase but decreased progressively to 65%. This suggested that under the latter conditions, vasodilation due to beta-adrenergic action on the myocardium and coronary vessels exceeds vasoconstriction by alpha-adrenergic action on the coronary vessel. Assuming the mean arterial pressure was constant, change in coronary blood flow (*CBF*) was estimated

tor tone in the control state has been demonstrated [5, 7, 20]. This suggested that the increase in CVR, resulting from alpha-adrenergic vasoconstrictor action of CSNA on the coronary vessel, was greater than the decrease in CVR due to a combined beta-adrenergic action of CSNA on the myocardium and coronary vessels. However, some investigators did not observe resting sympathetic vasomotor tone [4, 6]. In Fig. 5, when CSNA decreased to near-noise level or increased to about 275%, no sympathetic vasomotor tone could be found.

With vasomotor tone intact, a substantial portion (45%) of CVR resided in relatively large (>100 μm) coronary arterioles in the anesthetized open-chested cat [30]. Tonic sympathetic PCNA may have produced this large resistance by activation of alpha-adrenergic receptors [7, 8, 11, 12, 31], although there was a secondary myocardial metabolic vasodilator action [32] and probably beta-adrenergic vasodilator action [33–35].

In Fig. 5, CVR decreased from 100% to 65% in proportion to an increase in CSNA from 280% to 400%. This suggested that only when CSNA had increased to about 2–4 times the control value, did vasodilation (induced by beta-adrenergic action on the myocardium and coronary vessels) exceed and progressively predominate over the vasoconstriction induced by alpha-adrenergic action on the coronary vessels. During systemic hypotension, when MAP is reduced a critical level, vasodilation induced by beta-adrenergic action plays an important role in maintaining coronary blood flow, in addition to the autoregulatory mechanism.

Nonuniformity of Sympathetic PCNA and CSNA

Alpha- and beta-adrenergic receptors are distributed nonuniformly along the length of the coronary vessels [10, 11, 31, 36]. Alpha-adrenergic receptors predominated over beta receptors in the large coronary artery, whereas the opposite was true for the small coronary arteries [37, 38]. Coronary vasomotor fibers and their terminals are distributed nonuniformly [1, 3, 23, 24]. Such nonuniform distribution of the receptors and nerve terminals caused nonuniform vasomotor responses in the coronary segments [12, 36]. Nonuniform responses occurred, depending on the degree of nonuniformity of receptors and nerve terminals.

We found that there was temporal and spatial nonuniformity of sympathetic PCNA. The cardiac- and respiratory-related temporal nonuniformity of sympathetic PCNA may cause a fine cycle-to-cycle adjustment of contractile and elastic properties of the LAD coronary artery. The spatial nonuniformity of sympathetic PCNA, induced by an alteration of the number of active fibers, may regulate the number of microvessels utilized. Indeed, chemical sympathectomy increased the percentage of myocardial microvessels perfused, without changing average coronary flow [4].

Conclusion

In the present study, we measured pericoronary nerve activity (PCNA) of the LAD coronary artery in anesthetized, open-chest dogs, and found that there was cardiac- and respiratory-related PCNA. The majority of tonic PCNA originated from postganglionic sympathetic nerve fibers, and the magnitude of tonic PCNA was influenced by the degree of anesthesia. In contrast, there was no evidence of tonic parasympathetic PCNA. We concluded that sympathetic PCNA showed a temporal and spatial nonuniformity, which may have lead to temporally and spatially nonuniform vasomotion in the coronary arteries. The arrangement of the coronary arteries, in series and in parallel, determines the regulation of coronary vasomotor tone and the heterogeneous distribution of blood flow.

Sympathetic PCNA and cardiac sympathetic nerve activity (CSNA) were controlled by the baroceptor reflex during systemic hypertension and hypotension. During systemic hypotension, coronary vascular resistance (CVR) showed a biphasic change, i.e., increase and decrease, depending on the magnitude of CSNA.

Acknowledgments. This study was supported by grants from the Ministry of Education, Science, and Culture and from the Ministry of Health and Welfare of Japan.

References

1. Berne RM, Rubio R (1979) Coronary circulation. In: Berne RM, Sperelakis N, Geiger SR (eds) The cardiovascular system. Am Physiol Soc, Bethesda, pp 873–952 (Handbook of physiology, sect 2, vol 1)
2. Denn MJ, Stone HL (1976) Autonomic innervation of dog coronary arteries. J Appl Physiol 41: 30–35
3. Gerova M (1982) Autonomic innervation of the coronary vasculature. In: Kalsner S(ed) The coronary artery. Croom Helm, London, pp 189–215
4. Acad BA, Weiss HR (1988) Chemical sympathectomy and utilization of coronary capillary reserve in rabbits. Microvasc Res 36: 250–261
5. Brachfeld N, Monroe RG, Gorlin R (1960) Effect of pericoronary denervation on coronary hemodynamics. Am J Physiol 199: 174–178
6. Chilian WM, Boatwright RB, Shoji T, Griggs DM Jr (1981) Evidence against significant resting sympathetic coronary vasoconstrictor tone in the conscious dog. Circ Res 49: 866–876
7. Holtz J, Mayer E, Bassenge E (1977) Demonstration of alpha adrenergic coronary control in different layers of canine myocardium by regional myocardial sympathectomy. Pflugers Arch 372: 187–194
8. Gerova M, Barta E, Gero J (1979) Sympathetic control of major coronary artery diameter in the dog. Circ Res 44: 459–467
9. Hamilton FN, Feigl EO (1976) Coronary vascular sympathetic beta-receptor innervation. Am J Physiol 230: 1569–1576
10. Heusch G, Deussen A, Schipke J, Thamer V (1984) 1- and 2-α-adrenoceptor-mediated vasoconstriction of large and small canine coronary arteries in vivo. J Cardiovasc Pharmacol 6: 961–968
11. Kelley KO, Feigl EO (1978) Segmental α-receptor-mediated vasoconstriction in the canine coronary circulation. Circ Res 43: 908–917
12. Malindzak GS Jr, Kosinski EJ, Green HD, Yarborough GW (1978) The effects of adrenergic stimulation on conductive and resistive segments of the coronary vascular bed. J Pharmacol Exp Ther 206: 248–258
13. Reid JVO, Ito BR, Huang AH, Buffington CW, Feigl EO (1985) Parasympathetic control of transmural coronary blood flow in dogs. Am J Physiol 249: H337–H343
14. Rinkema LE, Thomas JX Jr, Randall WC (1982) Regional coronary vasoconstriction in response to stimulation of stellate ganglia. Am J Physiol 243: H410–H415

15. DiSalvo J, Parker PE, Scott JB, Haddy FJ (1971) Carotid baroreceptor influence on coronary vascular resistance in the anesthetized dog. Am J Physiol 221: 156–160
16. Ito BR, Feigl EO (1985) Carotid baroreceptor reflex coronary vasodilation in the dog. Circ Res 56: 486–495
17. Limet R, Fourny J, Kennedy JH (1975) Effects of the carotid sinus reflex upon coronary vessel tone. J Surg Res 19: 199–207
18. Mohrman DE, Feigl EO (1978) Competition between sympathetic vasoconstriction and metabolic vasodilation in the canine coronary circulation. Circ Res 42: 79–86
19. Powell JR, Feigl EO (1979) Carotid sinus reflex coronary vasoconstriction during controlled myocardial oxygen metabolism in the dog. Circ Res 44: 44–51
20. Vatner SF, McRitchie RJ (1975) Interaction of the chemoreflex and the pulmonary inflation reflex in the regulation of coronary circulation in conscious dogs. Circ Res 37: 664–673
21. Feigl EO (1975) Reflex parasympathetic coronary vasodilation elicited from cardiac receptors in the dog. Circ Res 37: 175–182
22. Feigl EO (1983) Coronary physiology. Physiol Rev 63: 1–205
23. Dolezel S, Gerova M, Gero J, Sladek T, Vasku J (1978) Adrenergic innervation of the coronary arteries and the myocardium. Acta Anat 100: 306–316
24. Dolezel S, Gerova M, Gero J, Sladek T, Vasku J (1980) Monoaminergic pathways to the coronary arteries and to the myocardium. Acta Anat 108: 490–497
25. Ninomiya I, Matsukawa K, Honda T, Nishiura N, Nabuchi A (1988) Effects of baroceptor reflex on cardiac and renal sympathetic nerve activity before and after atropinization in awake cats at rest. Jpn J Physiol 38: 491–506
26. Matsukawa K, Ninomiya I (1989) Anesthetic effects on tonic and reflex renal sympathetic nerve activity in awake cats. Am J Physiol 256: R371–R378
27. Vatner SF, Murray PA (1982) Reflex control of coronary arteries. In: Kalsner S (ed) The coronary artery. Croom Helm, London, pp 216–238
28. Ninomiya I, Irisawa A, Nisimaru N (1973) Nonuniformity of sympathetic nerve activity to the skin and kidney. Am J Physiol 224: 256–264
29. Ninomiya I, Nisimaru N, Irisawa H (1971) Sympathetic nerve activity to the spleen, kidney, and heart in response to baroceptor input. Am J Physiol 221: 1346–1351
30. Chilian WM, Eastham CL, Marcus ML (1986) Microvascular distribution of coronary vascular resistance in beating left ventricle. Am J Physiol 251: H779–H788
31. Ertl G, Fuchs M (1980) Alpha-adrenergic vasoconstriction in arterial and arteriolar sections of the canine coronary circulation. Basic Res Cardiol 75: 600–614
32. Macho P, Hintze TH, Vatner SF (1981) Regulation of large coronary arteries by increases in myocardial metabolic demands in conscious dogs. Circ Res 49: 594–599
33. Vatner SF, Hintze TH, Macho P (1982) Regulation of large coronary arteries by β-adrenergic mechanisms in the conscious dog. Circ Res 51: 56–66
34. Vatner DE, Knight DR, Homcy CJ, Vatner SF, Young MA (1986) Subtypes of β-adrenergic receptors in bovine coronary arteries. Circ Res 59: 463–473
35. Young MA, Vatner SF (1986) Regulation of large coronary arteries. Circ Res 59: 579–596
36. Turlapaty PDMV, Altura BM (1985) Heterogeneous distribution of adrenergic receptors in coronary arteries and their influence on coronary arterial tone. Microcirc Endothelium Lymphatics 2: 617–642
37. Bayer BL, Mentz P, Forster W (1974) Characterization of the adrenoceptors in coronary arteries of pigs. Europ J Pharmacol 29: 58–65
38. Zuberbuhler RC, Bohr DF (1965) Responses of coronary smooth muscle to catecholamines. Circ Res 16: 431–440

Humoral Control of the Coronary Circulation

GERALD A. KLASSEN[1]

Summary. Humoral control of the coronary circulation is reviewed. Endothelium acts as the gatekeeper to transduce humoral perturbations into vascular responses. Adenosine was chosen as an example of a humoral mediator which has potent vasomotor actions. It can stimulate the release of EDRF from endothelial cells or it can act directly on smooth muscle to induce vasodilation. These actions as well as vasoconstrictor activity of the coronary vessels can be probed by using a small dose of adenosine which excites damping oscillations of vasodilation and vasoconstriction. As well, adenosine can modify the responses of smooth muscle cells to catecholamines or peptides. Finally, these humoral responses are integrated throughout the coronary vasculature from epicardial artery to epicardial vein to control overall distribution of coronary blood flow.

Key words: Coronary circulation—Humoral control—Adenosine—EDRF—Endothelium

Introduction

Blood, through its fluid and solid elements, provides the wherewithal for myocardial function. Its importance can be demonstrated by interruption of the coronary circulation which results in loss of myocardial function within beats.

It has been proposed by Gregg [1] that the blood contained within the coronary circulation acts as a moveable plastic material which participates in ventricular shape changes and hence function (the garden hose model). As well, blood is the transporter of oxygen and metabolites required for energy production and structural integrity. Similarly, blood is the mode by which CO_2 and other end products of metabolism are removed. The key to comprehension of humoral control of the coronary circulation lies in understanding the function of the endothelium as it transduces humoral events to the coronary vasculature and the myocardium.

[1] Departments of Medicine, Physiology, and Biophysics, Dalhousie University, and Maritime Heart Centre, Victoria General Hospital, Halifax, N.S. B3H 2Y9, Canada

It has only recently been appreciated that the endothelium should be regarded as the "gateway" to the coronary circulation in that under physiological circumstances it is its responsive barrier which relates perturbations in the vascular system to the coronary vasculature and the myocardium. Blood can be considered to be both a transport media and a signalling organ [2] in that arterial blood contains the integrated products of all of the systemic venous circulatory systems as modified by the pulmonary circulation. These products when presented to the vascular endothelium of the coronary circulation exert specific effects depending upon the responses of this barrier. As well, endothelial cells can release chemicals into the blood [3] which may have effects downstream in the coronary circulation, or which, after leaving the coronary sinus and mixing in the right atrium, may be presented to the pulmonary and systemic circulations for subsequent endothelial-mediated responses. In addition it has been demonstrated that endothelial cells can respond to hemodynamic stress by changing shape, thereby altering their ambient stress [4]; or, by producing vasoactive chemicals [5] which in turn alter the geometry of coronary vessels, returning the vessel to a new vasomotor state. As new chemicals, which participate in this control system, appear in the literature with increasing frequency, this review will concentrate on only a few examples of how this system may function.

Anatomical Considerations

The endothelial lining of the coronary circulation consists of a contiguous continuous monolayer of flattened mononuclear cells. Anatomically they are characterized as being of the A/α type, meaning that they are tightly applied to each other without gaps between cells [6]. Endothelial cells are in communication with each other by tight junctions and by gap junctions. The basement membrane, consisting of Type IV collagen, forms a supportive structure for these delicate cells [7]. There are regional differences in these cells; that is, coronary arterial endothelial cells may differ from coronary venular cells [8] and arterial to venous differences in responses of these cells have been reported [9]. The luminal surface of these cells has an intense interaction with blood products, particularly platelets, through a variety of receptors. All materials which are transported to the vessel wall's media or to the myocardium must pass this barrier. Materials range from water molecules which pass the barrier easily and rapidly to large cells such as macrophages which may enter the vessel wall or myocardium in response to injury. The mechanisms involved in such diverse forms of transport will be variable and the required responses of the endothelial cell in regulating such transport will be multiple.

Bassingthwaighte et al. [10] have calculated that the surface area of the coronary endothelium is 500 cm^2/g of myocardium or approximately two square meters for the average heart. This surface is extremely active in transport of simple molecules such as adenosine [11]. As well, the luminal surface has receptors for lipoproteins [12], peptides [13], catecholamines [14], ions [15], etc., all of which trigger responses from these cells.

Physiological Considerations

The coronary endothelium may be regarded as having at minimum two principal functions. The first, already alluded to, is to act as a selective barrier or facilitative transporter between the blood and myocardium or vessel wall. The second is to regulate flow and pressure so as to accomplish the first. Given the multifactorial vasoactive components of the second function of the system (peptides, prostaglandins, amines, etc.) I have arbitrarily selected one vasoactive molecule to be a model for how vasoactive molecules in blood may act to control coronary vasomotion and hence vascular transport to the vessel wall and to the myocardium. The most studied of these is adenosine, hence its choice.

Adenosine—A Vasoactive Molecule

The vasoactivity of adenosine was early recognized by Szent-Györgyi in 1935 [16]. Subsequently, Berne [17] proposed that it has a major role in the regulation of coronary blood flow. This concept has both supporters and skeptics. Besides its vasoactive properties it is a known modulator of neural function [18].

Adenosine can be both taken up and produced by the coronary endothelial cells [19]. Figure 1 demonstrates the possible pathways involving this process. Adenosine produces vasodilation of coronary arteries both in the presence and absence of the endothelium. Endothelial cells are thought to possess at least two types of adenosine receptors, A_2 and a mixed type P_{1+2}; it is this latter type which is thought to regulate coronary blood flow through the release of endothelial-derived relaxing factor (EDRF) [19]. This compound, which may be NO, produces relaxation of vascular smooth muscle with coronary vasodilation. We have examined this response in the canine coronary circulation by measurement of gradients and have demonstrated that adenosine's major effects are on arterial resistance vessels, and on the post capillary vessels, with less effect on large artery and vein gradients [20].

A consideration of adenosine's effects on coronary vasomotion leads to demonstration of two separate processes. The first is the demonstration of coronary vascular reserve. Coronary vascular reserve [21] can be defined as the maximal capacity of a vessel to dilate. Traditionally adenosine or ischemia has been used to define this point. However, a consideration of the pressure-flow relationship of the coronary circulation demonstrates that the reserve is dependent upon several factors, particularly perfusion pressure. L'Abbate et al. [22] have demonstrated that if adenosine administration is prolonged, coronary flow continues to rise indicating that further dilation is possible. Therefore, coronary vascular reserve is not an absolute number but is determined by a complex interaction of events at the level of the endothelium.

A second component of endothelial control is the setting of coronary vasomotor tone so that coronary supply equals myocardial demand. This must involve a matching of these processes so that metablism of the myocardium is a regulator of coronary arteriolar tone. This is the origin of Berne's adenosine hypothesis.

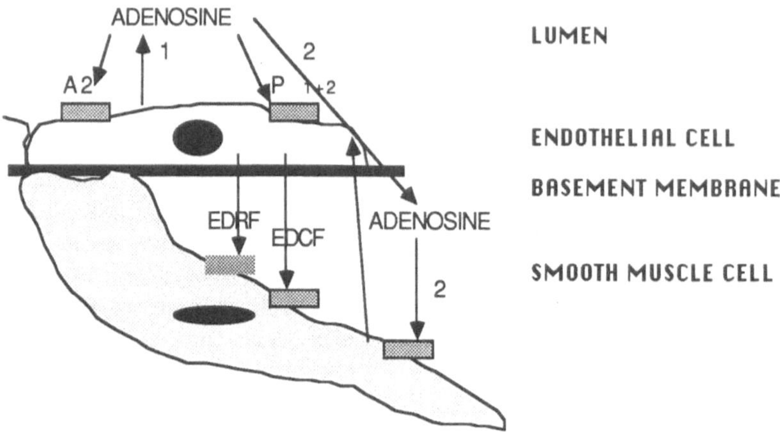

Fig. 1. The vascular-adenosine response unit: adenosine (*1*) in the vascular space binds to two types of receptors on the edothelial cell surface, A_2 (adenylate-cyclase complex) and P_{1+2} (phospholipase C). This latter receptor is thought to regulate coronary flow through release of EDRF (NO), endothelial-derived relaxing factor. As well, adenosine (*2*) can access vascular smooth muscle directly through a discontinuity in the endothelial layer and produce vasodilation. EDCF, endothelial-derived constricting factor (endothelin?) may also be released by the endothelial cell and participate in a "push-pull" fashion to regulate coronary tone dynamically. As well, both smooth muscle and endothelial cells may release adenosine into the vascular space

The endothelium, through gap junctions, has the capacity to act as an integrated responsive system. Evidence for this is not yet available. However, adenosine can be transported from the myocardium to the vascular space and thereby participate in the overall control of blood flow [5] (Fig. 2).

We have recently described a method of probing this control system using a technique described by Kenner and Ono [23]. They showed that a sudden injection (Δ function) of a vasodilator induces oscillation in flow. We studied this phenomenon in a dog model and produced oscillations (Fig. 3) with a frequency of approximately 3 per min in coronary flow. The oscillations demonstrate a vasodilator component (EDRF) and a constrictor component (possibly endothelin release).

The time constant for release of EDRF is thought to be short. Hence, when adenosine is injected as a bolus into the systemic circulation or into the coronary artery in a dose which is less than that which induces a sustained vasodilation, the endothelial cells rapidly take up adenosine and, via the P_{1+2} receptor, release EDRF. The result is a sudden increase in flow as vasodilation occurs. Apparently adenosine, or the process of vasodilation, also induces a slower-responding vasoconstriction which subsequently dominates, with coronary flow being reduced to minimal values. Two observations in our laboratory which may explain the mechanism of the constrictor response are: (1) that even smaller doses of adenosine may occasionally cause only coronary constriction or (2) if the injection is made into the descending aorta, constrictor responses may be observed. Therefore adenosine at a "low dose" may induce a reduction in coron-

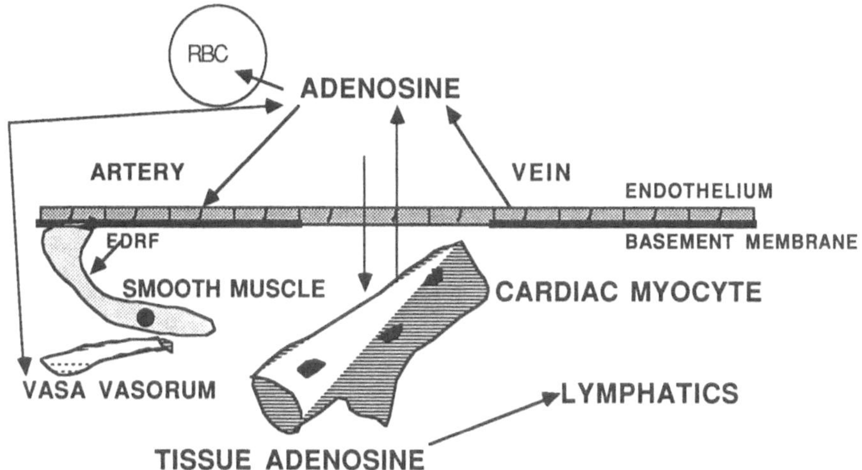

Fig. 2. An integrated coronary-myocardial adenosine system. Adenosine in the vascular lumen may be taken up by red cells, endothelial cells, myocytes, and vascular smooth muscle (? by vasa vasorum). It may also be released by these same tissues into the vascular space and into the lymphatics. Exchange between the myocardium and the vascular space probably is most extensive at the level of the capillaries; however, such exchange may be modified by the endothelium. A dynamic balance contributes to vascular regulation. However, adenosine is only one of many vascular regulatory compounds. *RBC*, red blood cell; *EDRF*, endothelial-derived relaxing factor

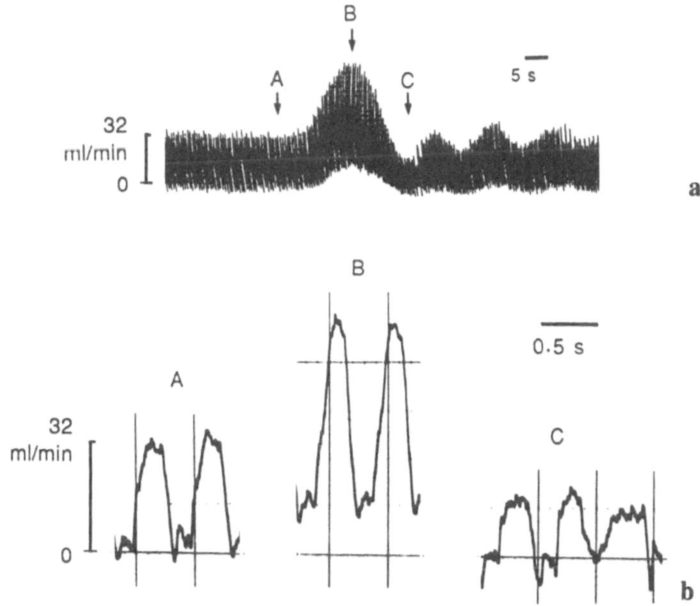

Fig. 3a, b. A typical coronary flow oscillation following a 3.7×10^{-7} mol/kg dose of adenosine into the left atrium. **a** Phasic coronary blood flow. *A* the time of injection, *B* peak hyperemic flow, and *C* the nadir of the first oscillation. **b** Phasic flow at these times for 2 cardiac cycles. The record in *C* is perturbed by an ectopic beat

ary flow, i.e., a paradoxical response. (It should be noted that adenosine always induces vasoconstriction in the kidney through a calcium channel dependent mechanism.) An alternative explanation might be the release of endothelin, a peptide which can be released by endothelial cells, and which is considered to be an effective constrictor of coronary arteries [24]. The release of this compound is known to have a longer time constant than EDRF. Therefore we can observe what has been recently called the *yin and yang* of coronary vasomotion [25]. Thus, through selecting the dose of a vasodilator we can induce oscillations in coronary flow in coronary vessel and thereby probe the vasoactive capacity of the endothelium. The use of this method to evaluate endothelial vasoactive capacity will depend upon the accurate, rapid response time measurement of coronary blood flow and upon being able to defind a dose in which a specified normal response can be observed.

Coronary Disease

It has been demonstrated that atherosclerosis alters the vascular response of coronary arteries [26]. It can do this without mechanical obstruction to flow. Presumably the atheroma interferes with both the mechanical capability of the vessel wall as well as being associated with disruption of the endothelial cell lining of the vessel. Furchgott and Zawadzki [27] have demonstrated that the vascular response to intraluminal acetylcholine changes from vasodilation to vasoconstriction when the endothelium has been interrupted. This has become a test to demonstrate the intactness of the endothelial layer [28]. Bassenge and Galle [29] have demonstrated that human low density lipoprotein (LDL) inhibits the release of EDRF.

Several classes of compounds have been identified which have the capacity to induce vasodilation or constriction of coronary vessels. The prostaglandins are known to be synthesized by the coronary vessels [30] and hence to act as autacoids to integrate vascular responses.

Peptides

The acidic polyamines [31] have been recognized to induce vasodilation of coronary vessels. This is though to occur through the mediation of EDRF as their responses depend upon an intact endothelium. We have demonstrated in Fig. 4 that the response to the peptide oxytocin can be modified by the presence of adenosine. When oxytocin is administered alone it has no effect upon coronary blood flow. However, when it is given within a minute after a bolus of adenosine it causes a vasodilator oscillation typical of that occurring after adenosine administration. Whether this response is common to other peptides is not known. Coronary vasomotor responses to peptides have a different time constant than EDRF—they are generally slow and prolonged. Analysis of such coronary responses is modified by consideration of their peripheral circulatory effects. The mechanism of action of peptides on the endothelial cell has been suggested by Ignarro et al. [32] as involving altered synthesis of EDRF.

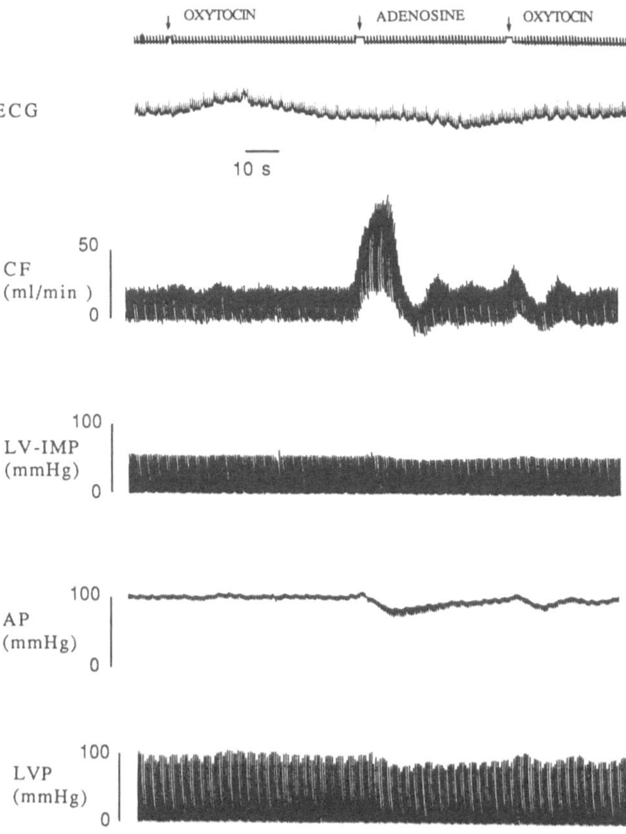

Fig. 4. Responses of ECG (electrocardiogram), CF (coronary blood flow), LV-IMP (left ventricular intramyocardial pressure), AP (aortic pressure), and LVP (left ventricular pressure) to 0.1 unit of oxytocin administered into the left atrium (*first arrow*), which produces no response. Following this a dose of 3.7×10^{-7} mol/kg of adenosine, also given into the left atrium, induces a flow oscillation. When the same dose of oxytocin is repeated an adenosine-like flow oscillation results

Catecholamines

The existence of α and β receptors in the coronary arteries has long been recognized [33]. More recently it has been demonstrated that the distribution of innervation may be heterogeneous within the coronary vasculature [34]. By measuring pressure gradients we have demonstrated that such variability exists [35]. The dominant control of the coronary arterial system is by the sympathetic system where its actions are primarily constrictive in nature. Chilian et al. [36] has demonstrated these differences in control of the epicardial versus the small coronary arteries. Such differences may act to facilitate both transmural and regional distribution of coronary blood flow as part of the integrated process of matching coronary supply to myocardial demand for oxygen delivery.

Another interaction has been reported by Hori et al. [37]. They have demonstrated that there is an interaction between α receptors in the coronary arteries and adenosine. It has become recognized that the myocardial and vascular responses to catecholamines can be modified by adenosine.

In the venous circulation the responses are again variable with different responses occurring at the level of the intramyocardial veins in contrast to the surface epicardial veins. We have demonstrated [38] that the epicardial veins are highly resistive, in that, associated with their shape changes during systole, resistance is reduced; in contrast, during diastole, resistance in these veins increases, which impedes the egress of blood. This results in surface vein flow being discontinuous or bolus in nature. During systole, blood in the high-capacitance intramyocardial vein is squeezed to the surface by contracting myocardium, propelling it towards the subsurface reservoirs of venous blood. As systole progresses, this cavitary venous blood is ejected into surface veins which open both in response to a higher venous pressure and to a change in their length as their epicardial chord shortens. This effect is maximal towards the end of systole when intramyocardial pressure is already decreasing. An element of venous blood momentum may assist in this process. Venous flow ceases during that phase of diastole when early diastolic filling of the ventricle is well on its way to completion and epicardial veins are stretched. Two forces tend to shut the epicardial vein during diastole: the fall in intramyocardial pressure with a resultant fall in venous pressure, and the increasing resistance of the epicardial vein as cavitary filling stretches the vein over the surface of the heart. This process is facilitated by the flatness of these veins and their lack of surface forces to keep them open. However, there is usually minimal flow during diastole; presumably collapse of these veins is not complete.

When sympathetic stimulation occurs, resistance in the epicardial venous system increases as it does in the epicardial arterial system. Thus, ejection of venous blood is retarded so that now more of it is ejected during diastole. This occurs even though intramyocardial pressure (IMP) is increased. This differential response suggests that the capability of epicardial veins to constrict is considerable. Coronary venous flow is as dependent upon humoral elements as is arterial flow.

Discussion

Most questions concerning humoral control of the coronary circulation remain unanswered by this analysis. These relate in part to understanding endothelial function. Do endothelial cells in the coronary circulation respond as a syncytium? If so, is there a hierarchy for such responses, and do local stimuli dominate over regional or integrated responses? What is the mechanism of endothelial-to-endothelial cell communications? What mechanisms other than EDRF permit signalling between endothelial cells and smooth muscle and cardiac muscle? Are there "feedback" systems between muscle and endothelial cells? What is the role of neural control in these interactions? To these questions must be added the questions of what diseases alter these processes or how drugs alter these functions. As well, how are the arterial and venous systems synchronized—is this endothelial dependent or is this related to myocardial function? Considerable research and modeling must be undertaken before the humoral control of the coronary circulation is understood.

Acknowledgment. This work was supported by a grant from the Nova Scotia Heart Foundation.

References

1. Gregg DE (1963) Effect of coronary perfusion pressure or coronary flow on oxygen usage of the myocardium. Circ Res 13: 497–500
2. Klassen GA, Zborowska-Sluis DT (1971) Carbon dioxide mediated glycolysis: a signal system. Proceedings of the International Union of Physiological Sciences, 25th International Congress, Munich, p 307
3. Gerlach E, Nees S, Becker BF (1985) The vascular endothelium: a survey of some newly evolving biochemical and physiological features. Basic Res Cardiol 80: 459–474
4. Davies PF (1989) How do vascular endothelial cells respond to flow? NIPS 4: 22–25
5. Busse R, Trogisch G, Bassenge E (1985) The role of endothelium in the control of vascular tone. Basic Res Cardiol 80: 475–490
6. Bennett HS, Luft JH, Hampton JC (1959) Morphological classifications of vertebrate blood capillaries. Am J Physiol 196: 381–390
7. Jaffe EA, Minick CR, Adelman B, Becker CG, Nachman R (1976) Synthesis of basement membrane collagen by cultured human endothelial cells. J Exp Med 144: 209–225
8. Schelling ME, Meininger CJ, Hawker JR Jr, Granger HJ (1988) Venular endothelial cells from bovine heart. Am J Physiol 254: H1211–H1217
9. De May JG, Vanhoutte PM (1982) Heterogeneous behavior of the canine arterial and venous wall: importance of the endotheium. Circ Res 51: 439–447
10. Bassingthwaighte JB, Yipinstoi T, Harvey RB (1974) Microvasulature of the dog left ventricular myocardium. Microvasc Res 7: 229–249
11. Bassingthwaighte JB, Sparks HV Jr, Chan IS, DeWitt DF, Gorman MW (1985) Modeling of transendothelial transport. Fed Proc 44: 2623–2626
12. Brown MS, Goldstein JL (1986) A receptor-mediated pathway for cholesterol homeostasis. Science 232: 34–47
13. Lynch DR, Snyder SH (1986) Neuropeptides: multiple molecular forms, metabolic pathways, and receptors. Annu Rev Biochem 55: 773–799
14. Vanhoutte PM, Verbeuren TJ, Webb RC (1981) Local modulation of adrenergic neuroeffector interaction in the blood vessel wall. Physiol Rev 61: 151–247
15. Zawadzki JV, Cherry PD, Furchgott RF (1980) Comparison of endothelium-dependent relaxation of rabbit aorta by A23187 and by acetylcholine. Pharmacologist 22: 271
16. Drury AN, Szent-Györgyi A (1929) The physiological action of adenine compounds with especial reference to their action on the mammalian heart. J Physiol (Lond) 68: 214–237
17. Berne RM (1980) The role of adenosine in the regulation of coronary blood flow. Circ Res 47: 807–813
18. Kuan CJ, Jackson EK (1988) Role of adenosine in noradrenergic neurotransmission. Am J Physiol 255 (Heart Circ Physiol 24): H386–H393
19. Nees S, Des Rosiers C, Böck M (1987) Adenosine receptors at the coronary endothelium: functional implications. In: Gerlach E, Becker BF (eds) Topics and perspectives in adenosine research. Springer-Verlag, Berlin Heidelberg, pp 454–467
20. Klassen GA, Armour JA, Garner JB (1987) Coronary circulatory pressure gradients. Can J Physiol Pharmacol 65: 520–531

21. Klassen GA (1989) Considerations of coronary flow reserve in man: methodology, utility and limitations. In: Vigorito C, Rengo F (eds) Evaluation of myocardial perfusion in man. Bibl Cardiol. Basel, Karger 44: 112–122
22. L'Abbate, Camici P, Trivella MG, Pelosi G, Davies GJ, Ballestra AM, Taddei L (1981) Time dependent response of coronary flow to prolonged adenosine infusion: doubling of peak reactive hyperaemic flow. Cardiovasc Res 15: 282–286
23. Kenner T, Ono K (1972) Analysis of slow autooscillations of arterial flow. Pflügers Arch 331: 347–356
24. Yanagisawa M, Kurihara H, Kimura S, Tomobe Y, Kobayashi M, Mitsui Y, Yazaki Y, Goto K, Masaki T (1988) A novel potent vasoconstrictor peptide produced by vascular endothelial cells. Nature 332: 411–415
25. Anonymous (1988) Yin and yang in vasomotor control. Lancet 2: 19–21 (editorial)
26. Freiman PC, Mitchell GG, Heistad DD, Armstrong ML, Harrison DG (1986) Atherosclerosis impairs endothelium-dependent vascular relaxation to acetylcholine and thrombin in primates. Circ Res 58: 783–789
27. Furchgott R, Zawadzki J (1980) The obligatory role of endothelial cells in the relaxation of arterial smooth muscle by acetylcholine. Nature 288: 373–376
28. Ludmer PL, Selwyn AP, Shook TL (1986) Paradoxical vasoconstrictors induced by acetylcholine in atherosclerotic coronary arteries. N Engl J Med 315: 1046–1051
29. Bassenge E, Galle J (1989) Low density lipoprotein (LDL) suppresses EDRF-mediated vasodilation by two mechanisms. FASEB J 3: A532
30. Moncada S, Vane JR (1979) Arachidonic acid metabolites and the interactions between platelets and blood vessel walls. N Engl J Med 300: 1142–1147
31. Ignarro LJ, Gold ME, Buga GM, Byrns RE, Wood KS, Chaudhuri G, Frank G (1989) Basic polyamino acids rich in arginine, lysine, or ornithine cause both enhancement of and refractoriness to formation of endothelium-derived nitric oxide in pulmonary artery and vein. Circ Res 64: 315–329
32. Ignarro LJ, Byrns RE, Buga GM, Wood KS (1987) Mechanisms of endothelium-dependent vascular smooth muscle relaxation elicited by bradykinin and VIP. Am J Physiol 253: H1074–H1082
33. Feigl EO (1983) Coronary physiology. Physiol Rev 63: 1–205
34. Holmgren S, Abrahamsson T, Almgren O (1985) Adrenergic innervation of coronary arteries and ventricular myocardium in the pig: fluorescence microscopic appearance in the normal state and after ischemia. Basic Res Cardiol 80: 18–26
35. Klassen GA, Armour JA, Garner JB (1989) The effects of propranolol, phentolamine and atropine on canine coronary vascular gradients. Can J Physiol Pharmacol 67: 140–151
36. Chilian WM, Layne SM, Eastham CL, Marcus ML (1989) Heterogeneous microvascular coronary α-adrenergic vasoconstriction. Circ Res 64: 376–388
37. Hori M, Kitakaze M, Tamai J, Koretsune Y, Iwai K, Iwakura K, Kagiya T, Kitabatake A, Inoue M, Kamada T (1988) α_2-Adrenoceptor activity exerts dual control of coronary blood flow in canine coronary artery. Am J Physiol 255 (Heart Circ Physiol 24): H250–H260
38. Hellenbrand WK, Klassen GA, Armour JA, Sezerman O, Paton B (1986) Autonomic nervous system regulation of epicardial coronary vein systolic and diastolic blood velocity as measured by a laser Doppler velocimeter. Can J Physiol Pharmacol 64: 1463–1472

Static and Dynamic Control of the Coronary Circulation

ISABELLE VERGROESEN[1], JENNY DANKELMAN[2], and JOS A. E. SPAAN[1,2]

Summary. To explain the characteristics of steady state control of the coronary circulation, many metabolic hypotheses have been formulated in the past. In this chapter, an oxygen-based model and an adenosine-based model are compared to the experimental findings in the literature. In these models the specific assumptions on cause and effect are made explicit by mathematical expressions. The comparison yielded a good fit for the oxygen-based model. For the adenosine model the requirement of the absence of a washout term was essential. Since experiments have shown washout of adenosine, making this latter requirement unrealistic, only the oxygen-based model was extended to the dynamic situation.

The major rate-limiting processes involved in the oxygen model are (1) the mixing of oxygen in the tissue and capillary space and (2) the adaptation of arteriolar resistance to a change in tissue concentration of oxygen. The predictions of this model in terms of response rates of the pressure/flow ratio were compared to experimentally obtained response rates to a sudden change in heart rate in anesthetized goats. The slow part of the experimentally found response was adequately predicted by the dynamic model, but the fast part of the response was not. Mechanical properties have to be incorporated in the model to explain the fast initial responses.

Key words: Regulation—Model—Coronary flow—Autoregulation—Metabolic flow adaptation—Adenosine—Oxygen—Time constants

Introduction

Steady state control of the coronary circulation under physiological conditions is characterized by autoregulation and metabolic flow adaptation [1–5]. Autoregulation can be defined according to Johnson [1] as "the intrinsic tendency of an organ to maintain constant blood flow despite changes in arterial perfusion pressure." When examining autoregulation of the heart, it is necessary to stress the fact that organ metabolism is assumed to be constant. Since the arterial

[1]Department of Medical Physics, University of Amsterdam, AMC, Meibergdreef 15, 1105 AZ Amsterdam, The Netherlands

[2]Laboratory of Measurement and Control, Faculty of Mechanical Engineering and Marine Technology, University of Technology Delft, Mekelweg 2, 2628 CD Delft, The Netherlands

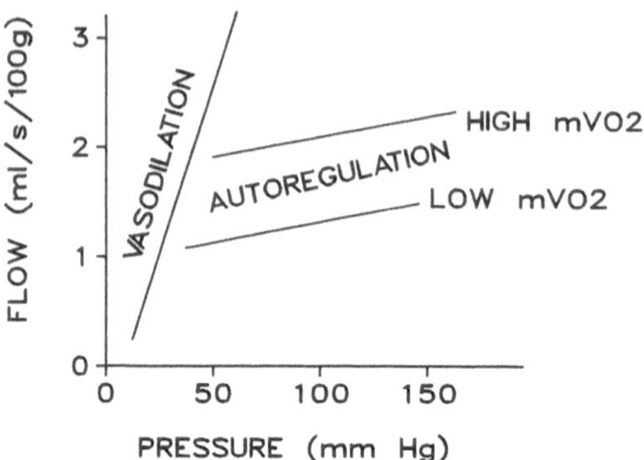

Fig. 1. Schematic representation of the steady state pressure/flow relations in the autoregulated coronary bed compared to this relation in the fully dilated coronary bed. At different MVO_2 levels, the autoregulation curves shift in parallel. Autoregulation is limited to a physiological range of coronary arterial pressures

pressure is under normal conditions both the perfusion pressure and the major determinant of the myocardial O_2 consumption [2], autoregulation of the heart can hardly be measured otherwise than within a coronary bed perfused from an external pressure source. In Fig. 1 the basic characteristics of autoregulation are explained. Early measurements showing this behaviour were published by Mosher et al. [3]. Myocardial oxygen consumption (MVO_2) is an important determinant of coronary flow as can be concluded from the measurements of Eckenhoff et al. [4]. This relation is named here "metabolic flow adaptation." Despite the direct effect of arterial perfusion pressure (P_a) on MVO_2 (the Gregg effect) Vergroesen et al. [5] showed recently that the steady state effects of both variables on coronary blood flow (CBF) were independent and linear over a wide range of P_a and MVO_2. In Fig. 1 the lack of interaction between metabolic flow adaptation and autoregulation is visualized by the parallel shift of the autoregulation curves at two levels of MVO_2. The influence of MVO_2 and P_a on CBF can be described by a linear relation:

$$CBF = b_0 + b_1 \cdot MVO_2 + b_2 \cdot P_a \qquad (1)$$

where b_0, b_1, b_2 are constants. Any hypothesis concerning the control of coronary flow should predict this linear relation in steady state.

In Fig. 2 the experimental variation between different animal experiments is depicted. CBF, P_a and MVO_2 were measured in a group of 7 dogs and two groups of 9 and 5 goats respectively. In the dogs and one group of goats (goat I, $n = 9$) the left main coronary artery was cannulated and perfused with a constant pressure source. MVO_2 was varied by HR variations and infusion of hexamethonium bromide or adrenalin. In the second group of goats (goat II) an occluder around the left main coronary artery was used to uncouple aortic pressure from coronary artery pressure (measured with a Herd-Barger catheter in the circumflex artery). Coronary blood flow was measured with an electromagnetic flow probe around the left anterior descending artery. MVO_2 was changed by pacing, or by changing the aortic pressure through the addition of blood.

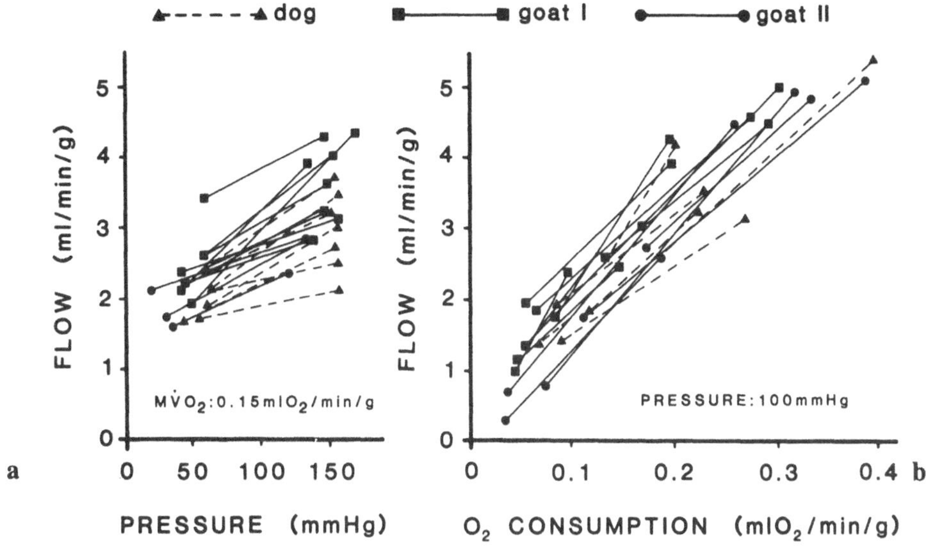

Fig. 2a, b. Plot of the regression lines found for the regression analysis of the linear relation (Eq. 1) to the experimental data from Vergroesen et al. [5]. **a** Regression of coronary arterial pressure to coronary blood flow (*CBF*) for an MVO_2 of 0.15 (ml $O_2 \cdot min^{-1} \cdot g^{-1}$) for each individual animal. **b** Regression lines of MVO_2 to flow at a P_a of 100 mmHg above P_w. To compare the animals with each other, the CBF was normalized to a peak reactive hyperemic flow of 10 ml/min per gram and MVO_2 to an arterial oxygen content of 10 ml 0_2/100 ml blood. P_a, arterial perfusion pressure; P_w, arterial pressure at zero flow

Since the effects of MVO_2 and P_a are independent, the measured CBF can be corrected to a standard MVO_2-value of 0.15 ml O_2/min per gram before plotting the autoregulation curves (Fig. 2a). In Fig. 2b the metabolic flow adaptation is plotted for each individual animal at a standard P_a of 100 mmHg. Each individual curve in Fig. 2a starts at the lowest P_a measured in that particular animal and ends at the highest P_a measured. The lowest and highest MVO_2 measured in each animal experiment influence the beginning and end of the metabolic flow adaptation curves in Fig. 2b in a similar manner. To compare the animals with each other the CBF was normalized to a peak reactive hyperemic flow of 10 ml/min per gram and the MVO_2 was normalized to an arterial O_2 content of 10 ml O_2/100 ml blood. For a discussion of this normalization procedure see Vergroesen et al. [5].

In the past many metabolic hypotheses were formulated in the literature concerning the regulation of coronary flow [6–8]. In particular, the interstitial oxygen tension or adenosine concentration were proposed as mediators between heart cells and arteriolar smooth muscle tone [7, 9, 10]. There is evidence both pro and con these mediators under physiological circumstances [2, 6, 7, 11, 12]. We have aimed at testing these hypotheses quantitatively by formulating models in which the specific assumptions on cause and effect are made explicit by mathematical expressions. The predictions of these models in terms of CBF, P_a, and MVO_2 can then be compared to the experimental results summarized above.

For the dynamic control of the coronary circulation, the phenomena are less well documented. Belloni and Sparks [13] measured the dynamic response of the coronary resistance to a step in heart rate. They found a half-time for the change of the coronary pressure/flow ratio of 14 s. During long diastoles the rate of metabolic vasoconstriction is still controversial. Some authors [14–16] reported a half-time in the order of 5 s and others [17, 18] found no vasoconstriction at all during the first 5 s of diastole. Measurements of the dynamic behaviour of the coronary flow control under different perfusion conditions have recently been presented by Dankelman et al. [19].

The static O_2 model was extended to a dynamic version in order to assess the influence of the rate of change of smooth muscle tone and the importance of flow itself on the feedback system. This model can predict the dynamic responses to a change in heart rate. The predictions of this model were compared to measurements under different perfusion conditions (constant flow perfusion versus constant pressure perfusion) [19].

Steady State Models

General Description of the Model

In Fig. 3 a schematic representation of a mass balance model is depicted. The model consists of three compartments representing myocytes, interstitium, and intramyocardial blood. The fluxes of the regulating substance are indicated by the arrows (F_1, F_2, F_3). The heart cell can consume or produce (F_1) the substance (S). The substance can be delivered or extracted by the blood (F_2). This process depends on the blood flow and the concentration of the substance in the blood

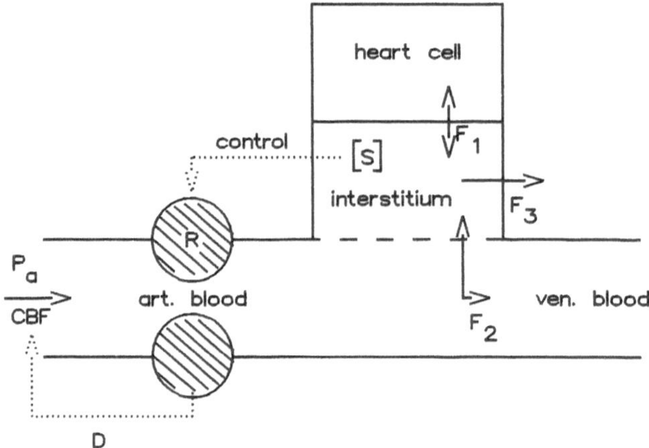

Fig. 3. Schematic representation of a mass balance model. $[S]$, concentration of any flow-regulating substance; P_a, the arterial pressure; *CBF*, the coronary blood flow; *R*, coronary resistance; *art*, arterial; *ven*, venous; F_1, F_2, and F_3 are the fluxes of the regulating substance in and out the interstitium. In steady state these fluxes are in balance with each other. *D* represents the adaptation of coronary flow to a change in arteriolar resistance

and interstitium. The flux F_3 is a concentration-dependent breakdown, concentration-dependent reuptake, or other concentration-dependent chemical process. The concentration of substance S determines the status of the vascular smooth muscle tone. These muscles determine the arteriolar resistance (R). D represents the influence of arteriolar resistance on coronary blood flow.

In the next section this general model will be specialized as oxygen-based model and as adenosine-based model.

Oxygen-based Model

If substance S is oxygen, then F_1 equals oxygen consumption of the myocyte cell, F_2 equals oxygen delivery by the blood, and F_3 is zero. In steady state F_1 equals F_2:

$$MVO_2 = 1.36 \cdot Hb \cdot CBF \cdot (S_a - S_v) \qquad (2)$$

where CBF = coronary blood flow, 1.36 is the oxygen-binding capacity of 1 g% hemoglobin (Hb) [20], and S_a and S_v are the arterial and venous saturation. The oxygen hypothesis assumes a linear relation between the coronary resistance, R, and the interstitial pO_2, with a minimal resistance R_0 as intercept:

$$R = R_0 + C_1 \cdot pO_2 = (P_a - P_w)/CBF \qquad (3)$$

where P_a equals coronary arterial pressure, and P_w the arterial pressure at zero flow. The interstitial pO_2 is assumed equal to the venous pO_2. Linearization of the hemoglobin-oxygen dissociation curve in the venous pO_2 region yields:

$$S_v = -0.1 + 0.024 \cdot pO_2 \text{ (pH 7.4, 37 °C)}. \qquad (4)$$

The combination of Eqs. 2, 3, and 4 yields:

$$MVO_2 = 1.36 \cdot Hb \cdot CBF \cdot \{S_a + 0.1 - 0.024/C_1 \cdot [(P_a - P_w)/CBF - R_0]\}. \qquad (5)$$

This equation can be rearranged to:

$$CBF = b_0 + b_1 \cdot MVO_2 + b_2 \cdot P_a \qquad (6)$$

where $b_0 = -0.024 \cdot P_w / [(S_a + 0.1) \cdot C_1 + 0.024 \cdot R_0]$
$b_1 = C_1 / \{(1.36 \cdot Hb) \cdot [(S_a + 0.1) \cdot C_1 + 0.024 \cdot R_0]\}$
$b_2 = 0.024 / [(S_a + 0.1) \cdot C_1 + 0.024 \cdot R_0]$.

Note that b_0 should be a negative figure, since P_w, S_a, C_1, and R_0 are positive numbers, and b_1 and b_2 should be positive figures.

Equation 6 gives a linear relation between CBF, MVO_2, and P_a as predicted from the experiments [5]. This means that the oxygen model predicts the linear shift of the autoregulation curves at different levels of MVO_2. The oxygen model fits the measured data. This means that although autoregulation and metabolic flow adaptation can both be effected by the same intermediate, the sensitivities characterizing both regulations are nevertheless independent.

Adenosine-based Models

Two adenosine models were formulated. The first is analogous to the oxygen model and is a mass balance of adenosine production, breakdown, and washout by blood flow. Actually it is the formalization of the adenosine hypothesis as formulated in many publications. The second one is derived in order to fit the experimental data and hence Eq.1.

In the first adenosine model F_1 equals oxygen supply/demand related adenosine production (Eq. 7). F_2 equals adenosine washout, and F_3 equals all adenosine concentration related processes, like chemical breakdown and reuptake by the heart cells or endothelial cells. F_1 can be expressed by:

$$\text{Adenosine production} = K_1 \cdot MVO_2/(CBF \cdot [O_2]_a) \tag{7}$$

with $[O_2]_a$ as arterial oxygen content and K_1 a constant. The adenosine loss is a combination of F_2 and F_3:

$$\text{Adenosine loss} = CBF \cdot [A] + K2 \cdot [A]. \tag{8}$$

In a steady state, adenosine production is in equilibrium with adenosine loss:

$$K_1 \cdot MVO_2/\ (CBF \cdot [O_2]_a) = CBF \cdot [A] + K_2 \cdot [A]. \tag{9}$$

In the adenosine hypothesis the vasoactivity of adenosine can be modelled as:

$$K_3 \cdot [A] = CBF/(P_a - P_w). \tag{10}$$

Equations 9 and 10 can be combined to:

$$K_1 \cdot [O_2]_{(a-v)}/[O_2]_a = (CBF^2 + K_2 \cdot CBF)/(K_3 \cdot (P_a - P_w)). \tag{11}$$

Rearranging yields:

$$CBF^2 + K_2 \cdot CBF = K_3 \cdot K_1 \cdot [O_2]_{(a-v)} \cdot (P_a - P_w)/[O_2]_a. \tag{12}$$

Equation 12 could not be fitted adequately to the experimental data of Vergroesen et al. [5]. For this adenosine model, the regulation of flow depends on interaction between P_a and $[O_2]_{(a-v)}$. This interaction ensures that the predicted autoregulation curves are nonlinear at constant MVO_2. However, linearity was found in experiments as reported by Vergroesen et al. [5]. Furthermore, two different levels of MVO_2 will give two nonparallel autoregulation curves. Such a parallel relationship was shown by the experimental results [5] (Fig. 1).

Adenosine-based Model Fitting the Experimental Data

With the model above, plausible relations between different quantities were hypothesized and then the predictions of the model were tested with respect to the experimental findings. Here we will work in reversed order. The question is to find relations constituting an adenosine model describing the experimental results. Then it will be tested whether these forced relations are plausible. The experimental results, with simplification of the constants to B_4 and B_5, can be written as:

$$CBF = B_4 \cdot MVO_2/[O_2]_a + B_5 \cdot (P_a - P_w). \tag{13}$$

Simple rearrangment yields:

$$B_4 \cdot MVO_2/\ ([O_2]_a \cdot CBF) = 1 - B_5 \cdot (P_a - P_w)/CBF. \tag{14}$$

The left hand side of this equation is similar to O_2 demand over O_2 supply, and the right hand side is a function of resistance. The [A]-to-resistance relation can be derived from linearizations of the adenosine dose response curve:

$$R = R_{max} - K_5 \cdot [A] = (P_a - P_w)/CBF \tag{15}$$

where R_{max} is a constant representing the virtually maximal resistance and K_5 is a constant. Combination of Eqs. 14 and 15 yields:

$$B_4 \cdot MVO_2/(CBF \cdot [O_2]_a) = 1 - B_5 \cdot (R_{max} - K_5 \cdot [A]). \tag{16}$$

Rearrangment yields:

$$B_4 \cdot MVO_2/(CBF \cdot [O_2]_a) = B_5 \cdot K_5 \cdot [A] + (1 - B_5 \cdot R_{max}). \tag{17}$$

Equation 17 forms a mass balance of adenosine, which was derived from the experimentally tested linear equation (Eq. 1.) and hence a priori is consistent with the experimental results of local coronary flow control. When comparing this Eq. 17 with the steady-state mass balance equations of the adenosine model formulated above (Eq. 9), one recognizes supply/demand-related adenosine production and an adenosine breakdown that is dependent on [A]-related processes, such as reuptake and chemical breakdown. However, no term in the right hand side of Eq. 17 contains CBF, meaning that significant washout of adenosine should not occur.

Discussion of the Adenosine Models

The required absence of a washout term in the second adenosine model provides a possibility for testing the model constraints. Is there significant adenosine washout under normal physiological circumstances? Adenosine washout has been reviewed by Feigl [2]. Adenosine has been measured in the effluent of Langendorff-perfused isolated heart preparations during hypoxia [21], and in similar preparations, which show autoregulation and metabolic flow adaptation [22–24]. Adenosine release in the coronary sinus has been detected during human angina, induced by rapid atrial pacing [25]. Adenosine washout has always been an argument in favour of adenosine as major regulating substance, yet on the basis of the experimental results in combination with the model analysis, significant washout of adenosine is now an argument against a major role of adenosine. One should note that practically no experiments on adenosine washout were carried out in normoxic, blood-perfused hearts. Such an experiment is complicated by the capacity of blood cells to inactivate adenosine.

Adenosine, according to the hypothesis [7, 22, 26], influences the coronary resistance during autoregulation and metabolic flow adaptation. As can be seen from the model analysis above, this hypothesis can be true in theory. However, in the last few years some elegant experiments have been performed in which all adenosine was removed from the interstitial space by adenosine deaminase [11, 12]. In these experiments the slope of the autoregulation curve was not influenced, indicating that the coronary resistance is not determined by the adenosine concentration in the interstitium. This makes it hard to believe that adenosine has an important role in local coronary flow control.

Conclusions from the Steady State Model Analysis

It is possible to formulate a model based on either tissue pO_2 or interstitial adenosine concentration that predicts the experimentally found relation between CBF, MVO_2, and P_a. This indicates that autoregulation and metabolic flow adaptation could be mediated by the same factor. Furthermore, a relatively simple mass-balance model can predict the experimental results. No complicated multi-factor models, with interactions between different mechanisms, are necessary to explain autoregulation and metabolic flow adaptation.

For the oxygen model the important parameter is the sensitivity of the coronary resistance for tissue pO_2 per se. This should be tested in isolated coronary arteries where no intermediate between pO_2 and the coronary resistance can influence the measurements. Venous pO_2, which has been used as indicator of tissue pO_2 [9, 27], is probably not the best choice for this purpose.

The first adenosine model formulated on the bases of hypotheses from the literature [7, 22] did not fit the experimentally found relation between CBF, MVO_2, and P_a. The second adenosine model derived by curve fitting from the experimental results lacked adenosine washout. Since adenosine washout has been demonstrated this means that adenosine is not responsible for flow control according to the generally assumed principles of adenosine production, breakdown, and washout. These conclusions are in accordance with the absence of effect in the adenosine deaminase experiments.

Dynamic Model

The general model presented above (Fig. 3) will now be modified to predict the dynamic control of the coronary circulation. The chain of information is the same as in Fig. 3, where this was explained in general terms considering a substance S. This dynamic model was only developed in terms of the oxygen hypothesis.

The following rate-limiting processes have to be dealt with. The first process is mixing of oxygen in tissue and capillary space. The second rate-limiting process is the adaptation of arteriolar resistance to a change in tissue concentration of oxygen.

When tissue and capillary space are assumed to be well-mixed compartments then the mixing process can be described by:

$$(Hb \cdot 1.36 \cdot K \cdot Vb + St \cdot Vt) \cdot (dpO_2/dt) = F_2 - F_1 \tag{18}$$

where Vb and Vt represent the volumes of blood and tissue, respectively; St is the solubility of oxygen in tissue; and K is the slope of the linearized oxygen dissociation curve. Equation 18 represents a process which is of first order, with a time constant, t_1, that equals:

$$t_1 = (Hb \cdot 1.36 \cdot K \cdot Vb + St \cdot Vt) \,/\, (Hb \cdot 1.36 \cdot K \cdot CBF). \tag{19}$$

The second rate-limiting process is the adaptation of arteriolar resistance to a change in tissue concentration of O_2. It is assumed that this process is of the first order and can be described by:

$$t_2 \cdot (dR/dt) + R_0 = C_2 \cdot pO_2 \quad (20)$$

where t_2 is the time constant of the process of adaptation of arteriolar resistance to a change in tissue pO_2, and C_2 is a constant. In the steady state Eq. 20 reduces to a linear relation between the arteriolar resistance R and pO_2.

The value for the time constant of mixing of oxygen is in the order of 4 s. The time constant for adaptation of coronary resistance to a change in tissue oxygen pressure has to be taken in the order of 20 s to fit the experiments [19].

The dynamic model can be elucidated by the schematic in Fig. 3. With constant pressure perfusion, coronary blood flow is allowed to change. This is indicated by the *arrow D* in Fig. 3. The change in coronary blood flow depends on the change in arteriolar resistance (R). In the case of constant flow perfusion, the perfusion pressure is the dependent variable. When flow is constant there is no effect of a change of arteriolar resistance on the oxygen supply. Hence, in this situation D is not active.

The model predicts that the response rate of the coronary resistance change is slower with constant flow perfusion compared to the response during constant pressure perfusion. The results of this simulation are given in Fig. 4b. The resistance change was normalized according to Dankelman et al. [19]. Normalization makes comparison of response rates found at the different perfusion conditions possible. The model predicted that the response rate of the coronary pressure/flow ratio depends on the level of pressure during constant pressure perfusion, but is independent of the flow level with constant flow perfusion [19].

The predictions above were experimentally tested in anesthetized goats [19]. In these experiments the response of the coronary pressure-flow ratio to a sudden change in heart rate was measured. In the dynamic situation this ratio is not equal to coronary resistance because of compliance effects [28, 29]. The normalized experimental response of the pressure-flow ratios are given in Fig. 4a. This figure shows that the response has a fast initial reversed response followed by a slow response to its end value. After the initial dip the response increased faster to its end value during constant pressure perfusion when compared with constant flow perfusion. The experiments also showed that, as predicted by the model, the response rate depends on the pressure level with constant pressure perfusion, and does not depend on the flow level with constant flow perfusion [19]. Independent of the type of perfusion the pressure/flow response to an HR step-up was about 10% faster compared to an HR step-down. This small but significant difference was not predicted by the model.

Discussion of the Dynamic Model

We have not yet modified the adenosine hypothesis into a dynamic model. In case one would like to do so, one should account for the major rate-limiting processes in the adenosine hypothesis. These rate-limiting processes might involve adenosine production as a result of the balance between O_2 supply and demand, mixing of adenosine in the interstitial space, and the reaction of the smooth muscle cells to the change in adenosine concentration. To construct a

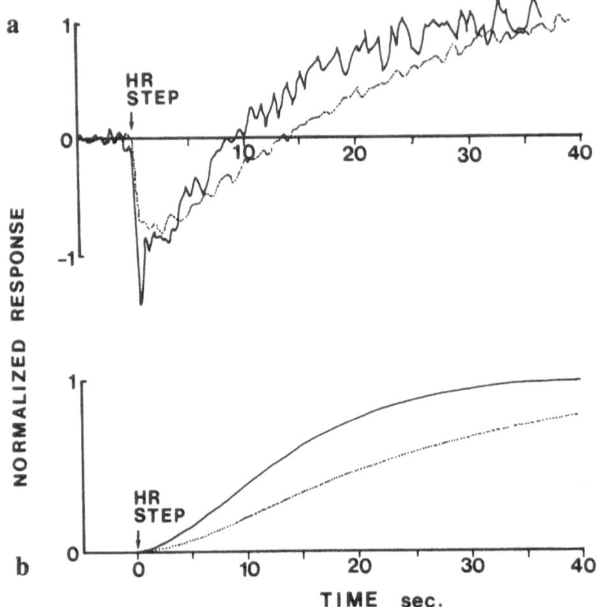

Fig. 4a, b. Normalized response of pressure-flow ratio to a step in heart rate. **a** Averaged results of experimentally obtained normalized responses of pressure-flow ratio to a heart rate step. **b** Model predictions of normalized pressure-flow ratio to a heart rate step. In the simulations the pressure-flow ratio is equal to coronary resistance. The *solid line* indicates the response with constant pressure perfusion, and the *dotted line*, with constant flow perfusion

dynamic adenosine model, the time dependency of these processes should be known, which is only partly the case at the moment.

The dynamic oxygen model predicts several experimental findings, especially the differences in rate of the slow component of the responses for both perfusion conditions (Fig. 4). However, in the experiments an initial dip is seen, which is not explained by our model. This reversed part of the response is the result of mechanical characteristics of the coronary circulation, such as compliance or elastance or pressure-dependent resistance. These properties were not incorporated in the model described here.

An explanation of this initial dip based on these mechanical properties can be thought as follows: an increase in heart rate results in an increase in time-averaged intramyocardial pressure on the coronary vessels leading to a squeezing of blood in the intramyocardial vessels. Because of the compliance of the intramyocardial blood vessels this becomes apparent in the initial dip. Besides this effect the intramyocardial vascular volume decreases due to the higher compression [14] which may lead to a permanent increase in the pressure/flow ratio [19]. This permanent increase can indeed be found when the coronary vessels are fully dilated. Such a description of the behaviour of the fully dilated coronary bed requires nonlinear elements such as pressure-dependent resistance and compliance [29] or a time-varying elastance [30].

The present model did not predict the significant difference in response time between a step-up and a step-down in HR. This difference is small (10%) compared to the difference in response rate between constant pressure perfusion and constant flow perfusion, which could be predicted by the present metabolic model. To explain the direction sensitivity a directional effect has to be assumed in the response of the vascular smooth muscle cells to a mismatch between O_2 supply and O_2 demand. A relative undersupply is more quickly corrected than a relative over-supply.

The present model can be further tested by comparing predictions of the model concerning a step in perfusion at constant HR (constant MVO_2) to experiments that measure this response in the anesthetized goat. Such experiments have been started in our laboratory and will provide additional information on the possibilities of a metabolic hypothesis in elucidating the dynamic behaviour of the regulation of the coronary flow and explaining the necessity of a directional element in the smooth muscle response.

Conclusions from the Dynamic Model Analysis

A dynamic model based on the oxygen hypothesis can explain the different response rates of the pressure/flow ratio with different perfusion conditions. This type of model is not capable of explaining the fast initial reversed response. Mechanical properties have to be incorporated in the model to explain this fast initial response.

General Conclusions

Simple models based on mass balance between O_2 supply and O_2 demand can explain the static phenomena in the coronary flow control, such as autoregulation and metabolic flow control. An important part of the dynamic behaviour of the pressure/flow response can be explained by a dynamic version of the same type of model. Incorporation of mechanical properties into the model is necessary to explain the initial transients in the response.

References

1. Johnson PC (1964) Review of previous studies and current theories of autoregulation. Circ Res 14,15(Suppl I): I–2–9
2. Feigl EO (1983) Coronary physiology. Physiol Rev. 63: 1–205
3. Mosher P, Ross J, McFate PA, Shaw RF (1964) Control of coronary blood flow by an autoregulatory mechanism. Circ Res 14: 250–259
4. Eckenhoff JE, Hafkenschiel JH, Landmesser CM, Harmel M (1947) Cardiac oxygen metabolism and control of the coronary circulation. Am J Physiol 149: 634–649
5. Vergroesen I, Noble MIM, Wieringa PA, Spaan JAE (1987) Quantification of O_2 consumption and arterial pressure as independent determinants of coronary flow. Am J Physiol 252 (Heart Circ Physiol 21): H545–H553
6. Rubio R, Berne RM (1975) Regulation of coronary flow. Prog Cardiovasc Dis 18: 105–122.
7. Berne RM (1980) Role of adenosine in the regulation of coronary blood flow. Circ Res 47: 807–813

8. Hilton R, Eichholtz F (1925) The influence of chemical factor on the coronary circulation. J Physiol 59: 413–425
9. Drake-Holland AJ, Laird JD, Noble MIM, Spaan JAE, Vergroesen I (1984) Oxygen and coronary vascular resistance during autoregulation and metabolic vasodilation in the dog. J Physiol 348: 285–299
10. Olsson RA, Khouri EM, Bedynek JL Jr, McLean J (1979) Coronary vasoactivity of adenosine in the conscious dog. Circ Res 45: 468–478
11. Dole WP, Yamada N, Bishop VS, Olsson RA (1985) Role of adenosine in coronary blood flow regulation after reductions in perfusing pressure. Circ Res 56: 517–524
12. Hanley FL, Grattan MT, Stevens MB, Hoffman JIE (1986) Role of adenosine in coronary autoregulation. Am J Physiol 250 (Heart Circ Physiol 19): H558–H566
13. Belloni FL, Sparks HV (1977) Dynamics of myocardial oxygen consumption and coronary vascular resistance. Am J Physiol 233 (Heart Circ Physiol 2): H34–H43
14. Vergroesen I, Noble MIM, Spaan JAE (1987) Intramyocardial blood volume change in first moments of cardiac arrest in anesthetized goats. Am J Physiol 253 (Heart Circ Physiol 22): H307–H316
15. Klocke FJ, Weinstein IR, Klocke JF, Ellis AK, Kraus DR, Mates RE, Canty JM, Anbar RD, Romanowski RR, Wallmeyer KW, Echt MP (1981) Zero-flow pressure and pressure-flow relationship during single long diastoles in the canine coronary bed before and during maximum vasodilation. J Clin Invest 68: 970–980
16. Eng C, Jentzer JH, Kirk ES (1982) The effects of the coronary capacitance on the interpretation of diastolic pressure-flow relationships. Circ Res 50: 334–341
17. Dole WP, Bishop VS (1982) Influence of autoregulation and capacitance on diastolic coronary artery pressure-flow relationships in the dog. Circ Res 51: 261–270
18. Dole WP, Bishop VS (1982) Regulation of coronary blood flow during individual diastoles in the dog. Circ Res 50: 377–385
19. Dankelman J, Spaan JAE, Stassen HG, Vergroesen I (1989) Dynamics of coronary adjustment to a change in heart rate in the anaesthetized goat, J Physiol 408: 295–312
20. Zijlstra WG, Assendelft van OW, Rijskamp A (1965) Oxygen capacity of normal human blood. Acta Physiol Pharmacol Neerl 13: 229
21. Edlund A, Fredholm BB, Patrignani P, Patrono C, Wennmalm Å, Wennmalm M (1983) Release of two vasodilators, adenosine and prostacyclin, from isolated rabbit hearts during controlled hypoxia. J Physiol 340: 487–501
22. Bardenheuer H, Schrader J (1986) Supply-to-demand ratio for oxygen determines formation of adenosine by the heart. Am J Physiol 250 (Heart Circ Physiol 19): H173–H180
23. Schrader J, Haddy FJ, Gerlach E (1977) Release of adenosine, inosine, and hypoxanthine from the isolated guinea pig heart during hypoxia, flow-autoregulation and reactive hyperemia. Pflügers Arch 369: 1–6
24. Wiedmeier VT, Spell LH (1977) Effects of catacholamines, histamine, and nitroglycerin on flow, oxygen utilization and adenosine production in the perfused guinea pig heart. Circ Res 41: 503–508
25. Fox AC, Reed GE, Glassman E, Kaltman AJ, Silk BB (1974) Release of adenosine from human hearts during angina induced by rapid atrial pacing. J Clin Invet 53: 1447–1457
26. Granger HJ, Shepherd AP Jr (1973) Intrinsic microvascular control of tissue osygen delivery. Microvasc Res 5: 49–72
27. Dole WP, Nuno DW (1986) Myocardial oxygen tension determines the degree and pressure range of coronary autoregulation. Circ Res 59: 202–215
28. Spaan JAE (1985) Coronary diastolic pressure-flow relation and zero flow pressure explained on the basis of intramyocardial compliance. Circ Res 56: 293–309
29. Bruinsma P, Arts T, Dankelman J, Spaan JAE (1988) Model of the coronary circulation based on pressure dependence of the coronary resistance and compliance. Basic Res Cardiol 83: 510–524
30. Krams R (1988) The effect of cardiac contraction on coronary flow. Introduction of a new concept. Thesis, Free Unversity of Amsterdam, Free University Press, Amsterdam

7. Coronary Circulation in Pathophysiological Conditions

Coronary Circulation in Myocardial Hypertrophy

ROBERT J. BACHE

Summary. Although mean myocardial blood flow has generally been reported to be normal or slightly greater than normal in animals with compensated left ventricular hypertrophy secondary to chronic pressure overload, these ventricles exhibit abnormalities of myocardial perfusion during stress. Thus, animals with left ventricular hypertrophy commonly demonstrate evidence of subendocardial under-perfusion during pacing-induced tachycardia or during exercise. It is likely that these transmural disparities of myocardial perfusion do result in subendocardial ischemia, as evidenced by myocardial production of lactate and development of systolic and diastolic dysfunction. Additional research is needed to determine whether such repetitive episodes of stress-induced subendocardial ischemia may cause the eventual development of left ventricular dilation and failure in hearts with chronic pressure overload.

Key words: Coronary blood flow—Myocardial blood flow—Left ventricular hypertrophy—Aortic stenosis—Hypertension

Introduction

Myocardial hypertrophy appears to be an appropriate response to chronic left ventricular pressure overload. As left ventricular wall thickness increases, systolic wall stress (force/unit area) is returned to normal or near normal levels despite an increased intracavitary systolic pressure [1, 2]. Although hypertrophy provides a useful mechanism by which systolic performance is maintained in the presence of an increased systolic load, clinical findings suggest that the pressure-overloaded, hypertrophied left ventricle has increased vulnerability to ischemia. Patients with severe concentric left ventricular hypertrophy secondary to aortic stenosis or arterial hypertension may develop exertional angina pectoris in the absence of occlusive coronary artery disease [3]. Such patients may exhibit electrocardiographic repolarization changes which are suggestive of ischemia either at rest or in response to exercise [4]. Pathologic studies have demonstrated that subendocardial fibrosis, suggestive of ischemic injury, may occur in the severely pressure-overloaded, hypertrophied left ventricle despite anatomically normal

---

Division of Cardiology, Department of Medicine, University of Minnesota Medical School, Box 338 UMHC, 420 Delaware Street S.E., Minneapolis, MN 55425, USA

coronary arteries [5]. These clinical findings, suggesting that myocardial hypertrophy may render the left ventricle at increased risk for developing ischemia, have stimulated interest in study of the coronary circulation in the presence of experimental myocardial hypertrophy.

Methods

Several laboratory models, including banding of the ascending aorta, surgical production of valvular aortic stenosis, and renovascular or perinephritic hypertension, have been used to produce experimental left ventricular hypertrophy [6–9]. The greatest degree of hypertrophy has been observed when banding of the ascending aorta is performed in young animals [6]. In dogs 6–9 weeks of age, the ascending aorta is constricted to produce a 20–25 mmHg peak systolic pressure gradient across the area of narrowing. The degree of left ventricular systolic overload progressively increases as normal body growth occurs in the face of fixed degree of aortic constriction. Because the systolic overload increases gradually, this model is capable of producing a greater degree of systolic overload and left ventricular hypertrophy than can be regularly achieved with other experimental models. Ascending aortic banding in young animals regularly produces an 80%–100% increase in left ventricular mass by the time the animals reach adulthood [6]. Ascending aortic banding can also be carried out in adult animals, but because the systolic overload is imposed in a single stage, the maximum tolerated systolic gradient across the area of constriction is generally only approximately 60 mmHg [10, 11]. This typically results in a less severe degree of hypertrophy, with approximately 50% increase in left ventricular mass within 6–8 weeks after aortic banding.

The advantage of ascending aortic banding for producing pressure overload is that it is technically easy to produce and it reliably results in substantial degrees of left ventricular hypertrophy. However, this experimental model is peculiar in that the aortic constriction is in a supracoronary location, resulting in elevated coronary perfusion pressure during systole (equal to left ventricular pressure). Coronary perfusion pressure in diastole is normal and equal to distal aortic pressure [6]. Experimental models for producing valvular aortic stenosis result in coronary perfusion pressures analogous to those seen in patients with aortic stenosis. Valvular aortic stenosis has been produced by plication of the noncoronary aortic cusp in young dogs to produce a modest systolic pressure gradient across the valve [7]. The degree of stenosis becomes progressively more severe with growth, and may result in an approximately 75% increase in left ventricular mass when the animals reach adulthood. Valvular aortic stenosis has also been produced in adult animals by plication of the non-coronary sinus of Valsalva proximal to the origin of the coronary arteries [12]. This technique results in mild to moderate degrees of left ventricular hypertrophy. Both techniques for producing aortic valve stenosis are substantially more difficult than ascending aortic banding, and the degree of left ventricular hypertrophy achieved is more variable.

A third experimental model of left ventricular hypertrophy secondary to chronic systolic overload involves producing arterial hypertension. Investigators have utilized either renal artery constriction or perinephritis to produce hypertension [8, 9]. These experimental models result in substantial arterial hypertension, and generally produce a 25%–50% increase in left ventricular mass within 6–8 weeks.

Results

Abnormalities of myocardial perfusion in the hypertrophied heart could be expressed in several different settings. If the abnormality were very severe, then values for mean myocardial blood flow might be decreased below normal even during basal conditions. Linzbach [13] predicted that with increasing severity of myocardial hypertrophy, coronary reserve would eventually become exhausted, leading to disturbances of myocardial nutrition. This might be associated with decreased myocardial blood flow rates even during resting conditions. If the perfusion abnormality were less severe, blood flow rates during basal conditions would remain normal, but the ability to increase flow (vasodilator reserve) in response to increasing metabolic requirements would be compromised. Finally, mean values for myocardial blood flow might not be abnormal, but a change in the transmural distribution of perfusion could occur, with selective underperfusion of the subendocardium.

Measurements of myocardial blood flow during basal conditions have generally yielded normal or slightly increased values for mean blood flow per gram of myocardium in animals with left ventricular hypertrophy secondary to chronic pressure overload [6–9, 14–17]. However, as shown in Fig. 1, when the transmural distribution of blood flow across the left ventricular wall has been examined with radioactive microspheres, the ratio of subendocardial/subepicardial perfusion is reduced in animals with hypertrophy. While in the normal heart, blood flow rates are greatest in the subendocardium, in animals with hypertrophy the highest blood flow rates tend to occur in the midwall of the left ventricle. The occurrence of left ventricular failure is marked by left ventricular dilation with increased systolic wall stress. In animals with aortic banding in which left ventricular failure developed, there was increased perfusion during basal conditions, so that mean myocardial blood flow at rest was significantly greater than in either normal dogs or dogs with similar degrees of compensated left ventricular hypertrophy [17]. The develpment of left ventricular failure was associated with further redistribution of blood flow away from the subendocardium, so that subendocardial/subepicardial flow ratios in animals with left ventricular failure were substantially less than in animals with compensated hypertrophy (Fig. 1). The increased mean myocardial blood flow rates suggested that increased systolic wall stress in the dilated, hypertrophied heart resulted in increased myocardial oxygen requirements even during resting conditions. The need for increased blood flow during resting conditions would impair the ability for further increases in blood flow to occur in response to increased cardiac activity.

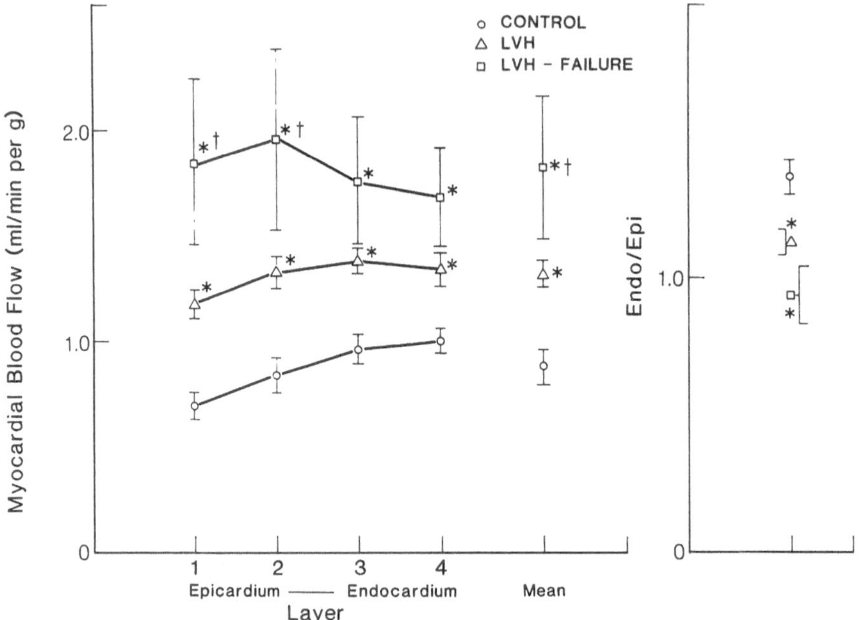

Fig. 1. Myocardial blood flow to four transmural layers from epicardium to endocardium in 16 control dogs, 16 dogs with compensated left ventricular hypertrophy produced by aortic banding (*LVH*), and 4 dogs with left ventricular hypertrophy and overt cardiac failure. Myocardial blood flow was significantly greater than normal in dogs with compensated left ventricular hypertrophy, and was further significantly increased in animals with left ventricular failure. $^{*}P<0.05$ compared with control. $^{\dagger}P<0.05$ compared with compensated LVH. From [17] with permission

Vasodilator Reserve

The increase of coronary blood flow in response to pharmacologic agents which produce maximal coronary vasodilation provides information regarding the ability of the coronary vasculature to respond to increased myocardial oxygen needs during periods of increased cardiac work. Maximum coronary blood flow rates have been reported for each of the experimental models of chronic pressure-overload left ventricular hypertrophy. Vasodilation has generally been produced with adenosine, which is known to cause maximal coronary vasodilation. In dogs with chronic left ventricular hypertrophy secondary to banding of the ascending aorta or arterial hypertension, maximal myocardial blood flow rates are generally not significantly different from those observed in animals with normal hearts [8, 18–20]. However, in both of these experimental models coronary perfusion pressure is abnormally increased, either by the presence of arterial hypertension or, in animals with aortic banding, because pressure in the proximal segment from which the coronary arteries arise is abnormally increased. Because of the increased coronary perfusion pressure, the computed minimum vascular resistance per gram of myocardium is increased in these two models of left ventricular hypertrophy [8, 18–20].

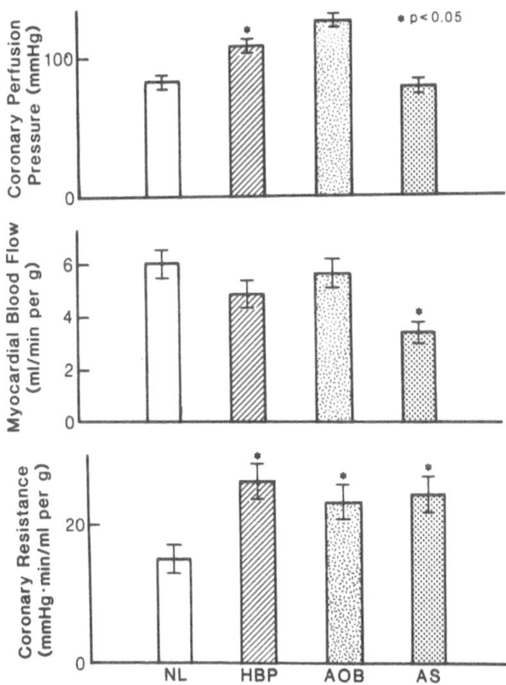

Fig. 2. Coronary perfusion pressure, maximal myocardial blood flow rates, and minimum coronary vascular resistance, during maximal coronary vasodilation with adenosine in normal dogs (*NL*) and 3 groups of animals with left ventricular hypertrophy produced by renovascular hypertension (*HBP*), banding of the ascending aorta (AOB), and experimental valvular aortic stenosis (*AS*). Minimum coronary vascular resistance was similar and significantly greater than normal in all 3 groups of animals with hypertrophy. However, the increased coronary perfusion pressure in HBP and AOB animals resulted in normal maximum myocardial blood flow rates. Since coronary perfusion pressure was not increased in AS animals, the increased minimum coronary resistance was expressed as an absolute decrease of myocardial blood flow during the adenosine vasodilation. From [25] with permission

When left ventricular hypertrophy results from valvular aortic stenosis, coronary perfusion pressure is not increased. In this situation, maximal coronary vasodilation with adenosine results in a subnormal maximum myocardial blood flow rate [7]. However, as shown in Fig. 2, the impairment of minimum coronary vascular resistance in left ventricular hypertrophy due to aortic valve stenosis in similar to that reported in left ventricular hypertrophy secondary to arterial hypertension or banding of the ascending aorta. In aortic valve stenosis, the presence of a normal coronary perfusion pressure causes the impairment of minimum coronary vascular resistance to be expressed as a reduction of maximal coronary blood flow rates. In contrast, maximum blood flow rates are not decreased in animals with hypertrophy due to arterial hypertension or supravalvular aortic stenosis, because the increased perfusion pressure compensates for the decreased minimum coronary vascular resistance. The finding of impaired minimum coronary vascular resistance in animals with valvular aortic stenosis is of importance, since it indicates that hypertrophy alone results in abnormalities of the coronary circulation, even when the coronary vessels are not exposed to increased pressure which might lead to hypertensive vascular changes in experimental models utilizing arterial hypertension or supravalvular aortic stenosis. Of interest is the finding in all of these experimental models that minimum coronary vascular resistance expressed for the entire left ventricle is not different in animals with hypertrophy than in normal animals. This suggests that the minimum cross-sectional area of the coronary vasculature does not increase as the myocardium undergoes hypertrophy.

Myocardial Blood Flow During Tachycardia

In the normal heart, increasing heart rates result in progressive increases of myocardial blood flow in proportion to the increased oxygen demands produced by tachycardia. Radioactive microspheres have been used to study the effects of heart rate on the pattern of perfusion across the left ventricular wall during tachycardia. In the normal heart, blood flow to the subendocardium is maintained greater than subepicardial flow despite the decreased diastolic perfusion interval associated with increasing heart rates. In dogs in which left ventricular hypertrophy was produced by renovascular hypertension or by ascending aortic banding, progressive increases of heart rate from 100 to 250 beats/min produced by cardiac pacing resulted in increases of mean myocardial blood flow which were not significantly different from normal [8, 21]. However, rapid pacing produced a redistribution of blood flow away from the subendocardium, which did not occur in normal hearts. Similarly, in dogs with left ventricular hypertrophy produced by experimental aortic valve stenosis, the ratio of subendocardial/subepicardial flow fell to 0.87 ± 0.15 at a heart rate of 250 beats/min, as compared with 1.21 ± 0.12 in normal animals ($P < 0.05$) [7].

Several lines of evidence suggest that the subendocardial perfusion abnormality observed during tachycardia in animals with left ventricular hypertrophy may result in myocardial ischemia. Bache et al. [22] measured aortic and coronary sinus lactate concentrations in normal dogs and in animals with left ventricular hypertrophy during pacing-induced tachycardia. In normal animals, increasing heart rates were associated with progressive increases in mean myocardial blood flow, while subendocardial flow was maintained equal to or greater than subepicardial flow. In contrast, pacing at 250 beats/min resulted in relative subendocardial underperfusion in 6 to 10 dogs with left ventricular hypertrophy secondary to ascending aortic banding. These 6 animals demonstrated myocardial lactate production, suggesting that the subendocardial perfusion abnormality produced by tachycardia resulted in myocardial ischemia. In contrast, none of the normal animals, and none of the 4 animals with hypertrophy that failed to show subendocardial underperfusion during rapid pacing, showed lactate production. Of interest was the finding that left ventricular end-diastolic pressure increased in animals with hypertrophy and subendocardial underperfusion during tachycardia, suggesting diastolic dysfunction [22]. Fujii et al. [23] examined left ventricular systolic and diastolic function during increasing heart rates produced by atrial pacing in normal dogs and in dogs with left ventricular hypertrophy produced by banding the ascending aorta. Normal animals tolerated pacing rates as fast as 240 beats/min without evidence of impairment of either systolic or diastolic function. In contrast, pacing-induced tachycardia resulted in significant depression of the mean velocity of circumferential fiber shortening during systole, as well as significant increases of diastolic myocardial stiffness with increases of end-diastolic pressure and wall stress in animals with hypertrophy. These data support the hypothesis that the subendocardial perfusion abnormality which is observed during pacing-induced tachycardia in animals with left ventricular hypertrophy does result in myocardial ischemia.

Myocardial Perfusion During Exercise

Several investigators have examined the response of myocardial blood flow to the physiologic stress of exercise in the chronically pressure-overloaded hypertrophied left ventricle. Most of these studies have utilized dogs in which banding of the ascending aorta was carried out in young animals to produce moderately severe degrees of hypertrophy. Holtz et al. [14] reported that subendocardial blood flow failed to increase in proportion to subepicardial flow during low-level treadmill exercise in animals with hypertrophy. Bache et al. [6] examined the response of myocardial blood flow during graded treadmill exercise in dogs with left ventricular hypertrophy secondary to banding of the ascending aorta. Myocardial blood flow increased in proportion to heart rate in both normal animals and animals with hypertrophy. However, heavy levels of exercise were associated with redistribution of perfusion away from the subendocardium in animals with hypertrophy but not in normal control animals. In animals with hypertrophy, the ratio of subendocardial to subepicardial blood flow fell to 0.72 ± 0.02 in the area of left ventricle that included the posterior papillary muscle, indicating relative subendocardial underperfusion. Animals with hypertrophy demonstrated evidence of diastolic dysfunction, with mean left ventricular end-diastolic pressure increasing from 8 ± 2 mmHg at rest to 37 ± 6 mmHg during heavy exercise. In contrast, left ventriclar end-diastolic pressure did not increase during exercise in normal animals.

The degree of subendocardial underperfusion during exercise in the hypertrophied heart may be influenced by diastolic coronary perfusion pressure. Unlike normal animals, diastolic aortic pressure does not increase during exercise in the presence of valvular or supravalvular aortic stenosis [6]. In contrast, in left ventricular hypertrophy due to arterial hypertension, aortic pressure undergoes a supranormal increase during exercise, resulting in substantially elevated diastolic coronary perfusion pressures. Bache and Vrobel [9] demonstrated that graded treadmill exercise resulted in similar increases of mean myocardial blood flow in normal dogs and in dogs in which left ventricular hypertrophy was produced by perinephritic hypertension. In contrast to animals with aortic banding, subendocardial flow was maintained equal to subepicardial flow in animals with hypertension [9]. During heavy exercise the subendocardial/subepicardial flow ratio was 1.31 ± 0.08 in normal animals, 1.07 ± 0.12 in animals with hypertensive hypertrophy, and 0.73 ± 0.08 in animals with hypertrophy produced by aor-

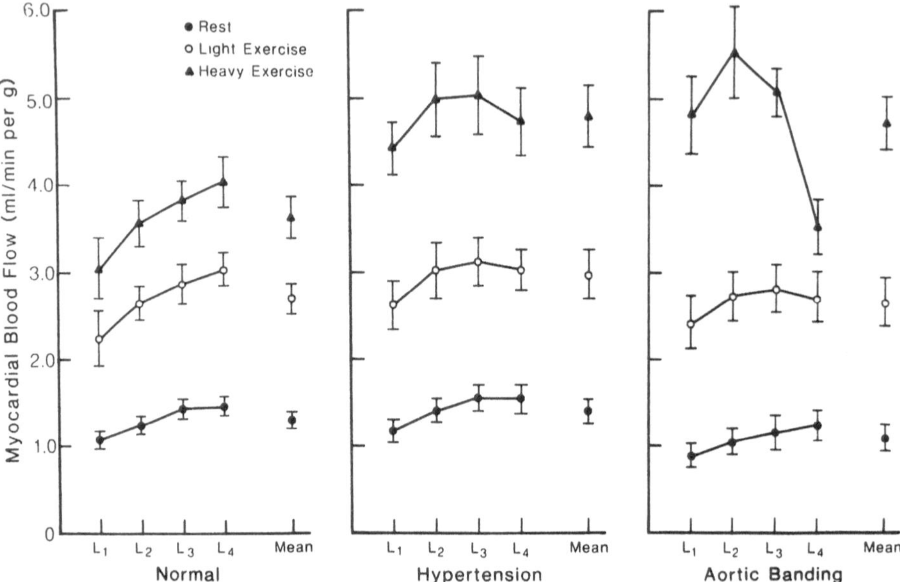

Fig. 3. Myocardial blood flow to 4 transmural layers from epicardium (L_1) to endocardium (L_4) in 7 normal dogs, 7 dogs with left ventricular hypertrophy secondary to perinephritic hypertension, and 9 dogs with left ventricular hypertrophy produced by banding the ascending aorta. Measurements were obtained during resting conditions and at 2 levels of treadmill exercise which increased heart rates to approximately 195 (light exercise) and 260 beats/min (heavy exercise). During heavy exercise, mean myocardial blood flow was significantly greater than normal in both groups of animals with hypertrophy. During heavy exercise subendocardial flow was maintained equal to subepicardial flow in animals with hypertension, but was substantially less than subepicardial flow in animals with aortic banding. From [24] with permission

tic banding (Fig. 3) [24]. These findings suggest that the increased diastolic coronary perfusion pressure associated with arterial hypertension is able to partially overcome the subendocardial underperfusion which occurs in the hypertrophied ventricle during heavy exercise.

Acknowledgment. This work was supported by U.S. Public Health Service Grant HL21872 from the National heart, Lung, and Blood Institute.

References

1. Hood WP, Rackley CE, Rolett E (1968) Wall stress in the normal and hypertrophied left ventricle. Am J Cardiol 22: 550–558
2. Gunther S, Grossman W (1979) Determinants of ventricular function in pressure overload hypertrophy in man. Circulation 59: 679–688
3. Goodwin JR (1973) Hypertrophic disease of the myocardium. Prog Cardiovasc Dis 16: 199–238
4. Wrobewski EM, Pearl FJ, Hammer WJ, Bove AA (1982) False-positive stress test due to undetected left ventricular hypertrophy. Am J Epidemiol 115: 412–417

5. Moller JH, Nakeb A, Edwards JE (1966) Infarction of the papillary muscle and mitral insufficiency associated with congenital aortic stenosis. Circulation 34: 87–91
6. Bache RJ, Vrobel TR, Ring WS, Emery RW, Anderson RW (1981) Regional myocardial blood flow during exercise in dogs with chronic left ventricular hypertrophy. Circ Res 48: 76–87
7. Alyono D, Anderson RW, Parrish D, Dai X, Bache RJ (1986) Effect of experimental canine left ventricular hypertrophy produced by valvular aortic stenosis on myocardial blood flow. Circ Res 58: 47
8. Muller TM, Marcus ML, Kerber RE, Young JA, Barnes RW, Abboud FM (1978) Effect of renal hypertension and left ventricular hypertrophy of the coronary circulation in dogs. Circ Res 42: 543–549
9. Bache RJ, Vrobel TR (1979) Effects of exercise on blood flow in the hypertrophied heart. Am J Cardiol 44: 1029–1033
10. Newman WH, Webb JB (1980) Adaptation of left ventricle to chronic pressure overload: Response to inotropic drugs. Am J Physiol 238 (Heart Circ Physiol 7): H134–H141
11. Bache RJ, Alyono D, Sublett E, Dai X (1986) Myocardial blood flow in left ventricular hypertrophy developing in growing and adult animals. J Physiol 251 (Heart Circ Physiol 20): H949–H956
12. Allard JR, O'Neill MJ, Hoffman JIE (1970) Valvar subcoronary aortic stenosis in dogs. Am J Physiol 236 (Heart Circ Physiol 5): H780–H784
13. Linzbach AJ (1960) Heart failure from the point of view of quantitative anatomy. Am J Cardiol 5: 370–382
14. Holtz J, Restroff W, Bard P, Bassenge E (1977) Transmural distribution of myocardial blood flow and of coronary reserve in canine left ventricular hypertrophy. Basic Res Cardiol 72: 286–292
15. Mittmann U, Bruckner UB, Keller EH, Kohler U, Vetter H, Waag KL (1980) Myocardial flow reserve in experimental cardiac hypertrophy. Basic Res Cardiol 5: 199–206
16. Rembert JC, Kleinman LH, Fedor JM, Wechsler AS, Greenfield JC, Jr (1979) Myocardial blood flow distribution in concentric left ventricular hypertrophy. J Clin Invest 62: 379–388
17. Parrish DG, Ring WS, Bache RJ (1985) Myocardial perfusion in the compensated and failing hypertrophied left ventricle. Am J Physiol (Heart Circ Physiol 18): H534–H539
18. Marcus ML, Mueller TM, Eastham CL (1981) Effects of short- and long-term left ventricular hypertrophy on coronary circulation. Am J Physiol 241 (Heart Circ Physiol 10): H358–H362
19. O'Keefe DD, Hoffman JIE, Cheitlin R,O'Neill MJ, Allard J, Shapkin E (1978) Coronary blood flow in experimental canine left ventricular hypertrophy. Circ Res 43: 43–51
20. Bache RJ, Vrobel TR, Arentzen CE (1981) Effect of maximal coronary artery vasodilation on transmural myocardial perfusion during tachycardia in dogs with left ventricular hypertrophy. Circ Res 49: 742–750
21. Vrobel TR, Ring WS, Anderson RW (1980) Effect of heart rate on myocardial blood flow in dogs with chronic left ventricular hypertrophy. Am J Physiol 239 (Heart Circ Physiol 8): H621–H627
22. Bache RJ, Arentzen CE, Simon AB (1984) Myocardial perfusion abnormalities during tachycardia in dogs with left ventricular hypertrophy: Metabolic evidence for myocardial ischemia. Circulation 69: 409–417
23. Fujii AM, Gelpi RJ, Mirsky I, Vatner SF (1988) Systolic and diastolic dysfunction during atrial pacing in conscious dogs with left ventricular hypertrophy. Circ Res 62: 462–470
24. Bache RJ, Dai X, Alyono D, Vrobel TR, Homans DC (1987) Myocardial blood flow during exercise in dogs with left ventricular hypertrophy produced by aortic banding and perinephritic hypertension. Circulation 76: 835–842
25. Bache RJ (1988) Effects of hypertrophy on the coronary circulation. Prog Cardiovasc Dis 31: 403–440

Provocation of Coronary Spasm and Its Pathophysiology: Experimental Studies

HITONOBU TOMOIKE and MOTOOMI NAKAMURA[1]

Summary. The present study was designed to develop an animal model of coronary spasm and to elucidate the mechanism and pathophysiology of coronary spasm. We used Göttingen miniature pigs as a model, because provoked coronary hyperconstriction in this animal often accompanied ischemic ECG-ST change 3–6 months after regional balloon denudation of the left coronary artery. Serotonin and histamine, but not phenylephrine, indomethacin, thiothromboxane A_2 or leukotrienes LTC_4 or LTD_4, provoked augmented constriction at the coronary segment(s) with intimal thickening. Topics discussed here include (1) the role of hypercholesterolemia and/or intimal thickening (geometric effect) on coronary spasm, (2) the mechanisms of arterial hypercontraction in the isolated perfused heart, and (3) the roles of endothelial and smooth muscle functions on the genesis of hyperresponsiveness to autacoids.

Key words: Coronary spasm—Göttingen miniature pig—Coronary arteriography—Histamine—Endothelium—Smooth muscle cell

Introduction

The amount of coronary blood flow follows tightly the oxygen requirement of the myocardium and is regulated at the level of small coronary arteries, or resistance vessels. Alpha-adrenergic stimulation constricts by 3.4% the epicardial coronary artery diameter, in vivo [1]. Such vasomotion along the epicardial vessel alters the total coronary resistance by only 5% [2]. Even in the isolated tube preparation of porcine intact coronary arteries, 120 m*M* KCl reduces the coronary external diameter by 35% ± 1% [3]. In other words, the epicardial coronary artery plays a role as a conduit under normal physiological conditions.

Angiographical and pathological studies have revealed that the main locus of ischemic heart disease is the epicardial coronary artery. Growing evidence indicates that dynamic changes in the severity of coronary stenosis may play a role in the genesis of various ischemic events, including variant angina [4], effort angina, myocardial infarction, or sudden death [5–9]. However, the mechanism and

[1]Research Institute of Angiocardiology and Cardiovascular Clinic, Faculty of Medicine, Kyushu University, 3-1-1 Maidashi, Higashi-ku, Fukuoka, 812 Japan

pathophysiology of coronary spasm have not been studied rigorously in experimental animal models.

We studied the functional and structural characteristics of regional coronary hyperconstriction in animal models. The following were prerequisites for the animal mode: (1) angiographically documentable regional coronary hyperconstriction, (2) transient coronary constriction, (3) evolution of myocardial ischemia distal to the site of a constricted coronary artery, and (4) repeatable provocation of coronary artery spasm. To produce regional coronary artery disease, we applied endothelial balloon denudation and cholesterol feeding in dogs [10] and miniature pigs [11], because these procedures were well-known to induce atherosclerosis at the lesion [12, 13]. Since evolution of myocardial ischemia due to coronary constriction was reproducible in the pig model [11], we studied the pathophysiology of experimental coronary artery spasm.

Provocation of Coronary Spasm in Göttingen Miniature Pigs

Characteristics of Animal Model

Göttingen miniature pigs were sedated by an intramuscular administration of ketamine hydrochloride, 12.5 mg/kg, and then anesthetized with an intravenous administration of sodium pentobarbital, 12.5 mg/kg. We made use mainly of coronary arteriography for documentation of spastic events, because this technique is the only available tool to show regional differences in vascular responsiveness to vasoactive stimuli, *in situ* [14]. The degree of luminal constriction was derived as a percentage of narrowing compared with the diameter after nitroglycerin 20 μg/kg i.v. The ECG was monitored to document ischemic changes associated with coronary artery spasm. Three to 6 months after endothelial denudation, coronary arteriography was performed before and after intracoronary or intravenous injections of vasoactive agents. Hypercontraction was specific to some autacoids such as histamine and serotonin but not to phenylephrine [15], indomethacin, thiothromboxane A_2 [16] or leukotrienes LTC_4 and LTD_4 [17].

Role of Cholesterol Level on Vasomotion

In 5 of 36 consecutive Göttingen miniature pigs, coronary spasm was repeatedly provoked by the intracoronary administration of histamine, and the left coronary arteries were examined histologically without endothelial denudation (group 1). Endothelial balloon denudation of the major branch of the left coronary artery was performed in 31 of 36 pigs 4–5 months of age by a 2F Fogarty embolectomy catheter. Five died during the procedure and the remaining 26 pigs were randomly assigned to one of 2 groups, one fed a 2% cholesterol-supplemented diet (group 2, $n = 13$) and one fed a regular low-cholesterol diet (group 3, $n = 13$) [18], because hypercholesterolemia has been repeatedly documented as one factor enhancing atherosclerosis [19] or altering vascular response to vasoactive substances [20–22].

In group 1, the content of total serum cholesterol was 51 ± 8 mg/dl. Three months after the coronary denudation, on a diet of a laboratory chow with or without 2% cholesterol, serum cholesterol levels had increased significantly from 57 ± 6 to 222 ± 27 mg/dl in group 2, but remained unchanged (48 ± 5 to 55 ± 6 mg/dl) in group 3.

Angiography revealed no significant coronary stenosis at the control state in all 3 groups. The percent luminal narrowing of the coronary diameter by histamine, 3 months after denudation, was 39% ± 3% (group 2) and 24% ± 2% (group 3) at the nondenuded site and 78% ± 3% (group 2) and 74% ± 4% (group 3) at the denuded site ($P < 0.01$ between nondenuded and denuded sites). Thus, the level of serum cholesterol, up to the level examined in the present study, did not play an important role in the evolution of coronary artery spasm.

Relationship Between Histamine-Induced Spasm and the Loci of Intimal Thickening

Determination of Spastic Site in Isolated Heart

In our previous studies, thickened intima was invariably noted at the spastic region [11, 15]. Thus, the following studies were designed to analyze rigorously the structural changes along the spastic coronary artery. After exsanguination, the left coronary artery of the isolated heart was cannulated, and perfused with physiological salt solution. Then, a barium-gelatin mixture was infused at a pressure of 90 mmHg, which provided coronary angiograms post-mortem to facilitate accurate determinations of the loci of histological specimens. After fixation with 20% formaldehyde solution, the left circumflex and left anterior descending coronary arteries were dissected serially at 5-mm intervals. Then, tissue samples were processed through paraffin, sectioned at 5–7 μm, and stained with hematoxylin and eosin or with the Weigert-van Gieson stain. Intimal thicknesses at the spastic site were 83 ± 10 μm, 80 ± 8 μm, and 71 ± 11 μm in groups 1, 2, and 3, respectively, values significantly larger than those of the nondenuded, contralateral coronary arteries: 10 ± 4 μm, 18 ± 3 μm and 13 ± 2 μm in thickness. We found no significant difference in intimal or medial thickness at the spastic site, among the 3 groups [18].

Role of Intimal Thickening on Coronary Constriction as a Geometric Effect

As the angiographical percent narrowing by histamine at the intact site was 27% ($n = 19$) in all 3 groups, the predicted percentage of histamine-induced narrowing at the spastic site, considering the geometric effects on coronary narrowing, was 33% ± 3%. The mean observed percent narrowing in the area with a thickened intima for all 3 groups was 79% ± 2% ($n = 19$), value significantly larger than that (33% ± 3%) predicted. In other words, the predicted value was only 42% ± 4% of the observed narrowing ($P < 0.01$). Thus, (1) the area with thickened intima had hyperresponsive characteristics with regard to histamine, (2) dynamic stenosis was attributed mainly to hypercontraction of the diseased coronary artery, and (3) the augmented constriction provoked by histamine was noted at the site with intimal thickening, spontaneously occurring (group 1) or mechanically induced (groups 2 and 3) [18].

Active Site on Augmented Coronary Vasomotion

When the degree of intimal thickening in micrometers and the percent luminal narrowing during a histamine challenge in vivo were compared, there was a close topologic correlation between the site of enhanced constriction and that of intimal thickening along the coronary artery. Focal and tubular luminal narrowings were observed arteriographically in 12 and 7 cases, respectively, and correlated well with the histologic distribution of intimal thickening such as focal and long-segmental localization. These lines of evidence suggested that coronary spasm was related to functional alterations of the coronary artery per se. However, the functional role of the endothelium, intima, and/or vascular smooth muscle at the site of thickened intima for the evolution of coronary spasm remained unclarified [18].

Coronary Hyperconstriction in Electrolyte-Perfused Isolated Heart

Factors Contributing to Augmented Vasoconstriction

Focal excessive narrowings of the epicardial coronary artery have been explained by the augmented transient contraction of the coronary artery [10, 11], platelet aggregation [23], geometric effects of the stenosed coronary artery [24], and injury of small resistance vessels [25]. As the role of the geometric effect on hyperconstriction and the site of spasm have been evaluated as just described, we determined the pharmacological characteristics of the spastic coronary artery in isolated heart to rule out the effects of blood constituents and neurohumoral factors [26].

After *in situ* provocation of coronary spasm by histamine, the heart was isolated and perfused with oxygenated Krebs-Henseleit solution under a constant perfusion pressure of 90 mmHg. After 90 min of equilibration, coronary angiograms on the arrested heart were taken before and after the infusion of vasoactive agents. The reduction of luminal coronary diameter of the isolated heart during continuous infusion of histamine ($10^{-5}M$) was $29\% \pm 4\%$ and $67\% \pm 3\%$ ($P<0.001$) in nondenuded and denuded areas, respectively. The constriction of the denuded areas in response to histamine was topologically the same in vivo and in vitro. The degree of focal constriction induced by histamine, defined as percent luminal stenosis from the mean diameter of the areas proximal and distal to the spastic site, was similar under in vivo (10 μg/kg i.c.) and in vitro ($10^{-5}M$) conditions. KCl (40 mM) or phenylephrine reduced the coronary artery diameters to a similar degree for denuded and for nondenuded coronary arteries. Thus, histamine is a specific vasoactive agent which produces coronary hypercontraction [26].

Characteristics of Histamine-Induced Hyperconstriction

The dose-response relation to histamine (10^{-7}–$10^{-5}M$) of the coronary diameter of the isolated heart was not influenced by pretreatment with the nerve transmitter blockers, guanethidine (3×10^{-6}M), atropine (10^{-6}M), and tetrodotoxin ($3 \times 10^{-7}M$). In Ca^{2+}-free solution with EGTA (2 mM), in which 118 mM KCl did not constrict the coronary diameter, histamine ($10^{-5}M$) reduced the coronary diameter by 17% ± 2% and 19 ± 1% (NS) in nondenuded and denuded areas, respectively. Mepyramine (pyrilamine), an H_1 blocker, abolished the constrictive response to histamine [26].

The present results suggest alterations in the influx of Ca^{2+} in face of H_1-receptor stimulation in vascular smooth muscle at the site of intimal thickening. Receptor function and the signal transduction mechanism may be an important factor for the genesis of augmented coronary contraction.

Vascular Strip Studies on Histamine-Induced Hypercontraction

Augmented Contraction by Histamine with or without Endothelium

The coronary arteries, for which coronary artery spasm was documented angiographically in vivo 3 months after endothelial denudation in 14 miniature pigs, were isolated. We sampled the specific segment of the coronary artery in referring to the in vivo angiograms and studied functional characteristics of the vascular segment at the spastic site [27].

Application of KCl (118 mM), acetylcholine (10^{-9}–$10^{-4}M$) and $PGF_{2\alpha}$ (10^{-8}–$3 \times 10^{-5}M$) produced similar tension at the respective doses in both the spastic and control coronary arteries. With increasing concentrations of histamine (10^{-8}–$3 \times 10^{-4}M$), the maximum tension was 136% ± 6% and 98% ± 4% of the tension produced by 118 mM KCl in the spastic and control vessels, respectively. The ED_{50} to histamine was significantly less in the spastic vessels. After mechanical removal of the endothelium, the tension generated during the cumulative administration of histamine was significantly larger in the spastic than in the control vessels. In addition, changes in tension after removal of the endothelium were larger ($P < 0.05$) in the control than spastic segment at $10^{-4}M$ histamine. Thus, histamine-induced hypercontraction may partly be explained by the functional alterations of medial smooth muscle cells [27].

Role of Endothelium on Coronary Relaxation

Since Furchgott and Zawadzki demonstrated the obligatory role of endothelial cells in the relaxation of arterial smooth muscle by acetylcholine [28], the endothelial cell has been extensively characterized, physiologically and pharmacologically. Damage to the endothelium [29] or atherosclerosis [30] has been reported to enhance the responsiveness of coronary vessels to vasoconstrictors.

The difference between the histamine-induced tension development seen in the presence and absence of endothelium was larger at the intact site than at the spastic site. Endothelium-related relaxation was also impaired at the spastic site in the case of serotonin. Thus, augmented responses to histamine in the smooth muscle along the spastic area were partly explained by the reduced level of endothelium-related relaxation. [27].

Perspectives

Histamine-induced coronary spasm in Göttingen miniature pigs is, so far, related to the presence of intimal thickening. However, the cause-effect relations of thickened intima to endothelial function and/or medial smooth muscle fibers remain to be elucidated. Vascular strip studies suggest the involvement of alterations of receptor function or signal transduction in medial smooth muscle in the evolution of coronary spasm. Although reduced relaxation to serotonin was noted in precontracted muscles, this evidence suggests a link to reduced production of endothelium-derived relaxing factor (EDRF), and/or reduced sensitivity of smooth muscle to EDRF.

Coronary spasm is only one step in a chain of reactions from the stimulus (trigger) to evolution of myocardial ischemia. There are many unknowns in the cascade of ischemic events, including triggers of spontaneous constriction, factors determining the degree and duration of myocardial ischemia, or whether coronary spasm leads to myocardial infarction. Recently, we found intramural hemorrhage at the site of severe coronary spasm [31]. Thus, the missing links from the spasm to myocardial ischemia may be tested in our animal model.

Acknowledgments. We thank Drs. Kenzo Tanaka, K. T. Lee and Noboru Tada for advice; Drs. Haruo Araki, Hikaru Suzuki, and Takeo Ito for discussions; Drs. Yoshiharu Kawachi, Hiroaki Shimokawa, Shozo Nabeyama, Yoshiharu Nagata, Akiko Suyama, Hideo Yamamoto, Yusuke Yamamoto, Kensuke Egashira, Yasuo Hayashi, Akira Yamada, Kazushige Nagasawa, Wateru Mitsuoka, and Shougo Egashira, Takeshi Kuga, and Hirofumi Tagawa for excellent work and discussions; Takayuki Kawasaki and Tadashi Kobayashi for technical assistance; and M. Ohara (Kyushu University) for comments on the manuscript.

This work was supported in part by Grants-in-Aid for Scientific Research (61440042, 62570396, 63870039, 63480231, 63440037, 01624006) from the Ministry of Education, Science, and Culture, Japan, and a grant for Research on Cardiovascular Disease (1-1) from the Ministry of Health and Welfare, Japan.

References

1. Ootsubo H, Tomoike H, Sakai K, Noguchi K, Takeshita A, Nakamura M (1984) Alpha adrenergic receptor activity of epicardial coronary artery in the anesthetized dog. Jpn Circ J 48: 596–601
2. Winbury M, Howe B, Hefner M (1969) Effect of nitrates and other coronary dilators on large and small coronary vessels: An hypothesis for the mechanism of action of nitrates. J Pharmacol Exp Ther 168: 70–95
3. Nagata Y, Araki H, Tomoike H, Nakamura M (1985) Vasoconstrictor agents correlatively alter diameter and tension development in isolated pig coronary arteries. Basic Res Cardiol 80: 210–217

4. Maseri A, Chierchia S (1982) Coronary artery spasm. Demonstration, diagnosis and consequences. Prog Cardiovasc Dis 25: 169–181
5. Oliva P, Potts D, Pluss R (1973) Coronary artery spasm in Prinzmetal angina: Documentation by coronary arteriography. N Engl J Med 288: 745–751
6. Epstein SE, Talbot TL (1981) Dynamic coronary tone in precipitation, exacerbation and relief of angina pectoris. Am J Cardiol 48: 797–803
7. Yasue H, Omote S, Takizawa A (1979) Circadian variation of exercise capacity in patients with Prinzmetal's variant angina: role of exercise-induced coronary artery spasm. Circulation 59: 938–948
8. Miller DD, Waters DD, Szlachcic J (1982) Clinical characteristics associated with sudden death in patients with variant angina. Circulation 66: 588–592
9. Maseri A, L'Abbate A, Baroldi G, Chierchia S, Marzilli M, Ballestra AM, Severi A, Parodi O, Biagini A, Distante A, Pesola A (1978) Coronary vasospasm as a possible cause of myocardial infarction. A conclusion derived from the study of "preinfarction" angina. N Engl J Med 299: 1271–1277
10. Kawachi Y, Tomoike H, Maruoka Y, Kikuchi Y, Araki H, Ishii Y, Tanaka K, Nakamura M (1984) Selective hypercontraction caused by ergonovine in the canine coronary artery under conditions of induced atherosclerosis. Circulation 69: 441–450
11. Shimokawa H, Tomoike H, Nabeyama S, Yamamoto Y, Araki H, Tanaka K, Ishii Y, Nakamura M (1983) Coronary artery spasm induced in atherosclerotic miniature swine. Science 221: 560–562
12. Ross R, Glomset JA (1976) The pathogenesis of atherosclerosis. N Engl J Med 295: 369–377, 420–425
13. Lee WM, Lee KT (1975) Advanced coronary atherosclerosis in swine produced by combination of balloon-catheter injury and cholesterol feeding. Exp Mol Pathol 23: 491–499
14. Tomoike H (1985) Animal models of coronary spasm and the pathophysiological events in regional vascular hypercontraction. Jpn Circ J 49: 101–107
15. Shimokawa H, Tomoike H, Nabeyama S, Yamamoto H, Ishii Y, Tanaka K, Nakamura M (1985) Coronary artery spasm induced in miniature swine: Angiographic evidence and relation to coronary atherosclerosis. Am Heart J 110: 300–310
16. Shimokawa H, Tomoike H, Nabeyama S, Yamaoto H, Nakamura M (1985) Histamine-induced spasm not significantly modulated by prostanoids in a swine model. J Am Coll Cardiol 6: 321–327
17. Tomoike H, Egashira K, Yamada A, Hayashi Y, Nakamura M (1987) Leukotrienes C_4 and D_4-induced diffuse peripheral constriction of swine coronary artery along with ECG-ST elevation: Angiographical analysis Circulation 76: 480–487
18. Egashira K, Tomoike H, Yamamoto Y, Yamada A, Hayashi Y, Nakamura M (1986) Histamine-induced coronary spasm in regions of intimal thickening in miniature pigs: Roles of serum cholesterol and spontaneous or induced intimal thickening. Circulation 74: 826–837
19. Faggiotto A, Ross R, Harker L (1984) Studies of hypercholesterolemia in the nonhuman primates. I. Changes that lead to fatty streak formation. Arteriosclerosis 4: 323
20. Heistad DD, Armstrong ML, Marcus ML, Piegors DJ, Mark AL (1984) Augmented responses to vasoconstrictor stimuli in hypercholesterolemic and atherosclerotic monkeys. Circ Res 54: 711–718
21. Yokoyama M, Henry PD (1979) Sensitization of isolated canine coronary arteries to calcium ions after exposure to cholesterol. Circ Res 45: 479–486
22. Rosendorff C, Hoffman JIE, Verrier ED, Rouleau J, Boerboom LE (1981) Cholesterol potentiates the coronary artery response to norepinephrine in anesthetized and conscious dogs. Circ Res 48: 320–329
23. Folts JD, Crowell EB, Rowe GG (1975) Platelet aggregation in partially obstructed vessels and its elimination with aspirin. Circulation 54: 365–370
24. MacAlpin RN (1980) Contribution of dynamic vascular wall thickening to luminal narrowing during coronary arterial constriction. Circulation 61: 296–301
25. Hellstrom HR (1982) The injury-spasm (ischemia-induced hemostatic vasoconstrictive) and vascular autoregulatory hypothesis of ischemic disease: Resistance vessel-spasm hypothesis of ischemic disease. Am J Cardiol 49: 802–810

26. Yamamoto Y, Tomoike H, Egashira K, Kobayashi T, Kawasaki T, Nakamura M (1987) Pathogenesis of coronary artery spasm in miniature swine with regional intimal thickening after balloon denudation. Circ Res 60: 113–121
27. Yamamoto Y, Tomoike H, Egashira K, Nakamura M (1987) Attenuation of endothelium-related relaxation and enhanced responsiveness of vascular smooth muscle to histamine in spastic coronary arterial segments from miniature pigs. Circ Res 61: 772–778
28. Furchgott RF, Zawadzki JV (1980) The obligatory role of endothelial cells in the relaxation of arterial smooth muscle by acetylcholine. Nature 288: 373–376
29. Brum JM, Sufan Q, Lang G, Bove AA (1984) Increased vasoconstrictor activity of proximal coronary arteries with endothelial damage in intact dogs. Circulation 70: 1066–1073
30. Ginsburg R, Bristow MR, Davis K, Dibiase A, Billingham ME (1984) Quantitative pharmacologic response of normal and atherosclerotic isolated human epicardial coronary arteries. Circulation 69: 430–440
31. Nagasawa K, Tomoike H, Hayashi Y, Yamada A, Nakamura M (1989) Intramural hemorrhage and endothelial changes in atherosclerotic coronary artery after repetitive episodes of spasm in X-irradiated hypercholesterolemic pigs. Circ Res 65: 272–282

8. Coronary Collaterals

Development of Coronary Collaterals

KONRAD W. SCHEEL[1]

Dedicated to Professor Kogo Onodera M.D. Sometimes we speak not each other's language, but we touch each other's soul.

Summary. This section begins with a brief description of the anatomical location of epicardial and intramural collaterals. The proposed mechanisms for collateral development are discussed in the context of presently existing experimental evidence. Thus, it is demonstrated that collateral growth occurs in response to an *ischemic stimulus* rather than a *pressure difference* between vascular beds. Data on the time course, collateral distribution, and coronary reserve in dog hearts after gradual coronary occlusion is presented. Collateral growth was investigated in hypertensive animals to address the hypothesis that intraluminal *pressure* per se promotes collateral growth. This experiment demonstrated that an elevated pressure does not induced collateral growth, and, since an elevated pressure also introduces an increase in tangential *wall stress*, this possible mechanism for inducing collateral growth was negated. Collateral growth occurred in experiments with coronary occlusion, severe chronic anemia, and after exercise in the presence of a compromised coronary circulation..However, exercise did not stimulate collateral growth in the normal heart. It appears that most experimental procedures seem to rule out mechanical factors in favor of ischemia-related factors as the primary stimulus for collateral growth. However, since decreased blood viscosity with anemia leads to an increased blood flow and increased vascular *shear forces*, this mechanism for stimulating collateral growth is not eliminated. It is further suggested that mechanical forces may be involved in limiting vascular growth once the need for oxygen is met. Evidence for the role of *adenosine* in endothelial *angiogenesis* is also presented.

Key words: Epicardial collaterals—Intramural collaterals—Ischemia—Pressure difference—Intraluminal pressure—Wall stress—Shear forces—Angiogenesis—Hypertension—Anemia—Exercise

Introduction

A coronary collateral vessel is an anastomosing branch or junction between two coronary blood vessels. It can exist between two major coronary vessels—intercoronary collaterals, or between branches of a coronary tree—intracoronary collaterals. Although the basic mechanism(s) for collateral development for either type of collateral may be identical, this section is primarily concerned with the development of intercoronary collaterals, simply referred to as coronary collaterals, or collaterals.

[1] Departments of Physiology and Medicine, Texas College of Osteopathic Medicine, 3500 Camp Bowie, Fort Worth, TX 76107, USA

Anatomical Considerations

Coronary collaterals can exist at the epicardial surface or intramurally, defined as "within the substance or boundary of the wall of any organ." The identification of intramural collaterals has led to further classifications by location of collaterals. Thus, collaterals found 2–3 mm below the epicardial surface have been termed subepicardial collaterals; and collaterals found below the endocardium as subendocardial collaterals. A term often used in describing the location of collaterals is "endomural." Although this term is not found in most medical dictionaries, some investigators use it synonymously with "intramural." We have also used the term "nonepicardial collaterals" to refer to all collaterals other than those at the epicardial surface. Collaterals with a diameter less than 100 μm [1] are labeled microvascular collaterals, and these can exist epicardially and/or intramurally.

The location and existence of collaterals is species-determined. In sheep or cattle no collaterals have been found [2]; in the pig [3, 4] and rhesus monkey [5] they are located subendocardially and intramurally; in the pony collaterals have been demonstrated in the subepicardial layer and intramurally [6]; while in the human heart epicardial, subendocardial, and intramural collaterals have been identified [7, 8]. The dog is thought to have primarily epicardial collaterals [9] and the nonepicardial component of collateral flow is reported to be insignificant [10]. However, in recent experiments from our laboratory, in which the epicardial collaterals were cauterized, it was demonstrated that 60% of the native collateral flow in the dog heart is from a nonepicardial source [11].

The adequacy of collateral flow, after coronary occlusion, in all of these species is not always sufficient to prevent myocardial ischemia or infarction, and the native endowment of collaterals is thought to be genetically determined [9].

Collateral growth has been most dramatically observed when a major coronary vessel occludes gradually. The developing collateral vessels create an alternate route of perfusion, bypassing the obstruction, and resupplying the original vessel distal to the occlusion. At times, however, intracoronary anastomoses develop from a site proximal to a site distal to the occlusion. It has not been clearly shown whether collateral growth occurs de novo, by expansion of preexisting collateral connections, or includes both possibilities. In the native dog heart small, but visible, collaterals can be observed on the epicardial surface interconnecting diagonal branches between the circumflex and left anterior descending coronary arteries. The observation that, after gradual coronary occlusion, interconnections in these same locations are found enlarged to a size approaching the dimensions of the parent vessel, has led to favoring the concept that collateral growth arises from preexisting connections. In the dog, which has a separate septal artery (in contrast to man), collateral growth between the septal artery and the occluded left circumflex coronary artery has been demonstrated. Since the septal artery courses deep within the intraventricular septum, collateral connections from this vessel constitute intramural connections [12].

The clinical concern with collateral growth is: (1) whether it can be sufficient to maintain the viability of the myocardium, (2) how adequately it can provide blood flow to the collateral-dependent myocardium under resting conditions and during periods of increased metabolic demand, (3) how rapidly it can proceed, and (4) what measures can be taken to enhance the growth of new or existing collateral vessels. For researchers to answer these questions, quantitative methods had to be developed.

Some Methods for Quantitation

The viability of the myocardium has been determined by histologically examining the magnitude of infarction or ischemia relative to the normal perfusion territory of the vessel [13]. A quantitative refinement to this method was to determine the amount of blood flow available relative to the resting or maximal (reserve) blood flow need. For these studies and in vivo interventions, the use of the radioactive microsphere method has yielded particularly useful information [14]. Because of the technical difficulties of resolving overlapping energy spectra, detection of very low blood flows, and because the spheres themselves can ultimately infarct the myocardium, this method has obvious limitations.

The intrinsic response of the coronary and collateral circulations can be strongly modified by the interplay between the heart and the peripheral circulation through neural, hormonal, and hemodynamic influences; in vivo results must always be cognizant of these heart-peripheral interactions. For this reason the isolated heart preparation has been developed for studies concerned with the intrinsic response of the heart. In advanced developments of this preparation, the major coronary vessels have been cannulated and perfused simultaneously, thus opening the way for investigations into hemodynamic interactions between the coronary vessels, more precise measurements of collateral flow, and quantitation of the rapidity of growth between the major coronary vessels. This preparation has also the advantage of allowing the coronary and collateral vasculature to be studied during maximal vasodilation of all coronary vessels, thus providing information on structural changes of vessels. To obtain the degree of vasodilation necessary for those studies, an equivalent dose of vasodilator in the in vivo preparation would result in peripheral vascular dilation with a precipitous fall in arterial pressure.

Future Investigations

Man, and as we have recently found, the dog, both possess a significant collateral network below the epicardial surface. Because nonepicardial collateral vessels are more subject to extravascular compressive forces, the interplay between collateral growth and the transmural damping effects of the myocardium on collateral flow requires further investigation.

Recent studies have also focused on research into the mechanisms involved with cellular and vascular proliferation and growth. These studies should ultimately result in answering the question on how collateral and vascular growth can be enhanced.

Mechanisms for Stimulating Collateral Growth

Research has been concerned with the stimulus for collateral growth in an attempt to distinguish between physical and biochemical factors. Thus, it has been proposed that:

(1). *Tangential wall stress* that arises when a vessel is exposed to maximum dilation or overstretching, due to ischemia and inflammatory reactions in the region, stimulates cellular proliferation and an increase in vascular diameter [15].
(2). A *pressure difference* between the patent and obstructed vascular trees gives rise to vascular growth [16].
(3). *Pressure per se* is a stimulant for vascular proliferation [17].
(4). Increased blood *flow velocity* precipitates collateral growth. A collateral vessel which connects between two patent coronary vessels has little or no blood flow passing through it, because the pressures at both ends of the collateral are identical. When one of the coronary vessels becomes occluded, a pressure drop across the collaterals occurs and blood flows; hence, blood flow velocity through the collateral is increased. When the shear forces along the wall of a vessel under these conditions are calculated, one finds that these are very much smaller than those produced by the hydrostatic pressure within the vessel [18]. On the other hand, findings that the endothelium responds to changes in blood flow by releasing endothelial-derived releasing factor, EDRF [19], may ascribe a unique function to the endothelium that may increase its sensitivity to changes in blood flow velocity or shear forces.
(5). The lack of *oxygen* supply to the vessel may be a stimulus.
(6). An accumulation of *metabolic end products* provides a stimulus for vascular growth.

There can be a further subdivision of this list when one considers that collateral growth can occur either in the local region of injury or remote or upstream from the impairment. For instance, epicardial collaterals develop in response to a lack of supply downstream, or at microcirculatory level of a vessel. Regional mechanisms could include tangential wall stress, lack of oxygen supply, and accumulation of metabolic end products. Pressure differences between vascular beds can occur at local and remote regions. An increased blood flow velocity upstream can be the result of vasodilation within the microcirculation.

Studies Concerned with the Mechanism for Stimulating Collateral Growth

Gradual left circumflex coronary occlusion. In order to distinguish between whether pressure difference or ischemia (considering the word "ischemia" to encompass both lack of oxygen and increased metabolic end products) provides the stimulus for vascular growth, one must experimentally produce one without the other. In experimental procedures, we produced an ischemic myocardium, by gradual coronary occlusion, and explored whether collaterals grew between any of two vascular beds in which a pressure difference was absent.

Gradual coronary occlusion was produced by the Ameroid method, which leads to total occlusion of the circumflex coronary artery within about 17 days after application of the device. (The Ameroid occluder is a device consisting of a slitted stainless steel ring filled with a casein "donut." When this device is slipped over the coronary artery, the casein absorbs body fluids and swells. Since it cannot expand outward, it expands inward and occludes the coronary artery.) Coronary and collateral blood flows were measured in 3 groups of dogs: control, and one and 3 months after application of the Ameroid device. After removal of the heart, a blood-perfused, maximally vasodilated, empty, beating, isolated heart preparation was used to study the coronary and collateral circulations. The left circumflex-, left anterior descending-, right-, and septal coronary artery were cannulated and blood flow was determined in each simultaneously. Collateral flow between each of these coronary vessels was measured by the retrograde flow method, which is considered to be an index of collateral flow. In this method, the cannulated vessel was disconnected from the perfusion system, allowed to bleed freely against atmospheric pressure, and blood was collected for 1 min.

The results of this study are shown in Fig. 1, and illustrate that collateral flows from the anterior descending-, right-, and septal coronary arteries to the occluded circumflex coronary artery were increased from control, demonstrating *directionality* of collateral growth [20].

Further inspection of Fig. 1 shows that during the first month after application of the Ameroid device, total collateral flow to the circumflex coronary artery had increased 10-fold (from 10 to 100.6 ml/min). Collateral growth continued, although at a slower rate, and stopped after reaching a plateau with an increase of 12 times the control magnitude. Dogs subjected to 5, 6, and 12 months of Ameroid occlusion demonstrated no significant increase in collateral growth beyond that seen at 3 months.

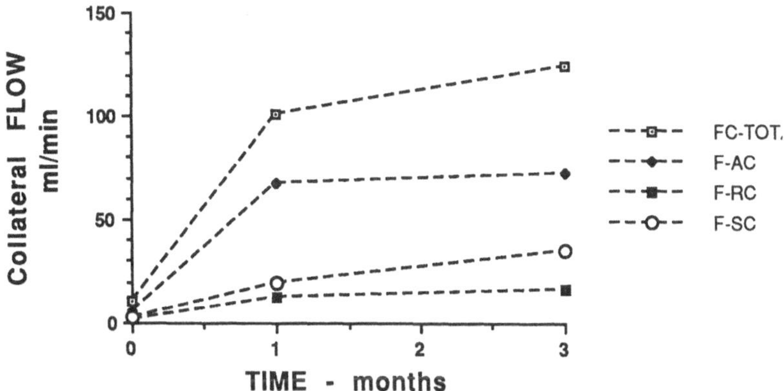

Fig. 1. Collateral development between the major coronary arteries in control hearts, and after 1 and 3 months of gradual circumflex coronary artery occlusion; *F-AC*, collateral flow from the left anterior descending- to circumflex coronary artery; *F-RC*, collateral flow from the right- to circumflex coronary artery; *F-SC*, collateral flow from the septal- to circumflex coronary artery

The total collateral flow to the circumflex coronary artery was distributed as shown in Fig. 2. Collateral flow was highest between the adjacent anterior descending and circumflex coronary arteries. However, because the septal artery adjoins the circumflex and left anterior descending coronary arteries within the intraventricular septum [21], the large collateral contribution from the septal artery was not surprising.

In Fig. 3 the data for collateral flow from the left anterior descending- to the right-; the septal- to the anterior descending-; and septal- to the right-, coronary arteries are plotted. It can be seen that although there was no pressure difference between these pairs of vessels during chronic circumflex occlusion, collateral growth between these vessels occurred [20, 22]. This suggests that myocardial

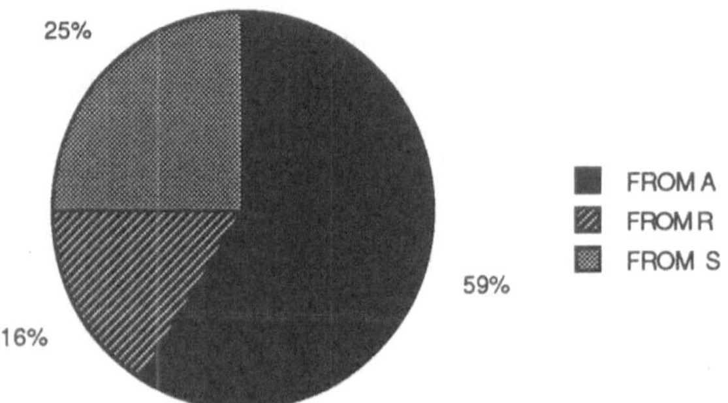

Fig. 2. Average collateral flow distribution to circumflex coronary artery after circumflex coronary artery occlusion in the dog. *FROM A* indicates the percentage of total collateral flow from the anterior descending- to the circumflex coronary artery. Similarly, *FROM R* and *FROM S* indicate the percentage of total collateral flow from the right- and septal-arteries to the circumflex coronary artery

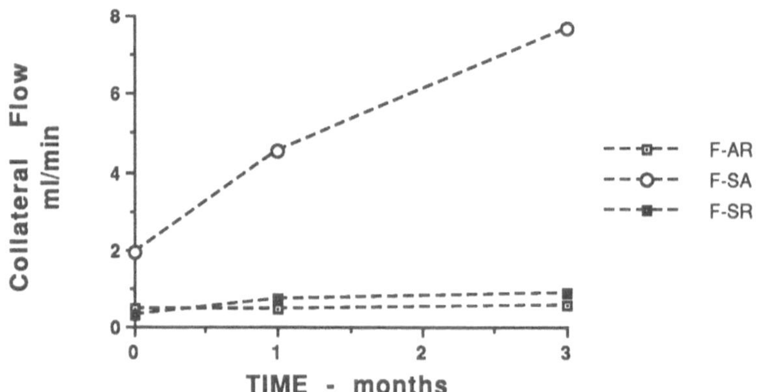

Fig. 3. Collateral flow between coronary vessels in control dogs, and after 1 and 3 months of gradual circumflex coronary artery occlusion. There was no pressure difference between these vessels; however, collateral growth between these vessels was significant. *F-AR*, *F-SA*, and *F-SR* indicate the collateral flows between the left anterior descending- and right- , septal- and left anterior descending- , and septal- and right coronary arteries

ischemia rather than a pressure difference between vessels is the stimulus for collateral growth.

Figure 4 illustrates the ratio of available collateral flow to that needed by the collateral-dependent coronary artery during resting conditions and during maximum metabolic demands. The resting flow needs were assumed to be 30 ml/min and the maximum flow, or flow reserve, was assumed to be 6 times control or 180 ml/min. It should be noted that the collateral flow measured during a retrograde flow measurement is not identical to the flow reaching the collateral-dependent vessel, because the collateral resistance is in series with the resistance of the collateral-dependent vessel, thus, increasing the total collateral resistance and reducing the collateral vessel's flow supply. For instance, if the collaterals grow to the extent that their resistance is identical to the resistance of the collateral-dependent vessel, the flow reaching the collateral-dependent vasculature is only about one-half or 50% of the flow determined during the retrograde flow measurement. During the retrograde flow measurement, collateral flow does not have to pass through the collateral-dependent bed, but only through the large coronary vessels, which have a very low resistance. When the ratio of control flow needs of the collateral-dependent vessel to collateral flow supply is equal to *1* (on the *left scale*, Fig. 4), collateral flow is sufficient to meet the resting needs.

The fact that the collaterals are in series with the collateral-dependent myocardium poses a serious reduction to reserve flow to this vascular bed, as can be seen from the *right scale* of Fig. 4. The figure demonstrates that after 3

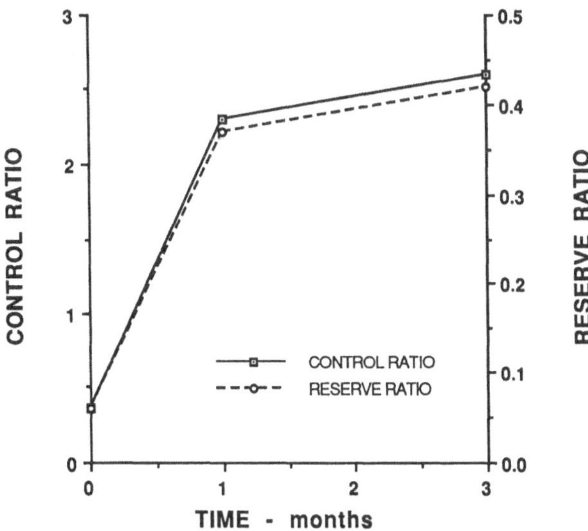

Fig. 4. Ratio of flow to collateral-dependent vessel relative to resting control needs and relative to needs during maximal metabolic demands (coronary reserve). Data shown are for normal dog hearts, and after 1 and 3 months of gradual circumflex coronary artery occlusion. The *left scale* refers to the ratio of flow to the collateral dependent vessel relative to resting control needs (*solid line*), while the *right scale* refers to the ratio of flow to the collateral-dependent vessel (*dashed line*) relative to that needed during maximum metabolic demands (reserve flow)

months of coronary occlusion, and when considerable collateral growth has occurred (about 1/10 of the original collateral resistance), the flow to the collateral-dependent vasculature can only obtain 45% of that needed during maximum metabolic demands. An additional impediment to flow for the collateral-dependent myocardium during maximum metabolic demands is that the vessel supplying the collaterals operates at maximum flow conditions, lowering the pressure at the origin of the collaterals due to a pressure drop across the large coronary vessels.

After coronary occlusion, blood flow to the collateral-dependent coronary artery is obtained from unobstructed coronary vessels. Because these vessels must not only supply their own needs, but those of the obstructed vessel, their coronary reserve is also diminished. Thus, the unobstructed vessels experience a "functional" decrease in coronary reserve flow rather than one determined by anatomical factors.

Gradual right coronary occlusion. The directionality of collateral growth was further emphasized in experiments in which the right coronary artery was equipped with an Ameroid device. In this experiment, an increased collateral flow to the right coronary artery from the left coronary arteries was seen [23]. However, the magnitude of collateral flow from the left to the right coronary artery was much less than that observed between the left circumflex and anterior descending coronary arteries after circumflex occlusion. Because the perfusion territory (grams of myocardium perfused) of the right coronary artery in the dog is much smaller than the perfusion territory of the circumflex coronary artery [21], it was hypothesized that collateral growth was sensitive to the magnitude of the ischemic stimulus [23].

Elevated intraluminal pressure. To test the hypothesis that pressure per se acts as a stimulus for collateral growth, the ascending aorta of dogs was banded. In this experimental preparation the coronary vessels were perfused at an elevated pressure for one month while peripheral pressure was normal. No increase in collateral conduction was observed. It should be noted that an increase in intraluminal pressure of the coronary arteries also produces an increase in tangential wall stress. Therefore, these experiments indirectly negated the hypothesis that tangential wall stress in the collateral vessels leads to collateral growth.

Severe chronic anemia. These experiments were conducted to investigate further the hypothesis that myocardial hypoxia causes collateral growth. After dogs were exposed to an average hematocrit of 11 vol. % for one month, the coronary and collateral conductances were determined in the isolated heart preparation with blood of normal hematocrit (40 vol. %). The data showed that both coronary and collateral conductances were increased [24]. The anemic dogs were normotensive, and since none of the vessels were occluded, there was no pressure difference between coronary vessels. Thus, the results of this study support the hypothesis that ischemia is the stimulus for not only intracoronary collateral but intercoronary collateral growth, and indicate further that, when the hypoxic stimulus is powerful enough, vascular growth is "turned on" globally throughout the heart. However, consideration must be given to the fact that during chronic anemia the blood viscosity is reduced and flow velocity in the coronary arteries was elevated. Thus, this experiment did not rule out the possibility that increased blood flow velocity induces growth of vessels.

Exercise and collaterals. The growth of collateral vessels reaches a plateau about three months after coronary occlusion. The following experiments were conducted to test whether this plateau can be raised by adding an ischemic stimulus. Four groups of research beagles were used. The first group was a control group; the second group was exercised daily at 50% of maximum oxygen consumption for two months; In the third group an Ameroid device was placed on the circumflex coronary artery for 3 months; and the fourth group was exposed to gradual circumflex coronary occlusion for 3 months, after which the dogs were exercised for 2 months, as for the dogs in the second group. The results showed that exercise alone did not cause collateral development. However, after 3 months of Ameroid application followed by an exercise program, collateral conduction doubled. Thus, collateral growth can be further stimulated by exercise-induced ischemia under conditions in which coronary reserve is diminshed [25].

The possible role of mechanical factors. To this point the mechanisms *for* stimulating collateral growth have been discussed. However, the question must also be raised, "What limits collateral growth from continuing indefinitely?"

The mechanical forces acting on the collateral vessel may provide some clues when one considers the following scenario. After an ischemic insult, the collateral vessel is passively stretched, increasing the lumen of the vessel but decreasing the wall thickness. Since the tangential wall stress on the vessel is determined by the ratio of radius to wall-thickness, the tangential wall stress is increased dramatically. Active growth of the collateral vessel, on the other hand, increases both the radius and the wall thickness [26]. During the ischemic phase of vascular growth, the increase in radius dominates over the increase in wall thickness, and wall stress remains high. When the vessel has increased in dimensions such that the ischemia is relieved, however, the growth in wall thickness occurs at a slightly higher rate than luminal growth, adjusting to the elevated intraluminal pressure. As this process continues, the ratio of lumen to wall-thickness returns to that in a normal vessel, wall stress returns to normal conditions, and vascular growth ceases. Thus, the mechanical forces acting on a collateral vessel may have a role in limiting vascular growth [18].

Angiogenesis. Angiogenesis or neovascularization is charaterized by migration and proliferation of endothelial cells. Investigations concerned with the growth of tumor cells have identified "tumor angiogenesis factors" which were demonstrated to stimulate capillary endothelial cell migration and proliferation [27, 28]. Growth factors have also been identified in other organs in which they possess vasoproliferative properties [29–31]. Because they are present in tissue not undergoing neovascularization, it has been hypothesized that a stimulus is necessary to activate the proliferative process. Hypoxia [32, 33], a direct role of oxygen [34, 35], and adenosine [36], have been implicated as the trigger substances for proliferation of endothelial cells. Since adenosine is released from the hypoxic myocardium it could act as the trigger for vascular proliferation. In recent experiments by Meininger et al. [37] vascular endothelial cells were cultured in vitro at 2% oxygen, and growth was more prolific than at 20% oxygen. The addition of adenosine to the medium further enhanced growth, but with the adenosine-receptor blocker, 8-phenyltheophylline, the proliferation caused by hypoxia was prevented. They concluded that the release of adenosine by hypoxic tissue may act as an angiogenic stimulus for growth of new vessels, but did not rule out the necessity for the presence of a growth factor.

Conclusion

The "Intelligent" Collateral

Collateral growth is not chaotic or undisciplined, as a cancer is, but seems to follow certain rules. Thus, collaterals: (1) *grow* in response to a need or stimulus, (2) display *directionality*; that is, collaterals grow primarily toward a region of the myocardium which is undersupplied or may experience an accumulation of metabolic end products, (3) matches growth to the *intensity* of the stimulus, and (4) "*knows*" that the myocardium downsteam is ischemic and institutes vascular growth upstream (epicardial collaterals), and (5) "knows" when to *stop* growing, thus distinguishing itself from a malignant tumor.

Acknowledgments. The author would like to thank Monica Scheel for assistance with the review of the literature, and Lynda Losen and Sue Williams for their help with typing of this manuscript.

References

1. Rhodin JAG (1967) The ultrastructure of mammalian arterioles and precapillary sphincters. J Ultrastruct Res 18: 181–223
2. Schaper W, Flameng W, DeBrabander M (1972) Comparative aspects of coronary collateral circulation. In: Bloor CM (ed) Comparative pathophysiology of circulatory disturbance. Plenum Publishing Corporation, New York, NY, pp 267–276
3. Schaper W, Jageneau A, Xhonneux R (1967) The development of collateral circulation in the pig and dog heart. Cardiologia 51: 321–335
4. Vastesaeger MM, Van Der Straeten PP, Friart J, Candaele G, Ghys A, Bernard RM (1957) Les anastomoses intercoronariennes telles qu'elles apparaissent a la coronarographie post-mortem. Acta Cardiol 12: 365–401
5. Buss DD, Hyde DM, Steffey EP (1983) Coronary collateral development in the rhesus monkey (*Macaca mulatta*). Basic Res Cardiol 78: 510–517
6. Amend JF, Carner HE, Fichtenbaum B (1980) Xeroradiographic observations of coronary arterial distribution in domestic ponies. Microvasc Res 20: 150–155
7. James TN (1970) The delivery and distribution of coronary collateral circulation. Chest 58: 183–203
8. Fulton WFM (1965) The coronary arteries. Charles C. Thomas, Springfield
9. Schaper W (1971) The collateral circulation of the heart. American Elsevier, New York, pp 29, 88
10. Harrison DG, Chapman MP, Christy JP, Marcus ML (1986) Studies of functional site of origin of native coronary collaterals. Am J Physiol 251: H1217–H1224

11. Scheel KW, Dautat GH, Williams SE (1990) Intramural coronary collateral flow in dogs. Am J Physiol 258 (Heart Circ Physiol 27) (to be published)
12. Wilson JL, Scheel KW (1989) Septal collateralization: demonstration of canine intramyocardial collaterals. Am J Anat 184: 62–65
13. Wilson JL, Scheel KW (1982) Myocardial infarction in dogs with acute and gradual occlusion of the circumflex or right coronary arteries. Anat Rec 204: 113–122
14. Schaper W (1979) The pathophysiology of myocardial perfusion. Elsevier/North-Holland Biomedical Press, Amsterdam, pp 13–42
15. Schaper W (1967) Tangential wall stress as a molding force in the development of collateral vessels in the canine heart. Experientia 23: 595–596
16. Gregg DE (1950) Coronary circulation in health and disease. Lea and Febiger, Philadelphia, p 198
17. Kattus AA, Major MC, Gregg DE (1959) Some determinates of coronary collateral blood flow in the open-chest dog. Circ Res 7: 628–642
18. Scheel KW, Fitzgerald EM, Martin RO, Larsen RA (1979) The possible role of mechanical stresses in coronary collateral development during gradual coronary occlusion. In: Schaper W (ed) Pathophysiology of myocardial perfusion. American Elsevier, New York, pp 489–518
19. Rubanyi GM, Romero JC, Vanhoutte PM (1986) Flow-induced release of endothelium-derived relaxing factor. Am J Physiol 250: H1145–H1149
20. Scheel KW, Rodriguez RJ, Ingram LA (1977) Directional coronary collateral growth with chronic circumflex occlusion in the dog. Circ Res 40: 384–390
21. Scheel KW, Ingram LA, Gordey RL (1982) Relationship of coronary flow and perfusion territory in dogs. Am J Physiol 243: H738–H747
22. Scheel KW, Wilson JL, Ingram LA, McGehee L (1980) The septal artery and its collaterals in dogs with and without circumflex occlusion. Am J Physiol 238: H504–H514
23. Scheel KW, Ingram LA (1981) Relationship between coronary collateral growth in the dog and ischemic bed size. Basic Res Cardiol 76: 305–312
24. Scheel KW, Williams SE (1985) Hypertrophy and coronary and collateral vascularity in dogs with severe chronic anemia. Am J Physiol 249: H1031–H1037
25. Scheel KW, Ingram LA, Wilson JL (1981) Effects of exercise on the coronary and collateral vasculature of beagles with and without coronary occlusion. Circ Res 48: 523–530
26. De Brabander M, Schaper W (1971) Quantitative histology of the canine coronary circulation in localized myocardial ischemia. Life Sci 10: 857–868
27. Weiss JB, Brown RA, Kumar S, Philips P (1979) An angiogenic factor isolated from tumors: a potent low-molecular-weight compound. Br J Cancer 40: 493–496
28. Zetter BR (1980) Migration of capillary endothelial cells is stimulated by tumor-derived factors. Nature 285: 41–43
29. D'Amore PA, Klagsbrun M (1984) Endothelial cell mitogens derived from retina and hypothalamus: Biochemical and biological similarities. J Cell Biol 99: 1545–1549
30. Gospodarowicz D, Cheng J, Lui GM, Baird A, Bohlent P (1984) Isolation of brain fibroblast growth factor with heparin—Sepharose affinity chromotography: identification with pituitary fibroblast growth factor. Proc Natl Acad Sci USA 81: 6963–6967
31. Schrieber AB, Kenney J, Kowalski J, Thomas KA, Gimenez-Gallego G, Rios-Candelore M, Di Salvo J, Barritault D, Courty J, Courtois Y, Moenner M, Loret C, Burgess WH, Mehlman T, Friesel R, Johnson W, Maciag T (1985) A unique family of endothelial cell mitogens: the antigenic and receptor cross-reactivity of bovine endothelial cell growth factor, brain-derived acidic fibroblast growth factor, and eye-derived growth factor II. J Cell Biol 101: 1623–1626
32. Knighton DR, Silver IA, Hunt TK (1981) Regulation of wound-healing angiogenesis-effect of oxygen gradients and inspired oxygen concentration. Surgery 90: 262–270
33. Kumar S, West D, Shahabuddin S, Arnold F, Haboubi N, Reid H, Carr T (1983) Angiogenesis factor from human myocardial infarct. Lancet 2: 364–368

34. Turek Z, Grandtner M, Kreuzer F (1972) Cardiac hypertrophy, capillary and muscle fiber density, muscle fiber diameter, capillary radius and diffusion distance in the myocardium of growing rats adapted to a simulated atmosphere of 3500 meters. Pflügers Arch 335: 19–28
35. Valdivia E (1958) Total capillary bed in striated muscle of guinea pigs native to the Peruvian mountains. Am J Physiol 194: 585–589
36. Hudlicka O, Tyler KR (1986) Angiogenesis: The growth of the vascular system. Academic Press, Orlando, pp 110–114
37. Meininger CJ, Schelling ME, Granger HJ (1988) Adenosine and hypoxia stimulate proliferation and migration of endothelial cells. Am J Physiol 225: H554–H562

A Conscious Animal Model for Studies of Coronary Collateral Blood Flow Dynamics

DEAN FRANKLIN[1], ATSUSHI MIKUNIYA[2], MASATOSHI FUJITA[3], MASAAKI TAKAHASHI[4], MICHAEL MCKOWN[1], DANIEL MCKOWN[1], JESSE HARTLEY[1], R. DUSTAN SARAZAN[1], and KOGO ONODERA[2]

Summary. We describe a new animal model and protocols for measurements of the rate of development of coronary collateral perfusion stimulated by repetitive episodes of regional, reversible myocardial ischemia in conscious dogs and for measurements of the integrated effects of interventions on collateral perfusion after development of collaterals. Examples of the results are illustrated and potential sources of errors are identified. We propose this model to complement the other *in vitro* and *in vivo* models for studies of coronary collateral dynamics in conscious animals.

Key words: Coronary collateral flow—Coronary collateral development—Regional myocardial ischemia—Regional myocardial function—Coronary collateral dynamics

Introduction

Recruitment of perfusion through coronary artery collateral channels is a primary natural defense mechanism for protection of myocardium jeopardized by stenosis or obstruction of a native coronary artery. Given the incidence of stenotic coronary artery disease, effective interventions for prophylactic enhancement of collateral recruitment have a potentially important role in preventive cardiology. We have developed a model with protocols that allow day-to-day measurements of collateral perfusion status in dogs during progressive development of collateral perfusion, and allow minute-to-minute measurements of the integrated effects of acute interventions, e.g., drugs, on collateral perfusion in dogs with mature collaterals.

Rationale

Coronary collateral perfusion channels may be schematically represented, for discussions, as a single resistive conduit interconnecting two native coronary blood vessels (Fig. 1). Assume the two natural coronary vessels to be the left

[1]Dalton Research Center, University of Missouri-Columbia, Columbia, MO 65211, USA
[2]Hirosaki University School of Medicine, Hirosaki, Japan
[3]Toyama Medical and Pharmaceutical University, Toyama, Japan
[4]Shizuoka Rohsai Hospital, Hamamatsu, Japan

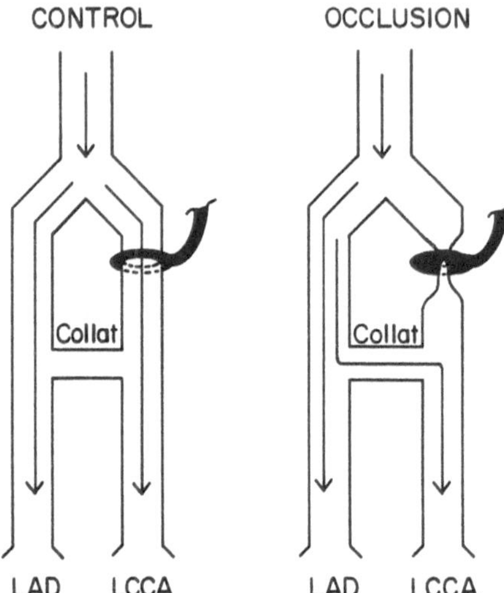

Fig. 1. Method for measurement of collateral (*Collat*) flow from the left anterior descending coronary artery (*LAD*) to the left circumflex coronary artery (*LCCA*) perfusion area. When the proximal LCCA is occluded, proximal LAD flow is the sum of flow to the LAD perfusion area and collateral flow to the LCCA perfusion area. Collateral flow is measured as the step decrease in LAD flow upon abrupt deflation of the LCCA occluder

circumflex coronary artery (LCCA) and the left anterior descending coronary artery (LAD). If the hydraulic resistances through the proximal portions of the LCCA and the LAD are negligible, the pressure gradient across the collateral channel will be zero and there will be no flow through the collateral conduit. However, if the LCCA is occluded, there will be a pressure gradient from the LCCA across the collateral channel, and blood will flow from the LAD through the collateral channel into the myocardium normally perfused by the LCCA. Under these conditions, flow through the proximal portion of the LAD will equal the sum of the flow to the myocardium normally perfused by the LAD and flow through the collateral channel into the myocardium normally perfused by the LCCA. Under ideal circumstances, i.e., where there is no change in vascular resistance in the LAD area, collateral flow is equal to the difference between LAD flow during the LCCA occlusion and the LAD flow prior to the occlusion. Cibulski et al. [1] used and validated this method for measurement of LAD to LCCA area collateral flow in the open-chested, anesthetized dog. He demonstrated that the LCCA occlusion did not modify LAD peripheral vascular resistance in his experiments. However, this ideal circumstance is not typically realized in the practical situation of a conscious animal. The LCCA occlusion is likely to cause a tachycardia which will increase the metabolic requirements and decrease the vascular resistance of the LAD area. Although the LAD flow during LCCA occlusion will be the sum of the flow into the LAD myocardium plus the flow through the collateral channels into the LCCA area, the variation in LAD area peripheral vascular resistance precludes the simple computation of collateral flow as the change in the LAD flow occurring when the LCCA artery is occluded.

We used a modification of Cibulski's method for measurement of collateral flow. After stabilization of the tachycardia and peripheral vascular resistances during the LCCA occlusion, the LCCA occlusion is abruptly released, i.e., the pressure gradient across the collateral perfusion channels is rapidly brought to zero. Flow through the collateral channel will abruptly cease and the resultant decrease in measured proximal LAD flow will equal the flow through the collateral perfusion channel prior to release of the occlusion. This rationale involves the assumption that LAD vascular resistance does not change instantaneously upon release of the occlusion. Note that under conditions where the collateral perfusion of the LCCA area is inadequate for the metabolic requirements, there will be a near-maximal physiological dilatation of the LCCA peripheral vasculature. Under these conditions the collateral channels will constitute the flow-limiting resistance. Thus, with a constant perfusion pressure, collateral flow measured in this way constitutes an index of the collateral hydraulic resistance and provides a basis for repeated daily, direct measurements of the extent of development of collaterals in response to a stimulus.

Measurement Implementation

Mongrel dogs of approximately 30-kg weight were used. After suitable anesthesia (pentobarbital) the heart was exposed by thoracotomy through the fifth left interspace, using sterile technique. Pulsed Doppler flow transducers were implanted around the proximal left anterior descending coronary artery (LAD) and around the proximal left circumflex coronary artery (LCCA). An externally inflatable occluder was implanted around the LCCA just distal to the flow transducer (Fig. 2). Pairs of 2-mm diameter, lensed piezoelectric crystals were implanted for measurement of subendocardial segment lengths in a circumferential plane in the regions perfused by the LCCA and the LAD. A Konigsburg pressure gauge was implanted within the left ventricle along with a catheter for direct

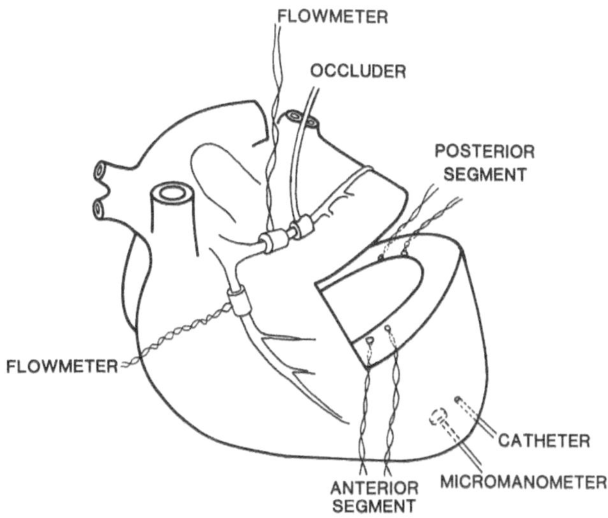

Fig. 2. Implanted instrumentation for measurements of LAD and LCCA flows, regional myocardial segment lengths and left ventricular pressure

pressure calibration. Where appropriate for the experiments, pacing electrodes were attached to the left atrium and to the epicardium near the right ventricular outflow tract. The chest was closed and evacuated using standard procedures. During the 2 weeks of recovery from surgery, the animal was conditioned to the experimental laboratory and trained to lie quietly on a rug on the laboratory floor. The experiments were initiated when the animal appeared healthy and vigorous at least 2 weeks postoperatively.

Collateral Development

Repeated episodes of regional myocardial ischemia constitute an adequate stimulus for the development of coronary collaterals in the dog and the pony [2–6]. Collateral perfusion was promoted in the dogs by occluding the LCCA for 2 min at hourly intervals 8–10 times each day. This pattern of stimulation was established from previous experiences that dogs do not appear to be aware of a 2-min LCCA occlusion (they commonly sleep through this procedure). However, if the occlusion is prolonged for much more than 3 min they commonly display signs of awareness. The one-hour interval was initially elected because of our observation that the modification in the shortening pattern of the ischemic segment following a 2-min occlusion (protodiastolic shortening) persisted for approximately 45 min.

Measurements were taken daily during the development of collaterals (Fig. 3). During the early LCCA occlusions there was a sustained holosystolic elongation of the collateral-dependent posterior segment, increased systolic shortening in the anterior segment, a tachycardia, and an elevation in LAD flow. Upon release of the LCCA occlusion there was a large and prolonged reactive hyperemia. There was typically a small but measurable step decrease in average LAD flow immediately upon release of the LCCA occlusion. In the same dog, after 160 two-min occlusion at hourly intervals, the LCCA occlusion resulted in only a transient reduction in posterior segment shortening, a transient increase in anterior shortening, and no tachycardia. LAD flow gradually increased during the LCCA occlusion. When the LCCA occlusion was released there was a small LCCA reactive hyperemia and a large step-reduction in LAD flow to near the preocclusion level. The step reduction in LAD flow immediately following release of the LCCA occlusion was measured and interpreted to be the collateral flow from the LAD donor artery to the LCCA perfusion area in both cases. The percentage error in measurement of collateral flow by this technique during the early stages of development is quite large since the step is comparable with spontaneous variations in LAD flow due to variations in heart rate, etc., in the conscious unmedicated dog. This difficulty is minimized by averaging of repeated measurements. In six dogs studied by Mikuniya et al. [7] using this technique, LAD to LCCA area collateral flow increased by 15 times from 0.6 ± 0.2 cm/s to 9.2 ± 2.8 cm/s during 212 ± 86 LCCA occlusions (mean ± SD).

Figure 4 is an example of the use of this technique to follow the day-to-day collateral development, in terms of the indirect indices of adequacy of collateral perfusion and LAD to LCCA area collateral flow as measured by the LAD step

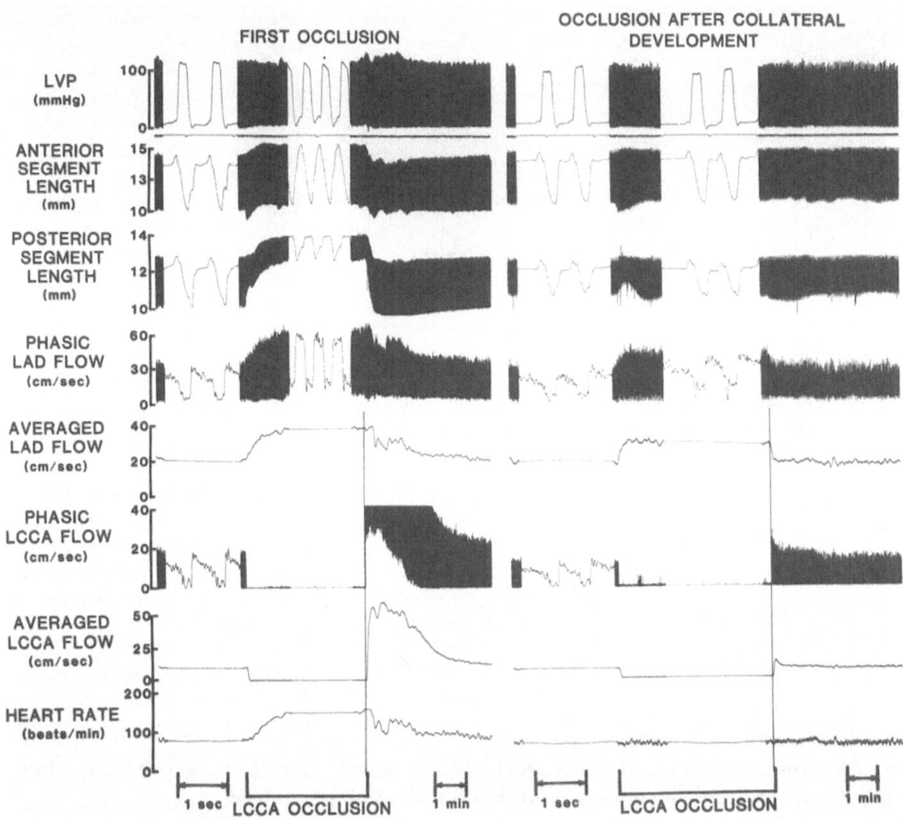

Fig. 3. Examples of measurements in a conscious dog prior to and after development of collaterals. *LVP*, left ventricular pressure; *LAD*, left anterior descending coronary artery; *LCCA*, left circumflex coronary artery

flow method. The measurements were taken during the last 2-min occlusion of each day, for 12 days. In six dogs so studied, there was a high correlation between blood flow debt repayment during reactive hyperemia and measured collateral flow [7]. There was a similar high inverse correlation between the percent reduction in collateral-dependent segment shortening during the occlusion and collateral flow, so long as the collateral flow exceeded approximately 30% of the resting metabolic requirements [7]. During the early stages of this experiment it is likely that the LCCA distal vasculature was widely dilated since there was clear ischemic reduction in function during the LCCA occlusion. Under these conditions it is likely that collateral flow is limited predominately by collateral vessel conductance. Therefore, the day-to-day measurement of collateral flow is an index of collateral conductance development. However, when the LCCA occlusion results in no sustained reduction in function, i.e., when collateral perfusion is adequate for the metabolic requirements, then collateral perfusion may be largely regulated by autoregulation of the distal LCCA vasculature. Under these conditions, although this approach may provide a reliable measure

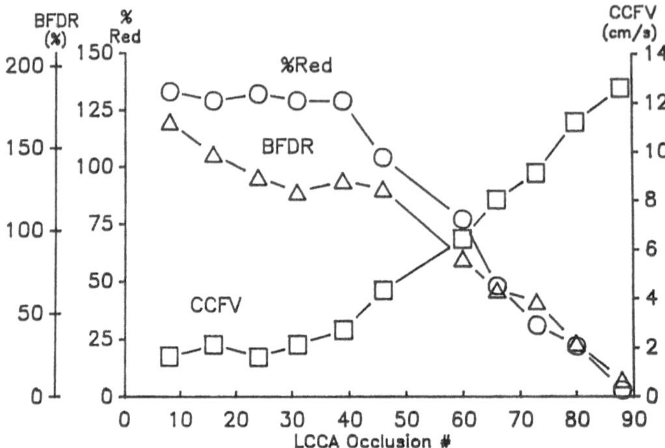

Fig. 4. Day-to-day measurements of direct and indirect indices of collateral perfusion in a dog during development of collaterals. The measurements were taken during the last 2-min (LCCA) occlusion each day for 12 days. *BFDR* is LCCA blood flow debt repayment following release of the occlusion. %*Red* is the percentage reduction in posterior segment systolic shortening during the occlusion as compared with the preocclusion shortening. *CCFV* is the derived LAD to LCCA area collateral flow

of collateral flow, it is no longer a reliable index of collateral perfusion capability. In other words, collateral perfusion reserve cannot be estimated when LCCA autoregulation predominately controls collateral flow.

The Integrated Effects of Interventions on Collateral Flow in Dogs with Well-Developed Collaterals

The flow of blood through coronary collateral blood vessels is determined by a host of interdependent factors. These include the pressure gradient across the collateral channels, the extent of dilation of the resistance vessels distal to the collateral channel, the state of dilation of the collateral vessels themselves, the intramyocardial compressing forces, the duration of systole and diastole, the extent of shortening, or lengthening, of the collateral dependent myocardium, the intraventricular pressure, and other factors. It is difficult to conceive of an experimental intervention to modify any one of these determinants of collateral perfusion without influencing the others in the conscious animal preparation. Schaper [8] recognized this difficulty inherent in conscious animal preparations and attributed the seemingly conflicting reports about the coronary effects of various vasoactive drugs to be due, in part, to the confounding actions of the various interdependent determinants of coronary and coronary collateral flow. Nevertheless, development of rational, clinically applicable strategies for treatment of patients with stenotic coronary artery disease is crucially dependent upon knowledge of the integrated (net) effects of interventions on collateral perfusion and upon regional myocardial function in the collateral-dependent myocardium. The following two illustrations from our current research will serve as examples of the use of the model to study the integrated effects of interventions on regional myocardial function and collateral flow in conscious dogs with well-developed collaterals.

Heart Rate

Since most interventions purported to modify coronary collateral flow are likely to have an indirect or direct effect on heart rate, the effect of heart rate as an intervention is important. The effects of heart rate in a dog with well-developed collaterals are illustrated in Fig. 5a–c. Collateral perfusion adequate for the metabolic requirements at rest was stimulated in this dog by the standard, repeated occlusion protocol. Subsequently, the repeated occlusions were continued but while the dog was running on a treadmill. This was continued until the LCCA occlusions during exercise at a modest level produced no sustained reduction in collateral-dependent segment shortening. The effects of pacing the ventricle at graded heart rates were then studied with the dog lying quietly on the floor.

During the LCCA occlusions at rest (unpaced) there was no deterioration of collateral-dependent, myocardial function and there was a negligible LCCA reactive hyperemia (Fig. 5a). LAD flow increased at the onset of the LCCA occlusion and decreased following the release. Note the difficulties in measurement of the step in LAD flow when there are beat-to-beat variations in heart rate, as in this case. This difficulty in measurement is minimized by averaging the measurements taken during repeated occlusions and releases.

When the heart was paced at 120 beats/min, LAD flow was elevated and increased further upon LCCA occlusion (Fig. 5b). There was a transient reduction in LCCA segment shortening at the onset of the LCCA occlusion. Upon release of the LCCA occlusion, there was an abrupt decrease in LAD flow to near preocclusion levels. The negative step in LAD flow upon release of LCCA occlusion was significantly larger at this heart rate than at the previous lower rate.

During pacing at 140 beats/min there was an additional increase in preocclusion LAD flow (Fig. 5c). During the LCCA occlusion there was a sustained reduction in segment shortening, i.e., collateral perfusion of the LCCA area was inadequate for the metabolic requirements at this heart rate. During occlusion, LAD flow gradually rose to a maximum and, upon release, decreased to a level above the preocclusion level. Both the increase in flow at the onset of occlusion and the decrease of flow upon release were less at this heart rate than at the previous lower heart rate. This combination of findings is strong evidence that collateral flow from the LAD to the LCCA perfusion area was decreased. Conceivably the decreased negative step in LAD flow was due to a persistent collateral perfusion pressure gradient following the LCCA release, i.e., collateral flow continued after release of the LCCA occlusion. This seems unlikely since the same phenomenon occurs with elevated heart rate and symmetrical dilatation using dipyridamole infusion.

From such studies we conclude that significant collateral perfusion reserve can

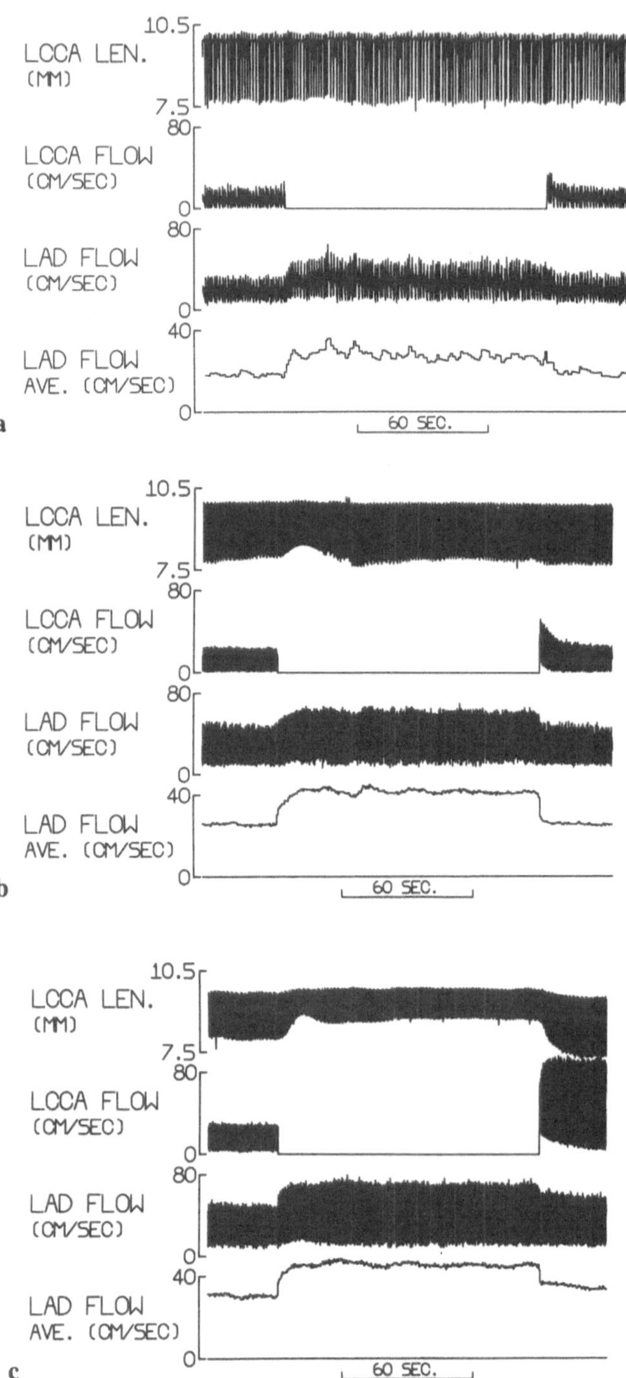

Fig. 5a–c. Digitized records of posterior segment length (*LCCA length*), phasic *LCCA flow*, *LAD flow*, and *LAD flow averaged* over each cardiac cycle during 2-min occlusion **a** without pacing, **b** paced at 120 beats/min, and **c** paced at 140 beats/min. *LCCA*, left circumflex coronary artery; *LAD*, left anterior descending coronary artery

be promoted by continuing the protocol of repeated episodes of regional myocardial ischemia. As heart rate is elevated, collateral perfusion increases until it becomes limited by collateral resistance. At heart rates above this level, the collateral-dependent myocardium becomes ischemic and collateral perfusion decreases, perhaps due to increased intramyocardial compressive forces associated with elevated diastolic pressure and increased segmental length.

Drug Interventions

A modification of the previously described protocol allows semicontinuous measurement of the extent and time-course of effects of an intervention on collateral flow. Note that in dogs with significant collateral perfusion reserve, the LCCA may be occluded indefinitely (Fig. 5a, b). The occlusion may be released briefly at any time for measurement of collateral flow in terms of the step in LAD flow. An example of use of this procedure for determining the effects of a drug on collateral perfusion and collateral-dependent regional myocardial function is shown in Fig. 6.

The effects of a bolus injection of nitroglycerin were studied in five dogs in which collateral perfusion adequate for the metabolic requirement at rest had been promoted by the repeated LCCA-occlusion protocol. Measurements of peak left ventricular pressure (LVP), heart rate (HR), LAD flow, and percentage systolic shortening in the collateral-dependent myocardium (Seg shrt) were taken prior to the occlusion and plotted as mean ± SE at the point on the x-axis marked "Pre occl." The LCCA was then occluded and the occlusion was released for 10s at 1-min intervals thereafter. Collateral flow (Coll flow) was measured at each release for five releases and the average measurements plotted on the graph at the point marked "Occl." Nitroglycerin (45 μg/kg) was injected into the left ventricle over a 30-s interval following the fifth release. Measurements were taken at 1-min intervals thereafter. Left ventricular pressure, heart rate, and segment shortening were measured just prior to each release. Collateral flow was measured as the step decrease in LAD flow upon each release. LAD flow to the LAD perfusion area was measured as the level to which LAD flow fell during the 10-s release. Measurements were made following the release of the last occlusion and are plotted at the point marked "Post occl."

Discussion

This model has a number of attractive features for studies of coronary collateral dynamics. It allows studies of the rate of development of collaterals in terms of the indirect indices of collateral perfusion, i.e., collateral-dependent myocardial function and postocclusion reactive hyperemic repayment, as well as the near-direct measurement of collateral flow from a major collateral donor artery. Thus, the model should be valuable in those studies of interventions purported to modify the rate and extent of development of coronary collaterals. The model allows studies of the effects of acute interventions purported to modify collateral flow in conscious animals with well-developed collaterals. Thus, the integrated

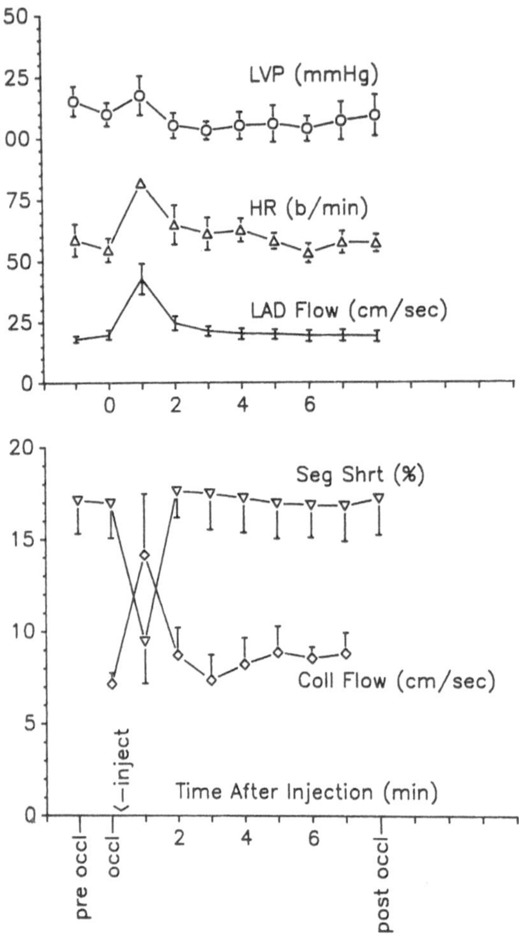

Fig. 6. Summary of effects of intraventricular bolus injection of nitroglycerin (45 μg/kg) in 5 dogs with well-developed collaterals (mean ± SE). *LVP*, left ventricular pressure; *b*, beats, *LAD*, left anterior descending coronary artery; *Seg shrt*, segment shortening (systolic, in collateral-dependent myocardium); *Coll*, collateral; *occl*, occlusion

effect of interventions, e.g., drugs, can be studied under conditions mimicking the clinical situation. This provides the opportunity for testing in the conscious animal those potentially valuable regimens of therapy suggested by the results from in vitro studies. It is likely that the collateral flow measurement technique reflects total LAD to LCCA perfusion area flow; i.e., collateral flow through both the low-resistance epicardial conduits as well as the high-resistance intramyocardial collaterals is measured. An additional feature of the model with great importance is the fact that each animal serves as its own control. The initial measurements provide an indication of the extent of postoperative preexisting collaterals with which subsequent measurements can be compared. In studies of the effects of intervention on well-developed collaterals, an effectively unlimited number of different interventions can be compared in the same animal. For example, current studies involve the evaluation of seven different drugs at graded dosages and heart rates. At a minimum this involves study of each drug at two different dosages, with and without pacing. Thus at least 28 individual experiments are conducted in the same dog. There is a 24-h interval between experiments to minimize residual effects. The order of the experiments is randomized. This capability for repeated comparative studies in the same animal alleviates many of the difficulties associated with comparisons among animals.

The model may be criticized on a number of bases. Note that the collateral perfusion measurement technique measures only that collateral flow from the LAD as a donor artery. This flow may or may not be proportional to total collateral flow from all donor arteries to the LCCA area. However, Mikuniya's demonstration that collateral flow as measured by this technique is highly correlated with the indirect indices of total collateral perfusion provides confidence in the generality of the findings [7]. The technique for measurement of collateral flow has not been calibrated against an independent technique in the conscious animal. Further studies of the situations under which the measurement technique may lead to serious error are needed. However, Cibulski's original validation of a very similar technique in open-chested anesthetized dogs provides confidence in the approach [1]. Under conditions where there is an extremely high flow through the LCCA during release, an LAD to LCCA perfusion pressure gradient may persist, resulting in persistent collateral flow during the LCCA release. This would result in an underestimation of collateral flow by the LAD flow step measurement. Preliminary pacing experiments with and without dipyridamole dilatation indicate that this potential error is likely to be small.

In summary, we have described a new animal model which allows studies of the rate of development of collateral perfusion and the effects of interventions on collateral-dependent myocardium and collateral flow in conscious unmedicated animals. We have described protocols for such studies and illustrated the application of these protocols. We have identified potential sources of error under certain experimental conditions. We propose this model to complement the other in vitro and in vivo models for studies in the important area of coronary collateral dynamics.

Acknowledgments. This research was supported in part by the U.S. Department of Health and Human Services, National Heart, Lung, and Blood Institute, Grant #HL32800, and by a gift of a cardiovascular data digital acquisition and analysis computer system from Po-Ne-Mah, Inc.

References

1. Cibulski AA Lehan PH, Hellems HK (1973) Myocardial collateral flow measurement in mongrel dogs. Am J Physiol 225: 559–565
2. Franklin D, McKown DP, McKown MD, Hartley JW, Caldwell WM (1981) Development and regression of coronary collaterals induced by repeated, reversible ischemia in dogs. Fed Proc 3: 603

3. Fujita M, McKown DP, McKown MD, Hartley JW, Franklin D (1987) Evaluation of coronary collateral development by regional myocardial function and reactive hyperemia. Cardiovasc Res 21: 377–384
4. Mikuniya A, McKown MD, McKown DP, Hartley JW, Franklin D (1987) Coronary collateral induction by repeated brief coronary occlusions: Validation by radioactive microspheres. Circulation 76(Suppl IV): 376
5. Yamamoto H, Tomoike H, Shimokawa H, Nabeyama S, Nakamura M (1984) Development of collateral function with repetitive coronary occlusion in a canine model reduces myocardial reactive hyperemia in the absence of significant coronary stenosis. Cir Res 55: 623–632
6. Rugh KS, Garner HE, Hatfield DG, Miramonti JR (1987) Ischemia induced development of functional coronary collateral circulation in ponies. Cardiovasc Res 21: 730–736
7. Mikuniya A, McKown MD, McKown DP, Hartley JW, Franklin D (1987) Relationships among coronary collateral flow, regional myocardial function and reactive hyperemia during coronary collateral development in conscious dogs. Fed Proc 46: 3014
8. Schaper W (1971) The collateral circulation of the heart. North-Holland, Amsterdam

9. Clinical Evaluation of Coronary Circulation

Coronary Reserve in the Human Coronary Circulation

MELVIN L. MARCUS, MICHAEL D. WINNIFORD, and JAMES D. ROSSEN[1]

Summary. Coronary flow reserve was initially measured in animals over 50 years ago. Recent technological advances have resulted in the extension of coronary reserve measurements to humans. Coronary flow reserve can be accurately measured using intracoronary Doppler and digital angiographic methods. Ultrafast computed tomography, positron emission tomography, and contrast echocardiography show future promise for the noninvasive measurement of coronary flow reserve. A host of factors, both related and unrelated to significant cardiac disease, can influence relative measurements of coronary flow reserve. Flow reserve determination is useful in several clinical settings.

Key words: Coronary circulation—Coronary vessels—Ischemic heart disease—Coronary vasodilators—Ventricular hypertrophy

Introduction

Measurements of coronary flow reserve were first made in animals decades ago by pioneering coronary physiologists including Louis Katz and Donald Gregg [1, 2]. More recently, measurements of coronary flow reserve in animals have been extensively explored by Gould and colleagues [3–6]. These studies have shown that in awake large mammals studied under resting conditions coronary flow can increase five- or sixfold above the resting value in response to a potent coronary dilator. Flow reserve is similar in the right and left ventricles but much lower in the atria [7, 8]. In addition, flow reserve is similar in the subepicardial and subendocardial layers of the left ventricle when measured under physiologic conditions. In small mammals such as the rat, flow reserve is much lower than in large mammals mainly because resting flow is markedly elevated [9, 10]. Minimal coronary vascular resistance in large and small mammals is remarkably similar [9, 10].

Studies in animals have shown that flow reserve is influenced by hemodynamic conditions, calcium channels antagonists, and various disease states such as severe pressure-induced left ventricular hypertrophy [11] and coronary obstructive lesions [3–6]. Thus, extensive investigation in animals has shown that coronary

[1]Department of Medicine and The Cardiovascular Center, University of Iowa Hospitals and Veterans Administration Hospital, Iowa City, IA 52240, USA

flow reserve measurements are very helpful in gaining insight into fundamental processes that influence the coronary circulation. In light of these observations, it was logical for clinical investigators to develop methods of assessing coronary flow reserve in individual coronary vessels of awake humans.

Methods of Measuring Coronary Flow Reserve in Humans

During the past several decades seven methodological approaches have been introduced for measuring flow reserve in humans (Table 1). None of these methods is ideal, but over the years there has been substantial improvement. The older methods, inert gas clearance and coronary sinus thermodilution, have substantial limitations. Gas clearance methods suffer from limited spatial resolution [12, 13], require prolonged steady-state conditions [12, 13], and seriously underestimate maximal coronary flow [14]. Several factors compromise the accuracy of the thermodilution method. Movement of the thermodilution catheter may lead to substantial changes in measured flow [15], particularly during respiration or interventions that result in large changes in heart size. Reflux of right atrial blood into the coronary sinus can lead to spurious overestimation of coronary blood flow by the thermodilution technique [16]. Finally, the coronary sinus thermodilution technique has not been adequately validated in subjects with coronary artery disease. These approaches were very popular prior to the mid-1980s but are now gradually being replaced by newer methods which have several advantages.

Two of the newer approaches—Doppler flow devices and digital subtraction angiography—are now well established and in use in many laboratories around the world. Our laboratory has been primarily involved in the development of the Doppler technique [7, 17–19]. Studies with the Doppler have shown that changes in coronary blood flow velocity correlate closely with changes in measured coronary flow (Fig. 1) and that the 3 French Doppler catheter (Fig. 2) does

Table 1. Methods of measuring coronary flow reserve in man

Older approaches
Inert gas clearance
Nonradioactive gases (helium, argon, nitrous oxide)
Radioactive gases (xenon-133)
Coronary sinus thermodilution
New approaches (established)
Dopper flow devices
Dopper epicardial suction probe
Coronary Doppler catheter
Digital subtraction angiography
New approaches in the development phase
Positron emission tomography
Ultrafast computed tomography
Contrast echocardiography

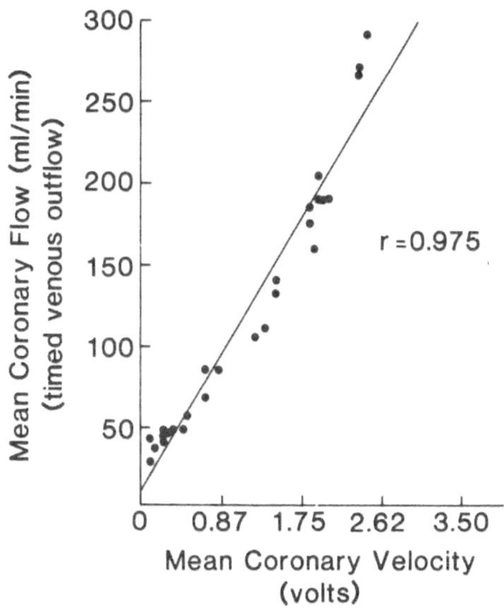

Fig. 1. Relationship between coronary flow measured by timed venous outflow from the coronary sinus and coronary velocity measured with the suction Doppler. These experiments were performed in a single dog. Coronary flow was varied over a broad range with various pharmacological maneuvers and severe hypotensive hemorrhage. Over a wide range of flows, there is a very close relationship between change in coronary blood flow velocity measured with the suction Doppler and change in timed venous outflow. These data indicate that changes in coronary cross-sectional area associated with the stimuli we employed do not significantly interfere with the correlation between coronary flow and coronary velocity. From [7] with permission of the American Heart Association

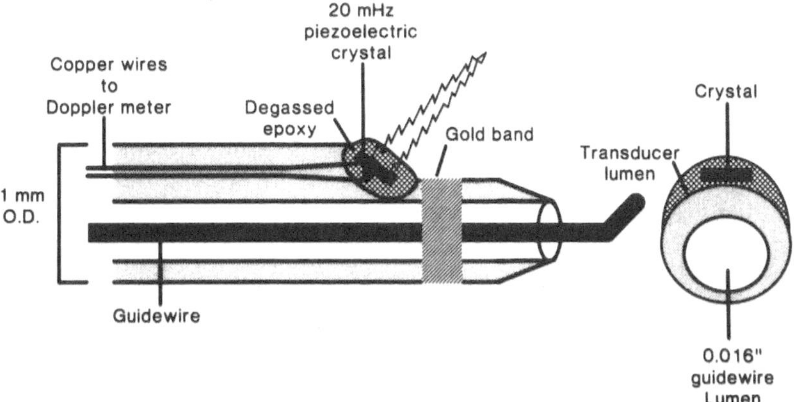

Fig. 2. Diagram of the 3 French coronary Doppler catheter. The Doppler crystal is placed on the side of the catheter and sends and receives Doppler information. *O.D.*, outside diameter. From [17] with permission of Prog Cardiovasc Dis

not produce detectable obstruction in conduit coronary arteries that are as large as the proximal coronary arteries in humans. Both the Doppler and digital subtraction angiographic methods have been the subject of extensive validation studes [7, 18, 20–22]. Both techniques can accurately detect large increments in coronary flow, and both yield flow reserve values in normal humans of 4–6 [17–22]. That is to say, coronary flow during maximal vasodilation is four to sixfold greater than resting flow. Both approaches have improved spatial resolution compared to the older methods. Digital subtraction angiography can measure coronary flow reserve in the perfusion territory of specific vessels, and the Doppler catheter can measure flow reserve in individual conduit coronary arteries. In addition, with both methods, multiple measurements of coronary flow reserve can be obtained safely in a single study. The digital subtraction angiography system is expensive and requires analysis off-line, whereas the equipment needed for Doppler studies is relatively modest in cost and the information is evaluable immediately.

Just as the Doppler and digital subtraction angiographic approaches share several advantages, both of them have three major disadvantages. They both require coronary ostial or intracoronary cannulation during cardiac catheterization, both measure relative changes in flow reserve as opposed to absolute flow, and neither method can measure coronary flow reserve in various transmural layers of the left ventricle.

The three newest methods of measuring coronary flow reserve in humans—positron emission tomography [23], ultrafast computed tomography [24, 25], and contrast echocardiography [26]—may potentially overcome the disadvantages of digital subtraction angiography and Doppler methods while maintaining their advantages. All three methods are "potentially" less invasive than the Doppler and digital subtraction angiographic methods and may ultimately require only intravenous injection of a radioactive tracer or a contrast agent. Most importantly, all three new approaches have the potential of measuring coronary flow reserve in various transmural layers of the left ventricle. This is a critically important requirement because many disease states can selectively impair flow reserve in the subendocardial layer of the left ventricle and have little or no effect on flow reserve in more superficial layers of the left ventricle. Recent studies conducted in our laboratory with ultrafast computed tomography are particularly encouraging with regard to the ability of this technology to separately measure coronary flow reserve in various transmural layers of the left ventricle [24].

It is our prediction that by the early 1990s one or more vastly improved method of measuring coronary flow reserve will be in broad use. Such a method will be relatively noninvasive, have good spatial resolution, and be able to assess absolute coronary flow reserve in individual transmural layers of the left ventricle in awake humans. At the present time, it would appear that ultrafast computed tomography may be first to satisfy these stringent requirements.

Whenever flow reserve is measured with any of the methods described, two additional critical issues must be considered: the method producing coronary dilation and factors that influence the interpretation of coronary reserve measurements. These important issues will be briefly discussed.

Clinically Applicable Coronary Dilator Stimuli

In the 1960s and 1970s, most investigations of the coronary circulation in humans employed very weak coronary dilator stimuli such as pacing, mild exercise, hand-grip, the cold pressor test, and infusion of inotropic agents such as dobutamine or isoproterenol. In general, these stimuli increase coronary flow by less than threefold. Recently, much more potent coronary dilator stimuli have been employed. There are two classes of potent coronary dilator stimuli applicable to studies in humans—mechanical and pharmacological (Table 2). The mechanical stimulus that has been employed is transient coronary occlusion either by direct compression of a surface coronary vessel at cardiac surgery [7] or intraluminal obstruction with a percutaneous balloon catheter [27]. In the absence of significant coronary collaterals, transient coronary occlusion produces profound myocardial ischemia in the perfusion field of the occluded vessel. This profound ischemia engenders maximal coronary dilation. With abrupt release of the occlusion, perfusion pressure is restored to a maximally dilated bed and flow increases strikingly. This entire sequence is referred to as the reactive hyperemic response. Studies in our laboratory in humans have shown that this stimulus increases coronary blood flow by four to sixfold [7] in normal coronary vessels (Figs. 3, 4). Although transient coronary occlusion is a very useful potent coronary dilator stimulus during cardiac surgery or coronary angioplasty, it is obviously not suited for use in other settings. Even during angioplasty, the balloon catheter can produce partial obstruction and thereby artifactually decrease coronary flow reserve.

Pharmacologic approaches to producing maximal coronary dilation are applicable in a wider range of settings. Unfortunately, papaverine and iodinated contrast agents produce coronary dilation without major systemic hemodynamic effects only when they are given by intracoronary injection. Dipyridamole currently is the sole widely studied coronary dilator that is safe and effective when administered intravenously. The experience with intravenous and intracoronary adenosine as coronary vasodilators in humans is limited.

Although dipyridamole and radiographic contrast material produce coronary dilation, they are less efficacious than intracoronary papaverine. Neither agent reliably produces maximal coronary vasodilation, and dipyridamole has a prolonged onset and duration of response [19, 28]. Papaverine is a very useful pharmacologic dilator when intracoronary administration can be utilized. It produces

Table 2. Potent coronary dilator stimuli applicable to studies in humans

Mechanical
Intraoperative transient coronary occlusion
Coronary occlusion with a percutaneous balloon catheter
Pharmacological
Intracoronary papaverine
Intracoronary adenosine
Intracoronary iodinated contrast agents
Intravenous dipyridamole

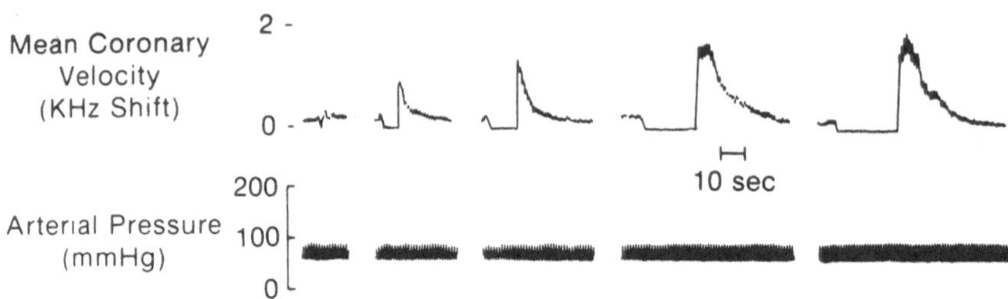

Fig. 3. Coronary reactive hyperemic responses obtained in a patient at the time of open-heart surgery. The suction Doppler was placed on the right ventricular branch of the right coronary artery. The patient had normal coronary arteries and normal ventricles. Progressively longer durations of coronary occlusion resulted in progressively greater reactive hyperemic responses. Maximum reactive hyperemic responses occurred when coronary occlusion was 20 s. Note that following the 20-s coronary occlusion there is a striking increase in coronary blood flow velocity indicating impressive coronary reserve in this patient. The coronary reactive hyperemic responses were not associated with cardiac arrhythmias or changes in arterial pressure. From [7] with permission of the American Heart Association

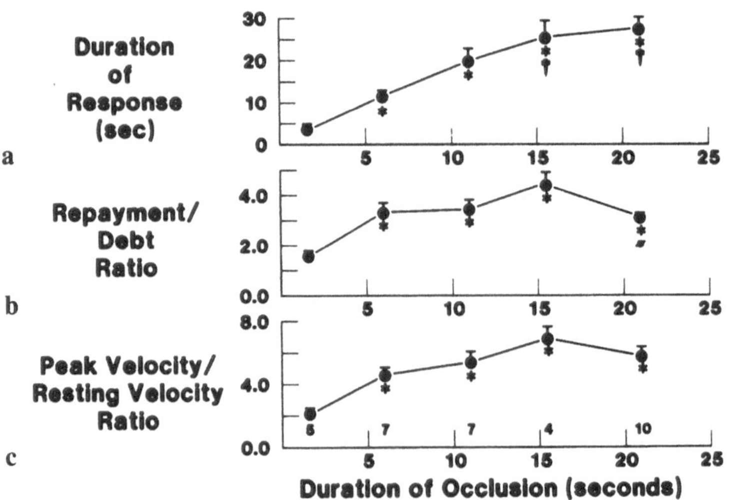

Fig. 4a–c. Quantitative characteristics of coronary reactive hyperemia in man. These measurements were obtained in patients at the time of open heart surgery who had angiographically normal coronary arteries and normal ventricles. The patients were having open heart surgery for conditions including atrial myoxma, mitral stenosis, and single vessel coronary disease. In patients with single vessel coronary disease, the coronary studied was angiographically normal. In these normal patients, maximum reactive hyperemic response occurred with an occlusion of 20-s duration. **c** Integers reflect the number of patients included in each of the subgroups. From [7] with permission from the American Heart Association

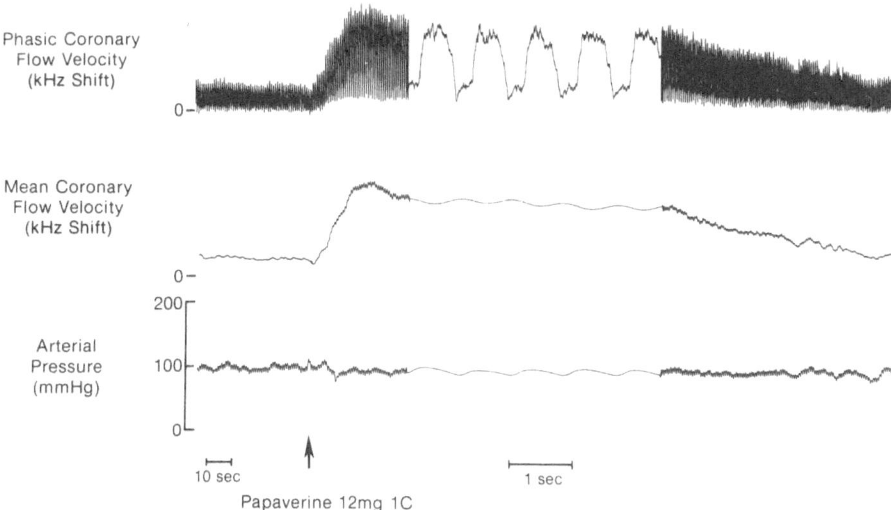

Fig. 5. Response of a normal coronary vascular system perfusing a normal ventricle to the intracoronary injection of papaverine. Papaverine produced a prompt rise in coronary blood flow velocity which persisted approximately 2 min. The injection of papaverine was associated with a minimal change in mean arterial pressure

maximal coronary dilation and the onset and duration of the response is sufficiently brief so that multiple measurements of coronary reserve can be made in a single patient (Fig. 5). Also, the effects of intracoronary papaverine on systemic hemodynamics are minor. The only significant adverse reaction is the development of transient non-sustained ventricular tachycardia or, rarely, ventricular fibrillation in patients who receive intracoronary papaverine [29].

Dipyidamole has received broad investigational use and has engendered wide interest in the cardiology community, though it is not commercially available in the Unites States at present. A substantial number of the patients who receive the standard dose of the drug do not achieve maximal coronary dilation [18, 19, 28]. Dipyridamole has both minor adverse side effects (nausea, vomiting, dizziness, flushing, mild chest pain) and significant adverse effects including severe hypotension and intense chest discomfort related to myocardial ischemia. Most importantly, when dipyridamole is administered to patients with unstable angina there is a small incidence of myocardial infarction and death.

Factors That Influence the Interpretation of Coronary Flow Reserve Measurements in Humans

The two currently dominant methods of assessing coronary flow reserve—digital subtraction angiography and the Doppler catheter—measure the relative change in flow from resting conditions to maximal coronary dilation. Neither approach measures minimal coronary vascular resistance. As a consequence, any factor that alters resting coronary flow (i.e., tachycardia) or maximal flow (i.e., severe

Table 3. Cardiac and non-cardiac factors that may influence measurements of coronary flow reserve

Hemodynamic alterations
Heart rate
Aortic pressure
Contractility
Preload
Non-cardiac disease states
Anemia
Polycythemia
Thyroid dysfunction
Febrile illness
Pharmacologic agents
Pharmacological agents that alter systemic hemodynamics
Vasodilators
Pressor agents
Positive or negative inotropic agents
Negative chronotropic agents
Pharmacological agents that directly effect the coronary circulation
Nitrates
Alpha-adrenergic agonists or antagonists
Cardiac diseases that affect flow reserve
Left ventricular hypertrophy
Syndrome X
Coronary collaterals
Myocardial infarction
Severe congestive heart failure
Coronary obstructive lesions

arterial hypotension) will decrease relative coronary flow reserve even though minimal coronary vascular resistance would be normal [30]. Unfortunately, there are a host of factors unrelated to significant cardiac disease that can influence relative measurements of coronary flow reserve. These conditions must be considered whenever relative coronary flow reserve measurements are interpreted (Table 3).

In addition to the non-cardiac factors that can influence coronary flow reserve, there are many cardiac diseases associated with decreases in coronary flow reserve and concomitant increases in minimal coronary vascular resistance (Table 3). These will be discussed in greater detail below.

Applications of Coronary Flow Reserve Measurements to Patient Care

There are at least three patient groups in which measurements of coronary flow reserve can influence diagnosis or clinical management: (1) patients with anginal chest pain and angiographically normal coronary vessels, (2) patients with coronary obstructions where the physiological significance of the obstruction is in doubt, and (3) patients who have had a major therapeutic intervention such as coronary bypass surgery or angioplasty and the effectiveness of the therapy is open to question.

Chest Pain and Normal Coronary Vessels

Approximately 10%–25% of patients undergoing coronary angiography for the evaluation of chest pain syndromes in the United States are found to have angiographically normal coronary vessels. A few such patients will have coronary spasm which may be demonstrable by ergonovine testing. An unknown fraction of patients with the syndrome of anginal pain and normal coronary arteries have decreased coronary flow reserve and myocardial ischemia as the cause of their symptoms. Several well described disease states can present in this manner, including Syndrome X, unrecognized cardiac hypertrophy of either ventricle, and diffuse coronary narrowing without focal obstruction. In these patients, direct assessment of coronary flow reserve would demonstrate impaired coronary vasodilator capacity and provide insight into the patient's clinical problems.

In our view, coronary flow reserve measurements should be performed in all patients with chest pain and normal coronary vessels, particularly those with evidence of ischemia on non-invasive testing. It is hoped that identification of patients with normal coronary arteries angiographically and depressed flow reserve will permit investigators to eventually develop effective therapeutic strategies in this patient group.

Coronary Obstructions of Intermediate Physiologic Significance

In patients with multivessel coronary disease, the physiologic significance of obstructions of intermediate severity is frequently in doubt because angiographically unidentifiable diffuse atherosclerosis is ubiquitous in such patients [31–33] (Fig. 6). Hence, even if the angiogram is analyzed with quantitative coronary angiographic techniques, the significance of these obstructions is often uncertain. This is less of a problem in patients with single vessel disease where diffuse atherosclerosis is less prevalent [34]. Whenever the physiologic significance of coronary obstructive lesion is in doubt, measurements of coronary flow reserve in the involved vessel can be extremely valuable. Interpretation of such flow reserve measurements must take into account the various factors that can influence coronary flow reserve independent of the coronary obstructive lesion being evaluated (Table 3).

In our view, direct measurements of coronary flow reserve in patients with obstructions of intermediate severity would improve the quality of patient care by allowing a better informed selection of patients for treatment with percutaneous transluminal coronary angioplasty or bypass surgery.

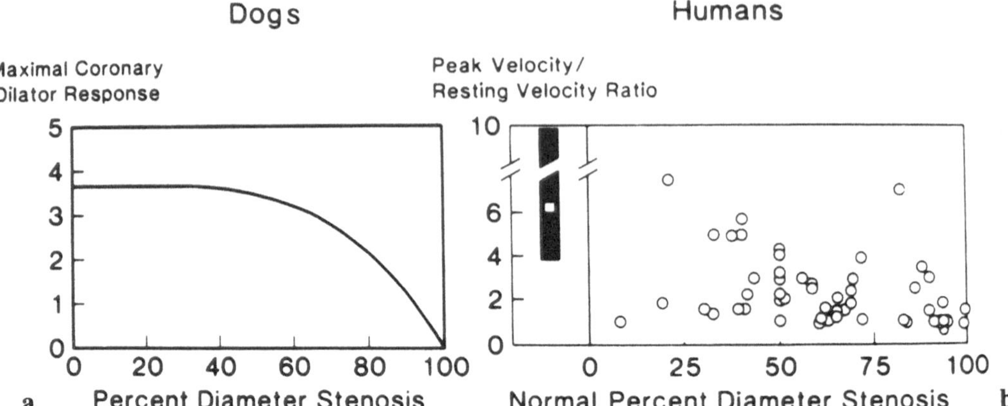

Fig. 6a, b. Relationship between percent diameter stenosis and maximal coronary dilator responses in **a** dogs studied by Gould et al. [3] and **b** humans studied by Marcus et al. [12], at the University of Iowa. **a** In dogs, with normal coronary arteries and normal ventricles who were subjected to progressive coronary obstruction with a short, relatively circular device, there was an orderly decrease in maximal coronary dilator response as percent diameter stenosis was increased. **b** In patients who were studied at the time of open heart surgery, percent diameter stenosis was assessed from visual interpretation of a coronary arteriogram and maximal coronary dilator responses from the peak velocity to resting velocity ratio after a 20-s coronary occlusion. In normal patients (*black bar*), the average peak velocity to resting velocity ratio was 6. In patients with coronary obstructive lesions, there was very poor correlation between percent diameter stenosis and the maximal coronary dilator response. These data indicate that percent diameter stenosis is a very poor index of coronary flow reserve. The patients included in these studies had multivessel coronary disease. From [11] with permission of McGraw-Hill

Efficacy of Treatment Interventions

In patients subjected to percutaneous transluminal coronary angioplasty or bypass surgery who develop recurrent symptoms suggestive of myocardial ischemia, it is critically important to evalute the efficacy of the therapy. In this clinical setting, determination of coronary anatomy alone does not always provide an explanation for the patient's symptoms. The diagnosis of clinically important restenosis or the presence of a significant obstruction at the bypass graft coronary anastomosis is often uncertain. Under such circumstances, direct measurement of coronary flow reserve in a bypass graft [35] or a native vessel following percutaneous transluminal coronary angioplasty [36] may yield critically important information that can influence the clinician's decision to proceed with further therapy. Making such therapeutic decisions on the basis of angiographic findings that are difficult to interpret should be discouraged in favor of obtaining a direct assessment of the physiologic significance of the involved vascular system.

Applications of Coronary Flow Reserve Measurements to Research

In the past few years, measurements of coronary flow reserve have been useful in a variety of research applications. For example, the coronary dilator effect of dipyridamole have been carefully examined and two new observations have come forth [18, 19, 28]. First, the time to maximal coronary dilator response after the standard dose of dipyridamole (0.56 mg/kg, i.v. over 4 min) is variable and may be as long as 9 min from the start of administration in some patients. Second, perhaps one-fourth of the patients given the standard dose of dipyridamole do not achieve maximal coronary dilation.

In the past, noninvasive tests for the diagnosis of coronary disease (treadmill exercise tolerance test, thallium scintigraphy) have been validated by comparing the test results to visual interpretations of the coronary arteriograms (percent diameter stenosis). Because percent diameter stenosis is such a poor gold standard [31–33], the true validation of these noninvasive tests remains in question. Recent studies from the University of lowa have compared the results of treadmill exercise tolerance tests and thallium scintigraphy to direct measurements of flow reserve and quantitative coronary angiographic analysis of lesion anatomy [37, 38]. These studies have demonstrated that these two widely applied noninvasive tests for coronary disease are far less sensitive than previously appreciated, and underline the importance of developing improved, noninvasive tests for the diagnosis of coronary disease and validating them by comparing the test results to a physiologic index of coronary stenosis severity.

Measurements of coronary flow reserve have also been helpful in evaluating effects of hypertrophy on the coronary circulation [11, 39]. Studies performed in patients with aortic stenosis (Fig. 7), aortic insufficiency, mitral insufficiency,

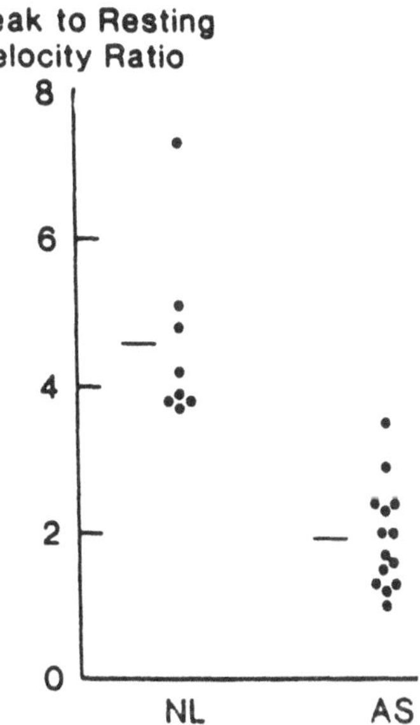

Fig. 7. Coronary flow reserve (peak-to-resting velocity ratio following a 20-s coronary occlusion) measurements made in normal patients (*NL*) and patients with aortic stenosis (*AS*) at the time of open heart surgery. The measurements were made in the left anterior descending coronary artery. In NL the left anterior descending perfuses a non-hypertrophied ventricle, whereas in As it perfuses a severely hypertrophied ventricle. All patients included in the study had angiographically normal coronary arteries. Coronary flow reserve measured by the peak-to-resting velocity ratio was strikingly decreased in AS. These data suggest that pressure-induced cardiac hypertrophy produces a profound decrement in coronary flow reserve. From [39] with permission of the New Engl J Med

and various congenital heart lesions have shown that all of these conditions are associated with a striking deterioration in coronary flow reserve in the hypertrophied ventricle. These measurements give strong support to the concept that cardiac hypertrophy of probably any cause (with the possible exception of thyrotoxicosis [40]), is associated with a profound decrement in coronary flow reserve. These data provide an explanation for why many patients with hypertrophy exhibit clinical evidence of myocardial ischemia which may include exercise-induced anginal pain, ischemic electrocardiographic responses during treadmill exercise, and decrements in ventricular function during exercise.

Finally, measurements of Doppler coronary flow velocity and flow reserve have been central in the identification of species-specific differences in the coronary circulation. The pulmonary inflation reflex produces large increases in coronary blood flow in dogs but is absent in humans [41]. Diltiazem and other calcium channel antagonists profoundly diminish maximal coronary flow following brief coronary occlusion or pharmacologic vasodilation in dogs, but do not significantly reduce pharmacologically stimulated coronary flow reserve in humans [42].

Conclusion

Measurements of coronary flow reserve in humans can now be relatively easily obtained during cardiac catheterization utilizing commercially available systems. Such information is useful in clinical decision making and indispensable in several research areas. The current methods of measuring coronary flow reserve are invasive, measure relative as opposed to absolute flow reserve, and do not allow an assessment of coronary flow reserve in individual transmural layers of the left ventricular myocardium. New approaches to measuring coronary flow reserve now in the developmental phase are likely to overcome these limitations. In the future, measurements of coronary flow reserve will play an ever-increasing role in the clinical evaluation and research study of patients with cardiac disease.

Acknowledgment. The original studies which form the basis for this review were supported by an Ischemic SCOR grant from NHLBI—HL 32295.

References

1. Katz LN, Linder E (1939) Quantitative relation between reactive hyperemia and the myocardial ischemia which it follows. Am J Physiol 126: 283–288
2. Shipley RE, Gregg DE (1944) The effect of external constriction of a blood vessel on blood flow. Am J Physiol 141: 289–296
3. Gould KL, Lipscomb K, Hamilton GW (1974) Physiologic basis for assessing critical coronary stenosis. Am J Cardiol 33: 87–94
4. Gould KL, Lipscomb K (1974) Effects of coronary stenoses on coronary flow reserve and resistance. Am J Cardiol 34: 48–55
5. Gould KL (1980) Dynamic coronary stenosis. Am J Cardiol 45: 286–292
6. Gould KL (1985) Quantification of coronary artery stenosis in vivo. Circ Res 57: 341–353

7. Marcus ML, Wright CB, Doty DB, Eastham C, Laughlin D, Krumm P, Fastenow C, Brody M (1981) Measurements of coronary velocity and reactive hyperemia in the coronary circulation of humans. Circ Res 49: 877–891
8. White CW, Holida MD, Marcus ML (1986) Effects of acute atrial fibrillation on the vasodilator reserve of the canine atrium, Cardiovasc Res 20: 683–689
9. Wangler RD, Peters KG, Laughlin DE, Tomanek RJ, Marcus ML (1981) A method for continuously assessing coronary velocity in the rat. Am J Physiol 241: H816–H820
10. Peters KG, Wangler RD, Tomanek RJ, Marcus ML (1984) Effects of long-term cardiac hypertrophy on coronary vasodilator reserve in SHR rats. Am J Cardiol 54: 1342–1348
11. Marcus ML (1983) The coronary circulation in health and disease. McGraw-Hill, New York (Chapter 13; Effects of cardiac hypertrophy on the coronary circulation)
12. Rowe GG (1968) Pharmacology of the coronary circulation. Annu Rev Pharmacol Toxicol 285–306
13. Klocke FJ (1976) Coronary blood flow in man. Prog Cardiovasc Dis 19: 117–166
14. Cannon PJ, Dell RB, Dwyer EM Jr (1972) Measurement of regional myocardial perfusion in man with ^{133}xenon and a scintillation camera. J Clin Invest 51: 964–977
15. Bagger JP (1984) Coronary sinus blood flow determination by the thermodilution technique: Influence of catheter position and respiration. Cardiovasc Res 19: 27–31
16. Mathey DG, Chatterjee K, Tyberg JV, Lekven J, Brundage B, Parmley WW (1978) Coronary sinus reflux. A source of error in the measurement of thermodilution coronary sinus flow. Circulation 57: 778–786
17. White CW, Wilson RF, Marcus ML (1988) Methods of measuring myocardial blood flow in humans. Prog Cardiovasc Dis 31: 79–94
18. Wilson RF, Laughlin DE, Ackell PH, Chilian WM, Holida MD, Hartley CJ, Armstrong ML, Marcus ML, White CW (1985) Transluminal, subselective measurement of coronary artery blood flow velocity and vasodilator reserve in man. Circulation 72: 82–92
19. Wilson RJ, White CW (1986) Intracoronary papaverine: an ideal coronary vasodilator for studies of the coronary circulation in conscious humans. Circulation 73: 444–451
20. Vogel R, LeFree M, Bates E, O'Neill W, Foster R, Kirlin P, Smith D, Pitt B (1984) Application of digital techniques to selective coronary arteriography: use of myocardial contrast appearance time to measure coronary flow reserve. Am Heart J 107: 153–164
21. Cusma JT, Toggart EJ, Folts JD, Peppler WW, Hangiandreou NJ, Lee C-S, Mistretta CA (1987) Digital subtraction angiographic imaging of coronary flow reserve. Circulation 75: 461–472
22. Cohen MD, Field WR, Hodgson JM (1987) Correlation of digital angiographic with Doppler estimation of coronary flow reserve. J Am Coll Cardiol 9(Suppl A): A161
23. Bergmann SR, Fox KAA, Geltman EM, Sobel BE (1985) Positron emission tomography of the heart. Prog Cardiovasc Dis 28: 165–194
24. Weiss RM, Hajduczok ZD, Marcus ML (1988) A new Cine CT algorithm for quantitation of myocardial perfusion. Circulation 78(Suppl II): II-398
25. Rumberger JA, Feiring AJ, Higgins CR, Lipton MJ, Ell SR, Marcus ML (1987) Use of ultrafast Computed Tomography to quantitate regional myocardial perfusion: a preliminary report. J Am Coll Cardiol 9: 59–69
26. Armstrong WF (1986) Assessment of myocardial perfusion with contrast enhanced echocardiography. Echocardiography 3: 355–370
27. Zijlstra F, Reiber JC, Juilliere Y, Serruys PW (1988) Normalization of coronary flow reserve by percutaneous transluminal coronary angioplasty. Am J Cardiol 61(1): 55–60
28. Wilson RF, White CW (1988) Serious ventricular dysrhythmias after intracoronary papaverine. Am J Cardiol 62: 1301–1302
29. Rossen JD, Simonetti 1, Marcus ML, Winniford MD (1989) Coronary dilation with standard dose dipyridamole and dipyidamole combined with handgrip. Circulation 79: 566–572

30. Klocke FJ (1987) Measurements of coronary flow reserve: defining pathophysiology versus making decisions about patient care. Circulation 76: 1183–1189
31. White CW, Wriht CB, Doty DB, Hiratzka LF, Eastham CL, Harrison DG, Marcus ML (1984) Does the visual interpretation of the coronary arteriograms predict the physiological significance of a coronary stenosis? N Engl J Med 310: 819–824
32. Harrison DG, White CW, Hiratzka LF, Doty DB, Barnes DH, Eastham CL, Marcus ML (1984) The value of lesion cross-sectional area determined by quantitative coronary angiography in assessing the physiological significance of proximal left anterior descending coronary arterial stenoses. Circulation 69: 1111–1119
33. McPherson DD, Hiratzka LF, Lamberth WC, Brandt B, Hunt M, Kieso RA, Marcus ML, Kerber RE (1987) Delineation of the extent of coronary atherosclerosis by high-frequency epicardial echocardiography. N Engl J Med 316: 304–309
34. Wilson RF, Marcus ML, Drews TA, White CW (1987) Prediction of the physiologic significance of coronary arterial lesions by quantitative lesion geometry in patients with limited coronary artery disease. Circulation 75: 723–732
35. Wilson RF, White CW (1987) Does coronary artery bypass surgery restore normal maximal coronary flow reserve? Circulation 76: 563–571
36. Wilson RF, Johnson MR, Marcus ML, Aylward PEG, Skorton DJ, Collins S, White CW (1988) The effect of coronary angioplasty on coronary flow reserve. Circulation 77: 873–885
37. Talman CL, Aylward PE, White CW, Wilson RF, Kirchner PT, Rezai K, Marcus ML (1986) Inability of planar thallium scintigraphy to predict 50% decrease in coronary vasodilator reserve. J Am Coll Cardiol 7: A5
38. Simonetti I, Rossen JD, Winniford MD, Talman CL, Hollenberg M, Marcus ML (1988) Computerized treadmill does not improve the accuracy of the exercise electrocardiogram in patients on antianginal therapy (abstract). Circulation 78 (suppl II): II–109
39. Marcus ML, Doty DB, Hiratzka LF, Wright CB, Eastham CL (1982) Decreased coronary reserve—a mechanism for angina pectoris in patients with aortic stenosis and normal coronary arteries. N Engl J Med 307: 1362–1367
40. Chilian WM, Wangler RD, Peters KG, Marcus ML, Tomanek RJ (1985) Thyroxine-induced left ventricular hypertrophy in the rat. Anatomical and physiological evidence for angiogenesis. Circ Res 57: 591–598
41. Wilson RF, Marcus ML, White CW (1988) Pulmonary inflation reflex: its lack of physiological significance in coronary circulation of humans. Am J Physiol (Heart Circ Physiol) 24: H866–871
42. Rossen JD, Simonetti I, Marcus ML, Braun P, Winniford MD (1989) The effect of diltiazem on coronary flow reserve in humans. Circulation 80: 1240–1246

Flow Dynamics in Coronary Stenosis

Fumihiko Kajiya[1], Katsuhiko Tsujioka[1], Yasuo Ogasawara[1], Shuji Matsuoka[1], and Takashi Fujiwara[2]

Summary. The pressure loss across coronary artery stenosis is mainly caused by two factors: viscous friction loss by stenotic segments and separation loss by the flow separation after (and before) stenosis. When the percentage diameter stenosis increases, the coronary flow reserve decreases. In addition to percentage stenosis, the length of the stenosis and the number of stenotic segments affect the hemodynamic significance of stenosis. The eccentric type stenosis is not only mechanically compliant, but it also changes the area of stenosis with the alteration of vasomotor tone. Accordingly, pressure-dependent and vasomotor tone-dependent diameter change should be taken into consideration in the compliant stenosis. Following our intraoperative measurements of blood flow velocities, the flow separation and flow disturbance were observed in the poststenotic region in human coronary arteries. The systolic-to-diastolic flow ratio was high in stenotic coronary arteries, while the ratio was normalized after bypass grafting in most patients.

Key words: Coronary artery stenosis—Friction and separation pressure-loss—Percentage stenosis—Compliant stenosis—Systolic-to-diastolic flow ratio

Introduction

Detailed assessment of flow dynamics in coronary artery stenosis is essential for understanding of the pathophysiology of obstructive coronary disease. In the presence of moderate to severe coronary stenosis, a significant pressure-loss develops across the stenosis, and that causes reduction in coronary flow reserve and effective coronary flow to the myocardium. In the clinical situation, percentage luminal narrowing read from coronary arteriogram is often used as the gold standard for describing stenosis severity. However, several problems have arisen with this conventional evaluation method because various factors may contribute to generation of the pressure-loss across the coronary stenosis. Model experiments provide potent information on the mechanism of the pressure loss across the stenosis, although they cannot completely simulate a diseased human arterial system. Information on human coronary flow dynamics is limited, but some promising techniques are now contributing to the evaluation of the physiologic significance of coronary stenosis in the human.

Department of Medical Engineering and Systems Cardiology[1] and Department of Thoracic Surgery[2], Kawasaki Medical School, 577 Matsushima, Kurashiki, 701-01 Japan

Quantitative Description of Stenosis Severity and Percent Narrowing

Young et al. have investigated flow dynamics in arterial stenosis, through an extensive series of model experiments [1, 2] and animal experiments using the femoral artery [3], and they have found that the pressure drop, ΔP, across a stenosis can be expressed by:

$$\Delta P = \frac{K_v \mu}{D} V + \frac{K_t}{2}\left(\frac{A_0}{A_1} - 1\right)^2 \rho |V| V + K_u \rho L \frac{dV}{dt} \tag{1}$$

where A_0 is the area of the tube remaining unobstructed, A_1 is the minimum cross-sectional area of the stenosis, D is the diameter of the unobstructed tube, K_v, K_t, K_u are experimentally determined coefficients, L is length over which the pressure drop is measured, t is time, V is instantaneous velocity in the unobstructed tube (average over the cross section), ρ is fluid density, and μ is fluid viscosity. Young and co-workers indicated that the first source of the pressure drop across the stenosis is the viscous loss given by the first term on the right of the equation. The second source is the nonlinear loss due to convergence and divergence of the fluid (second term), and the third source, inertia of the fluid (third term). Since the viscosity and density of blood do not usually vary significantly in the circulation, these two factors can be considered to be relatively constant. Thus the pressure drop across the stenosis is determined mainly by the geometry of the stenosis along with the instantaneous velocity and its phasic change.

Gould [4] investigated the relation between diastolic pressure gradient across the stenosis (ΔP) and blood velocity (V) measured by an ultrasonic Doppler flowmeter in the canine coronary artery, and he adopted a simplified equation:

$$\Delta P = FV + SV^2$$

$$\left\{F = \frac{8\pi\mu L}{A_1} \cdot \frac{A_0}{A_1},\ S = \frac{\rho K}{2}\left(\frac{A_0}{A_1} - 1\right)^2\right\} \tag{2}$$

where F is the coefficient of viscous friction loss, and S is the coefficient of separation loss. The inertial term related to phasic variation of the flow was not included in this equation because diastolic flow is less phasic, and the phasic nature of the flow through a moderate to severe stenosis will be dampened. Gould demonstrated that the pressure gradient across a coronary stenosis was expressed by this equation in the conscious animal under resting conditions during diastole. As indicated by these equations, the pressure loss across the coronary stenosis is mainly related to the geometry of the stenosis, including percentage narrowing, absolute area (diameter), length, and shape. The absolute area (diameter) and the length mainly contribute to viscous friction coefficient F, and the percentage narrowing and the shape mainly relate to separation coefficient S (Fig. 1). At high flow, separation loss will account for a greater proportion of the total pressure drop than viscous loss, since the separation loss is generated as the square of the velocity.

The relation between percentage diameter stenosis and flow has been well analyzed in canine coronary circulation. In resting dogs, about 85% narrowing in diameter is required to reduce coronary flow [5, 6], when vasomotor tone is

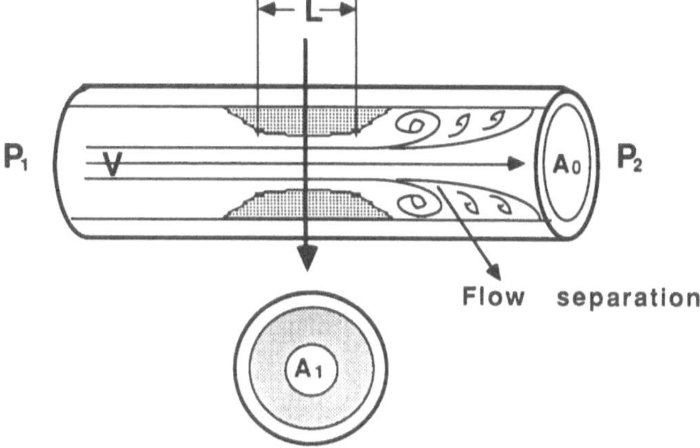

Fig. 1. Sources of pressure loss across stenosis. The absolute area (A_1) and the length (L) mainly contribute to viscous friction loss, and the percentage narrowing A_1/A_0 mainly relates to separation loss. P_1, proximal pressure; P_2, distal pressure; V, blood velocity; A_1, minimum cross-sectional area of stenosis

present. However, when the coronary vascular bed is dilated maximally, flow begins to decrease with a stenosis of 30%–45% of the diameter. Since coronary flow reserve is the flow difference between the presence and absence of the vasomotor tone, the flow reserve reduces with an increase in the percentage diameter stenosis. When the percentage diameter stenosis becomes greater than 85%–90%, a so-called critical stenosis, the flow reserve will be greatly decreased. Although the percentage diameter stenosis is one of the major determinants of flow reserve, other geometrical factors such as the absolute diameter of the stenosis, stenosis length and distribution, and the shape of the stenosis should be taken into account, as will be discussed.

Geometrical Factors and Pressure Loss

Although physicians give little consideration to the length of the coronary stenosis, the length was found to alter the hemodynamic significance of coronary stenosis, in both model and animal experiments [7–10]; i.e., lengthening of the stenosis increased the pressure loss across a stenosis, and the fall in coronary flow reserve. This length-effect on the pressure loss was included in the coefficient F in Eq. 2. Multiple stenotic lesions in a single vessel, which may confuse the prediction of the severity of a coronary stenosis, are frequently found in patients with coronary artery disease (Fig. 2b). Model experiments done by Seeley and Young [7] indicated that if the distance separating two stenoses was sufficiently long, the pressure drop could be estimated by a simple summation of the pressure drops due to the two individual stenoses. However, as the distance between the stenoses decreased, there occurred an interaction between the two stenoses, and the overall pressure drop was less than the sum of the two. Gould and Lipscomb [6], and Feldman et al. [9] also indicated in dog experiments that the resistances of coronary stenoses in series were additive, although they did not test the influence of the distance between the two stenoses. As indicated by both model and animal experiments, when the coronary stenoses exist in series, the hemodynamic significance is not generally determined solely by the dominant or most severe lesion. In the clinical setting, however, the importance of these serial lesions is frequently belittled.

The shape of the stenosis may influence its hemodynamic severity. However, some model experiments [7, 11] indicated that detailed geometry, such as eccentricity and the shape of the inlet and the outlet of the stenosis, did not significantly affect the pressure drop across the stenosis when the percentage narrowing was severe. Therefore, in the patients with hemodynamically significant stenosis, the detailed geometry of the stenosis may have little effect on its severity.

Compliant Stenosis

For many years, the pathophysiological evaluation of coronary stenosis has been done under the assumption that the geometry of coronary stenosis is fixed. Recently, however, several investigators suggested that this assumption is not always valid and that the dynamic nature of the coronary stenosis should be considered. For example, some pathological studies of postmortem coronary artery segments [12, 13] demonstrated that human coronary stenoses were frequently eccentric, i.e., the atheroma frequently failed to involve all of the circumference of the vessel (Fig. 2c). This type of stenosis can act as "compliant stenosis" in the sense that the geometry of stenosis is changeable. Logan [14] examined an effect of perfusion pressure on stenotic resistance by using fresh postmortem coronary artery segments with concentric or eccentric stenoses. The resistance to flow in a human artery with eccentric stenosis was inversely related to the perfusion pressure, although the resistance of concentric stenosis was independent of the perfusion pressure. Schwartz et al. [15] found a similar phenomenon in the canine coronary artery *in situ*. The increase in stenotic resistance induced by distal vasodilation was explained by the pressure-dependent resistance change of the stenotic portion [16–18], although the mechanism of the increase in resistance is still controversial. The eccentric type stenosis is not only mechanically compliant, but it may also vary in area due to the change in vasomotor tone of the stenotic segment itself. Alteration of stenosis severity due to this mechanism was demonstrated during infusion of nitrate, and exercise [19, 20]. Another important cause of the pressure loss is the dilatation of the vascular segment immediately distal to the stenosis, i.e., the separation loss expressed by the coefficient S in Eq. 2. Gould [4] indicated that a dilatation of the artery just distal to the stenosis induced more severe relative percentage stenosis and a greater pressure-loss due to the flow separation. Thus, for the analysis of so-called "compliant" stenosis, it is necessary to take account of three factors: (1) pressure-dependent diameter change of the stenotic portion, (2) vasomotor-tone-dependent diameter change of the stenotic segment, and (3) change in the relative percentage stenosis due to the dilatation or narrowing of the segment adjacent to the stenosis. These factors affect the drop in blood pressure through changes in friction and /or flow separation.

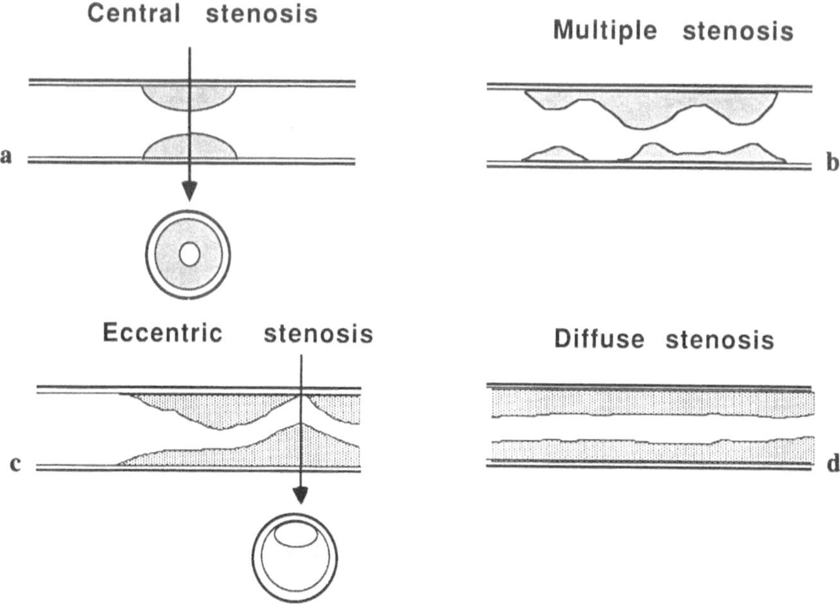

Fig. 2a–d. Types of coronary stenoses

Intraoperative Assessment of Stenotic Coronary Flow

For many years, assessment of coronary flow dynamics in human beings has been limited. Recently, a new and promising technique has been developed that makes it possible to measure instantaneous coronary velocity and coronary reactive hyperemia in humans. Hartley and Cole [21] developed high-frequency (20-MHz) ultrasound pulsed Doppler velocimetry, and Marcus and his colleagues measured phasic blood velocity and reactive hyperemia during cardiac surgery by using a modification of Hartley's circuit and by introducing a unique suction-cup probe [22, 23]. They found that the percentage area or percentage diameter stenosis given by conventional or quantitative coronary angiography did not accurately predict the coronary flow reserve (coronary reactive hyperemic response) [24]. Furthermore, they demonstrated that the degree of the reactive hyperemic response correlated significantly with the minimal cross-sectional area of the stenosis [25]. It is considered that some part of the reactive hyperemic response during operation may be confused by several factors including operation procedure, anesthesia, drugs, coronary perfusion pressure, and the open pericardium. Therefore, variation of these factors among individuals may contribute in part to scattering of data points in their study. They explained that the percentage narrowing is invalid for describing the stenosis severity, in that it underestimates the hemodynamic severity when the arteriosclerosis diffusely involves a coronary artery. Since the coronary angiogram images only internal (luminal) outlines of the vessels, diffuse coronary narrowing (Fig. 2d) cannot be expressed by the percentage narrowing. In addition, the compliant

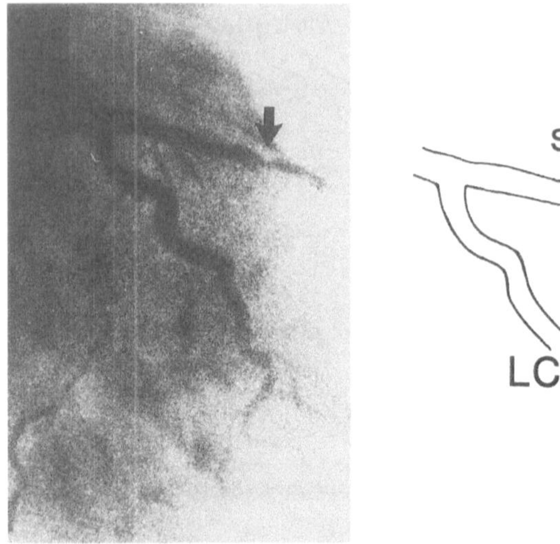

Fig. 3. An example of the coronary angiogram in a patient (53-year-old female) with an eccentric stenosis. *LAD*, left anterior descending coronary artery; *LCX*, left circumflex coronary artery

stenosis will cause a poor correlation between percentage stenosis and hemodynamic severity. Furthermore, one angiographic projection will not accurately predict the severity of the abstruction. Accordingly, computer-based coronary angiography has been introduced for quantitative assessment of coronary artery stenosis, and there are many reports available for demonstrating the validity of the method for the evaluation of the severity of stenosis. The optimal index describing physiological significance of the coronary stenosis, however, has yet to determined.

More recently, we developed a new type of 20-MHz ultrasound Doppler velocimeter as described in this volume by Fujiwara et al. (pp. 325–334). This system makes it possible to measure an instantaneous velocity profile across the vessel and velocity spectrum within a sample volume in real time, which gives valuable information for more detailed understanding of stenotic flow dynamics [26, 27]. We performed intraoperative measurements of velocity in the human coronary artery by using this system with a specially designed holder probe. Figure 3 shows a coronary angiogram in a patient with an eccentric coronary artery stenosis (90% luminal narrowing). In this case we observed a significant reverse flow, indicating flow separation near the outer free wall over a poststenotic portion close to the stenosis, in mid-diastole (Fig. 4) [28]. This observation is an important finding which demonstrates the existence of flow separation in a stenotic human coronary artery. Velocity spectra over the poststenotic portion were also analyzed, and we observed that broad spectra, indicating flow disturbance, extended to a relatively distal portion. These findings indicate substantial pressure-loss in the poststenotic portion due to flow separation and flow disturbance.

We also evaluated the change in the phasic nature of native coronary artery

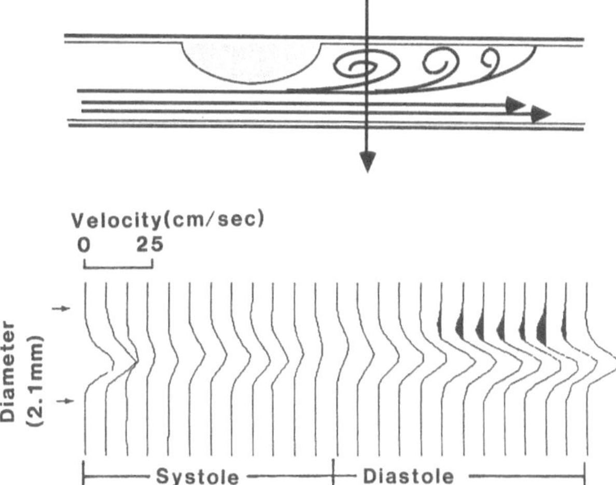

Fig. 4. The poststenotic velocity profile for the same patient as in Fig. 3 (*bottom*), and an illustrative drawing of the flow field in the poststenotic region. (*top*). Reverse flow (*dark shading*) was clearly noted in mid-diastole, indicating the existence of flow separation

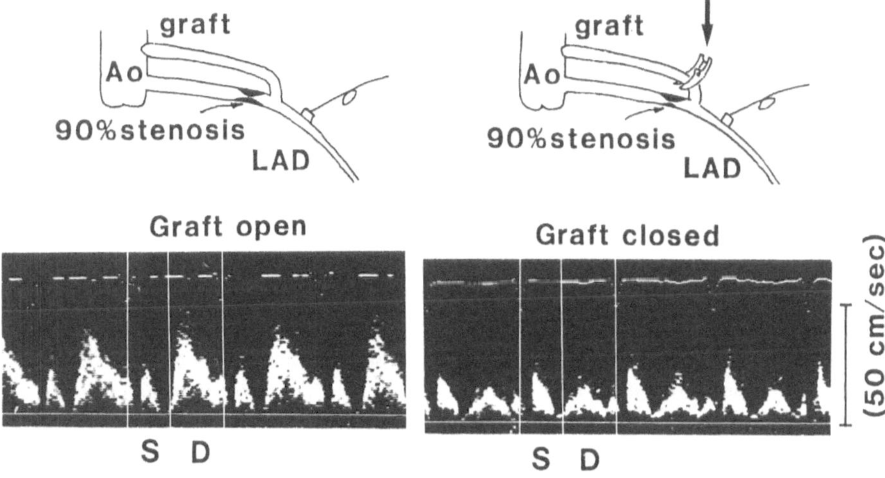

Fig. 5. Comparison of blood velocities in a distal LAD beyond a bypass insertion before and during transient occlusion of the graft. It should be noted that the transient graft occlusion caused a marked reduction in the diastolic velocity component. *Ao*, aorta; *LAD*, left anterior descending coronary artery; *S*, systole; *D*, diastole. From [27] with permission of the American Heart Association

flow before and after bypass grafting [26–28]. Figure 5 shows phasic velocities in the left anterior descending artery (LAD) distal to the graft anastomosis before and after a transient graft-occlusion in a patient with 90% narrowing of diameter. The phasic blood velocity in the LAD before the graft occlusion showed a diastolic-predominant pattern, which indicates good myocardial perfusion through the graft. After the graft occlusion, the diastolic flow component be-

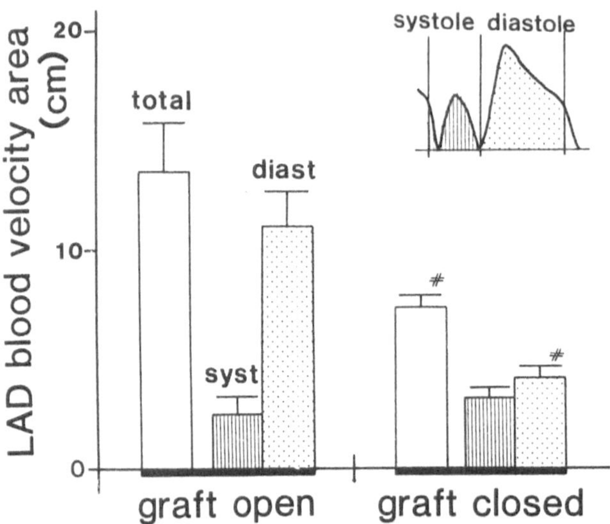

Fig. 6. Comparison of the LAD blood velocity area throughout the cardiac cycle (*total*), in systole (*syst*), and in diastole (*diast*) before and during transient occlusion of the graft. Mean ± SE. #Statistical significance ($P < 0.05$) before and during transient graft occlusion. *LAD*, left anterior descending coronary artery. From [28] with permission of the American Heart Association

came very small whereas the systolic flow component was almost unchanged (Fig. 6), indicating poor myocardial perfusion before the bypass grafting.

An increase in the systolic-to-diastolic flow ratio induced by the introduction of an artificial stenosis has been reported by some investigators [5, 29, 30]. Although the mechanism of such a change in phasic nature of coronary flow is not well understood, it may be mainly due to the change in relative importance of a stenotic resistance versus distal (intramyocardial) vascular resistance between systole and diastole. During diastole, since the distal vascular resistance is low, the stenotic resistance would mainly limit the flow. But during systole, in which the distal vascular resistance becomes high due to myocardial compression, the stenotic resistance will be much less important in limiting the flow than during diastole. Therefore, introduction of the stenosis to the proximal coronary artery (graft closing) may cause a relatively large reduction of the diastolic flow, while there is a smaller change in the systolic flow. This may be the main reason for the change in systolic-to-diastolic flow ratio. Some other factors could possibly affect the phasic nature of flow. Collateral circulation, change of vasomotor tone, and change of myocardial contractility should be taken into account, although these factors may contribute relatively little to the change in the systolic-to-diastolic flow ratio. In total coronary artery occlusion or near-total occlusion, however, the diastolic flow component diminished greatly, as discussed before, during graft clamping, but a systolic reverse flow appeared. The net flow during graft occlusion was negligible. This to-and-fro blood velocity waveform with total coronary artery occlusion may be attributable to compliance of the epicardial coronary artery and collateral flow. If the amplitude of the to-and-fro waveform is high enough, there might be some nutritional flow in the vascular bed.

References

1. Young DF, Tsai FY (1973) Flow characteristics in models of arterial stenosis: 1. Steady flow. J Biomech 6: 395–410
2. Young DF, Tsai FY (1973) Flow characteristics in models of arterial stenosis: 1. Unsteady flow. J Biomech 6: 547–559
3. Young DF, Cholvin NR, Roth AC (1975) Pressure drop across artificially induced stenosis in the femoral arteries of dogs. Circ Res 36: 735–743
4. Gould KL (1978) Pressure-flow characteristics of coronary stenosis in unsedated dogs at rest and during coronary vasodilation. Circ Res 43: 242–253
5. Gould KL, Lipscomb K, Hamilton GW (1974) Physiologic basis for assessing critical coronary stenosis. Am J Cardiol 33: 87–94
6. Gould KL, Lipscomb K (1974) Effects of coronary stenosis on coronary flow reserve and resistance. Am J Cardiol 34: 48–55
7. Seeley BD, Young DF (1976) Effect of geometry of pressure losses across models of arterial stenosis. J Biomech 9: 439–448
8. Hillis WS, Friesinger GC (1976) Reactive hyperemia: An index of the significance of coronary stenosis. Am Heart J 92: 737–740
9. Feldman RL, Nichols WW, Pepine CJ, Conti CR (1978) Hemodynamic effects of long and multiple coronary arterial narrowing. Chest 74: 280–285
10. Feldman RL, Nichols WW, Pepine CJ, Conti CR (1978) Hemodynamic significance of the length of coronary arterial narrowing. Am J Cardiol 41: 865–871
11. Mates RE, Gupta RL, Bell AC, Klocke FJ (1978) Fluid dynamics of coronary artery stenosis. Circ Res 42: 152–162
12. Vlodaver Z, Edwards JE (1971) Pathology of coronary atherosclerosis. Prog Cardiovasc Dis 14: 256–274
13. Freudenberg H, Lichtlen PR (1981) The normal wall segment in coronary stenosis—a postmortem study. Z Kardiol 70: 863
14. Logan SE (1975) On the fluid mechanics of human coronary artery stenosis. IEEE Trans Biomed Eng BME-22: 327–334
15. Schwartz JS, Carlyle PF, Cohn JN (1980) Effect of coronary artery stenosis. Circulation 61: 70–76
16. Schwartz JS, Carlyle PF, Cohn JN (1979) Effect of dilatation of the distal coronary bed on flow and resistance in severely stenotic coronary arteries in the dogs. Am J Cardiol 43: 219–224
17. Walinsky P, Santamore WP, Weiner L, Brest AN (1979) Dynamic changes in the hemodynamic severity of coronary artery stenosis in a canine model. Cardiovasc Res 13: 113–118
18. Santamore WP, Walinsky P (1980) Altered coronary flow responses to vasoactive drugs in the presence of coronary arterial stenosis in the dog. Am J Cardiol 45: 276–285
19. Brown BG, Bolson E, Peterson RB, Pierce CD, Dodge HT (1981) The mechanics of nitroglycerin action: Stenosis vasodilation as a major component of the drug response. Circulation 64: 1089–1097
20. Brown BG, Lee AB, Bolson EL, Dodge HT (1984) Reflex constriction of significant coronary stenosis as a mechanism contributing to ischemic left ventricular dysfunction during isometric exercise. Circulation 70: 18–24
21. Hartley CJ, Cole JS (1974) An ultrasound pulsed Doppler system for measuring blood flow in small vessels. J Appl Physiol 37: 626–629
22. Marcus M, Wright CB, Doty D, Charles E, Laughlin D, Krumm P, Fastenow C, Brody M (1981) Measurement of coronary velocity and reactive hyperemia in the coronary circulation of humans. Circ Res 49: 877–891
23. Wright CB, Doty DB, Eastham CL, Marcus ML (1980) Measurements of coronary reactive hyperemia with a Doppler probe. J Thorac Cardiovasc Surg 80: 888–897
24. White CW, Wright CB, Doty DB, Hiratzka LF, Eastham CL, Harrison DG, Marcus ML (1984) Does visual interpretation of the coronary arteriogram predict the physiologic importance of a coronary stenosis? N Engl J Med 310: 819–824

25. Harrison DG, White CW, Hiratzka LF, Doty DB, Barnes DH, Easthan CL, Marcus ML (1984) The value of lesion cross-sectional area determined by quantitative coronary angiography in assessing the physiologic significance of proximal left anterior descending coronary arterial stenosis. Circulation 69: 1111–1119
26. Ogasawara Y, Hiramatsu O, Kagiyama M, Tsujioka K, Tomonaga G, Kajiya F, Yanashima T, Kimura Y (1984) Evaluation of blood velocity profile by high frequency ultrasound pulsed Doppler velocimeter by a multigated zerocross method together with a Fourier transform method. IEEE Trans Comput Cardiol: 447–450
27. Kajiya F, Ogasawara Y, Tsujioka K, Nakai M, Goto M, Wada Y, Tadaoka S, Matsuoka S, Mito K, Fujiwara T (1986) Evaluation of human coronary blood flow with an 80 channel 20 MHz pulsed Doppler velocimeter and zero-cross and Fourier transform methods during cardiac surgery. Circulation 74(Suppl III): 53–60
28. Kajiya F, Tsujioka K, Ogasawara Y, Wada Y, Matsuoka S, Kanazawa S, Hiramatsu O, Tadaoka S, Goto M, Fujiwara T (1987) Analysis of flow characteristics in poststenotic regions of the human coronary artery during bypass graft surgery. Circulation 76: 1092–1100
29. Furuse A, Klopp EH, Brawley RK, Gott VL (1975) Hemodynamic determinations in the assessment of distal coronary artery disease. J Surg Res 19: 25–33
30. Ball RM, Bache RJ (1976) Distribution of myocardial blood flow in the exercising dog with restricted coronary artery inflow. Circ Res 38: 60–66

Characteristics of Coronary Artery Blood Velocity Waveforms in Aortic Stenosis and Regurgitation

TAKASHI FUJIWARA[1], TATSUKI KATSUMURA[1], and FUMIHIKO KAJIYA[2]

Summary. In twelve adult patients with pure aortic stenosis (AS) and pure aortic regurgitation (AR), we measured blood velocity waveforms in the left anterior descending coronary artery (LAD) before cardiopulmonary bypass and after aortic valve replacement. The coronary artery blood velocity waveform in AS was characterized by: (1) a decreased systolic velocity component (frequently reversed) and (2) a slowly increasing rate of diastolic inflow. On the other hand, the characteristics of the velocity waveform in AR were: (1) a relatively large systolic velocity component and (2) a rapidly decreasing diastolic flow wave. After valve replacement both for AS and AR, the velocity waveform returned to an almost normal pattern, indicating that the valve replacement induces beneficial effects on the myocardial inflow.

Key words: Aortic stenosis—Aortic regurgitation—Valve replacement—High-frequency ultrasound pulsed Doppler velocimeter

Introduction

Angina pectoris is a common clinical symptom in patients with aortic valve diseases in the presence of normal coronary arteries [1, 2]. Earlier clinical studies have demonstrated the decreased coronary flow reserve in patients with aortic valve diseases [3–5]. The causative factor responsible for the development of myocardial ischemia may be an imbalance between oxygen demand and supply: the mechanism of angina pectoris in aortic stenosis (AS) has been explained by a combination of a decreased myocardial oxygen supply due to reduced coronary perfusion pressure, and an increased oxygen demand due to increased wall tension and prolonged ejection time [2, 3]. On the other hand, the mechanism of aortic regurgitation (AR) has been explained by a low coronary perfusion pressure during diastole, and increased oxygen demand due to an elevated end-diastolic volume [1]. Pathologically, AS and AR cause the pressure-induced and the volume-induced left ventricular hypertrophy, respectively, which may also contribute to the reduced coronary reserve in patients with these diseases.

Department of Thoracic Surgery[1] and Department of Medical Engineering and Systems Cardiology[2], Kawasaki Medical School, 577 Matsushima, Kurashiki, 701-01 Japan

Phasic properties of the blood velocity waveforms of left coronary arteries, whose normal patttern is diastolic-predominant, may be affected by the hemodynamic changes caused by the aortic valve diseases. It is likely that the changes in the blood velocity waveforms closely relate to the insufficiency of myocardial blood perfusion in both AS and AR. Accordingly, we evaluated characteristics of velocity waveforms of the left coronary artery in patients with aortic stenosis and aortic regurgitation using a high-frequency multi-channel pulsed Doppler velocimeter during cardiac surgery before and after aortic valve replacement [6, 7].

Materials and Measurement Technique

In adult patients with pure AS (6 patients) or with pure AR (6 patients), we measured blood velocity waveforms in the left anterior descending coronary artery (LAD) before cardiopulmonary bypass and after aortic valve replacement. The absence of coronary artery stenosis was confirmed by coronary angiography in all patients. The velocity was measured by our 20-MHz multichannel pulsed Doppler velocimeter. A specially designed probe holder made of silicon rubber was used to place the transducer at a 60° angle on the LAD without dissection of the vessels (Fig. 1). Also, the operator followed cardiac motion carefully with his fingers. The Doppler signals were analyzed by a fast Fourier transform (FFT) or a zero-cross method, both in real time.

Coronary Artery Blood Velocity Waveform in Patients with Aortic Stenosis

Figure 2 shows a representative tracing of the LAD velocity waveform in a patient with AS before (Fig. 2a) and after (Fig. 2b) the aortic valve replacement. The velocity waveform before the valve replacement was characterized by (1) a

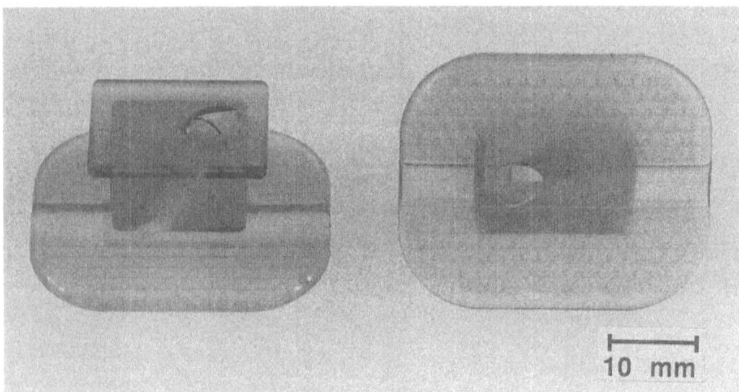

Fig. 1. Photograph of the specially designed probe holder that was used to place the transducer on the native epicardial coronary artery

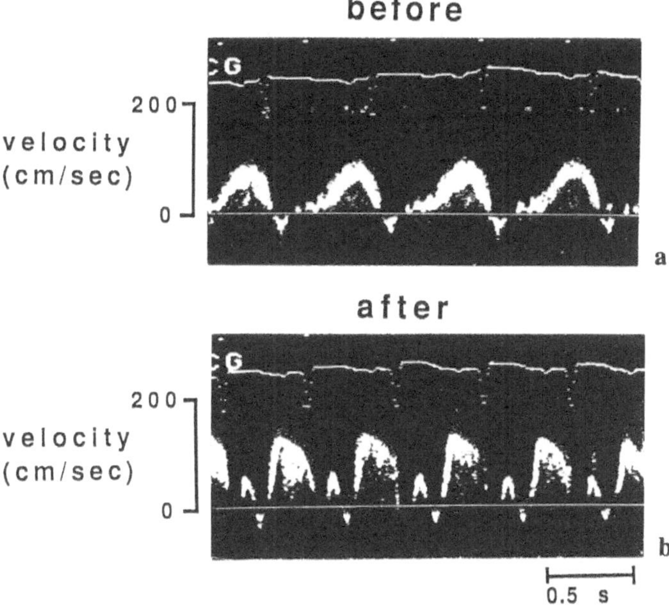

Fig. 2a, b. A representative left anterior descending coronary artery (LAD) velocity waveform in a patient with aortic stenosis before the aortic valve replacement (**a**) and that after valve replacement (**b**). The velocity waveforms are shown by the fast Fourier transform (FFT) display. From [7] with permission of the Society of Thoracic Surgeons

decreased systolic flow component with a first-half systolic reverse flow, and (2) a slowly increasing rate of early diastolic inflow. After the valve replacement, the first-half systolic reverse flow disappeared, but an end-systolic reverse flow appeared in most patients. The time to the diastolic peak flow from the onset of diastole (T_{DPV}) was significantly shortened by the valve replacement (about 50%), while the value of T_{DPV} had no correlation with heart rate. The peak diastolic velocity increased by about 30% after the valve replacement.

To indicate the temporal relation between the systolic reverse flow and the aorta/left ventricular pressure gradient, their simultaneous tracings are displayed in Fig. 3. It was noted that the timing of the systolic reverse flow before aortic replacement coincides well with that of the maximum negative aorta/left ventricular pressure gradient, while after the valve replacement, the peak systolic reverse flow occurred at the dicrotic notch of the aortic pressure in human AS. Carroll and Falsetti [8] have observed a reverse flow angiographically in their AS patients. Recently, Matsuo et al. [9] have measured blood velocities in the ostia of the left and right coronary arteries in conscious patients with AS and AR, by a Doppler flowmeter catheter. They observed a decreased systolic flow wave with an increased diastolic flow wave in AS. Although they did not report the velocity waveform in detail, the decreased systolic flow may include a reverse flow. Reports obtained from animal experiments, e.g., artificially reduced acute AS [10], and dogs with congenital subaortic stenosis [11], support our observation of the systolic reverse flow in human AS, although the systolic dip was not reversed in

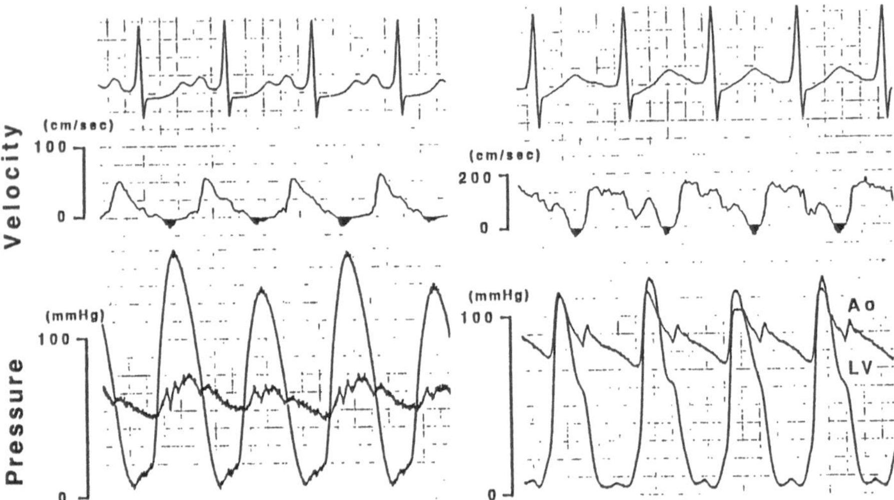

Fig. 3. A typical example indicating the temporal relationship between the systolic reverse flow and pressure waveforms in the ascending aorta (*Ao*) and the left ventricle (*LV*) before (*left*) and after (*right*) the aortic valve replacement. From [7] with permission of the Society of Thoracic Surgeons

some of animal experiments when the maximum aorta/left ventricular pressure gradient seemed to be small.

Although the fundamental characteristic of the velocity waveform, i.e., the diastolic predominancy, was shared by AS patients and normal subjects, as indicated by Marcus et al. [4], a substantial difference in the diastolic flow waveform was noticed. The "slowly increasing rate of diastolic inflow" may be caused by an increased resistance due to the preceding augmented systolic extravascular compressive force, and/or a decreased filling rate of blood into intramyocardial vessels due to the impaired myocardial relaxation. A high correlation coefficient between T_{DPV} and the aorta/left ventricular pressure gradient might support the hypothesis of the increased early diastolic resistance, due to the augmented systolic vascular compressive force. Pathophysiologically, the delayed peak velocity in diastole may cause a decrease in intramyocardial inflow, and thus subendocardial ischemia, especially when the heart rate and/or intramyocardial pressure increases.

Characteristics of Blood Velocity Waveform in Human Aortic Regurgitation

Figure 4 shows a representative tracing of the velocity waveform in a patient with severe AR before and after aortic valve replacement. The upper tracing is a velocity waveform obtained by the zero-cross method near the central region of the vessel, and the lower tracing is the velocity waveform by the FFT method.

Before the valve replacement, a large systolic flow with a high peak velocity was observed, while the diastolic flow component was small and blood velocity decreased rapidly after a peak formation in early diastole. Matsuo et al. [9] also

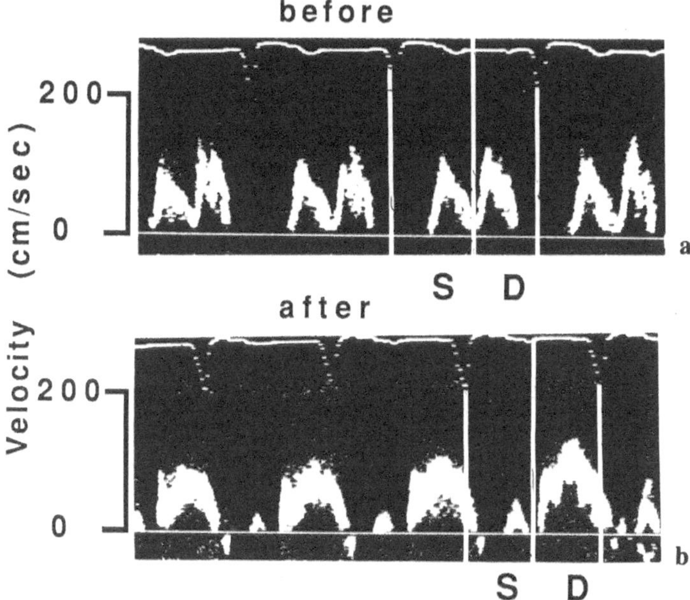

Fig. 4a, b. A representative left anterior descending coronary artery (LAD) velocity waveform in a patient with aortic regurgitation before the aortic valve replacement (**a**) and that after the valve replacement (**b**). *S*, systole; **D**, diastole. From [6] with permission of the Society of Thoracic Surgeons

observed a markedly increased systolic flow and a decreased diastolic flow. Thus, the coronary artery velocity waveform in severe aortic regurgitation was characterized by an augmented systolic flow component and a relatively small and rapidly decreased diastolic velocity waveform, which distinctly contrasted with the normal coronary artery flow waveform. After the valve replacement, a remarkable increase in the coronary flow throughout diastole had been achieved, while the systolic component decreased significantly (Fig. 5).

Experimental studies of mechanically induced acute AR demonstrated an increase in systolic flow and a marked reduction of diastolic flow [12–16]. In end-diastole, the flow was frequently reversed. These findings are consistent with our observations, except for the end-diastolic reverse flow. However, this reverse flow might be masked in our study by the cut-off of low frequency Doppler signals. Falsetti et al. [13] recognized an increase in the systolic-to-diastolic flow ratio and a significant decrease in the inflow into the subendocardial layer in severe AR, although the total coronary flow remained essentially unchanged. The abnormal increase in systolic blood flow may be attributable to insufficient flow into subendocardial layers, since effective blood perfusion into subendocardial muscle is supplied in distole, and systolic blood flow perfuses mainly subepicardial muscle of the left ventricle [17]. The timing and magnitude of the systolic flow may depend on the instantaneous epimyocardial wall stress. The inflow into the epicardial coronary arteries owing to the low diastolic pressure (causing a pressure-dependent capacitance change) and the high acceleration of aortic flow may also be attributable to the augmentation of systolic coronary flow.

Another factor responsible for myocardial ischemia in AR is a rapid decrease

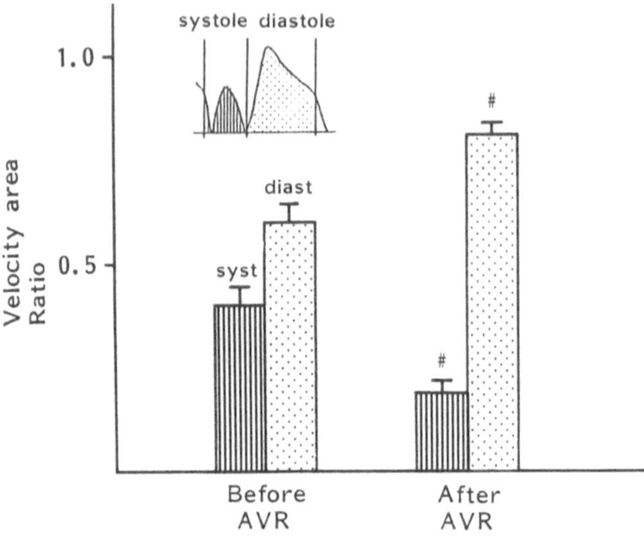

Fig. 5. Comparison of the velocity area ratios [systolic area/total area (*syst*), and diastolic area/total area (*diast*)] in arotic regurgitation *before* and *after* the aortic valve replacement (*AVR*). Mean ± SE. # $P < 0.05$ vs before AVR

in diastolic blood velocity waveform after its peak formation. This rapid decrease in the diastolic flow may be caused by a rapid fall of diastolic aortic pressure and also by an elevation of left ventricular end-diastolic pressure. Pathophysiologically, bradycardia in AR, i.e., a relatively long diastole, may not result in a beneficial effect on myocardial inflow; rather it may, worsen the coronary inflow. It has been noted clinically that a number of patients with AR developed symptoms of angina at rest, although they tolerated an exercise relatively well [18]. The angina at rest in AR may be mainly explained by bradycardia. Judge et al. [19] reported that an increase in heart rate, by means of pacing, reduced the regurgitant flow associated with decreases in the left ventricular end-diastolic pressure and volume, and a rise in aortic diastolic pressure. This may indicate the beneficial effect of moderate tachycardia on coronary arterial inflow.

In our study, immediately after the aortic valve replacement, the coronary blood velocity waveform changed from the systolic-predominant pattern to the diastolic-predominant pattern. This finding implies the achievement of effective myocardial blood flow immediately after the aortic valve replacement. Folts et al. [20] measured the right coronary blood flow in patients with AR by an electromagnetic flowmeter with an isolation of the artery. They observed an increase in the coronary flow by about 55%, along with a decrease in the systolic-to-diastolic flow ratio after the aortic valve replacement. The significant increase in coronary blood flow after the valve replacement in humans may be partly caused by an increase in cardiac output resulting from the disappearance of regurgitation and by a hyperkinetic state of left ventricular wall motion. However, the results of animal experiments are not necessarily consistent with the human data, although the cause of these differences has not yet been clarified.

Concluding Remarks

The blood velocity waveforms in the LAD were studied in patients with aortic stenosis (AS) and aortic regurgitation (AR) using our high-frequency multichannel pulsed Doppler velocimeter during cardiac surgery. The coronary artery blood velocity waveform in AS was characterized by a decreased systolic flow component with a first-half systolic reverse flow and a slowly increasing rate of early diastolic inflow. The coronary artery blood velocity waveform in AR was characterized by a relatively large systolic flow component, and a small and rapidly decreasing diastolic flow wave. After valve replacement both for AS and AR, the velocity waveforms changed to a pattern similar to the normal coronary artery velocity waveform. These results suggest that the operation for AS and AR induces beneficial effects on the myocardial inflow immediately after the valve replacement.

References

1. Graboys TB, Cohn PF (1977) The prevalence of angina pectoris and abnormal coronary arteriograms in severe aortic valvular disease. Am Heart J 93: 683–686
2. Basta LL, Raines D, Najjar S, Kioschos JM (1975) Clinical, haemodynamic, and coronary angiographic correlates of angina pectoris in patients with severe aortic valve disease. Br Heart J 37: 150–157
3. Fallon EL, Elliot WC, Gorlin R (1967) Mechanisms of angina in aortic stenosis. Circulation 36: 480–488
4. Marcus ML, Doty DB, Hiratzka LF, Wright CB, Eastham CL (1982) Decreased coronary reserve—a mechanism for angina pectoris in patients with aortic stenosis and normal coronary arteries. N Engl J Med 307: 1362–1367
5. Marcus ML (1983) The coronary circulation in health and disease. McGraw-Hill, New York, pp 285–306
6. Fujiwara T, Nogami A, Masaki H, Yamane H, Kanazawa S, Matsuoka S, Yoshida H, Katsumura T, Ogasawara Y, Kajiya F (1988) Coronary flow characteristics of left coronary artery in aortic regurgitation before and after aortic valve replacement. Ann Thorac Surg 46: 79–84
7. Fujiwara T, Nogami A, Masaki H, Yamane H, Matsuoka S, Yoshida H, Fukuda H, Katsaumura T, Kajiya F (1989) Coronary flow velocity waveforms in aortic stenosis and the effects of valve replacement. Ann Thorac Surg 48: 518–522
8. Carroll RJ, Falsetti HL (1976) Retrograde coronary artery flow in aortic valve disease. Circulation 54: 494–499
9. Matsuo S, Tsuruta M, Hayano M, Imamura Y, Eguchi Y, Tokushima T, Tsuji S (1988) Phasic coronary artery flow velocity determined by Doppler flowmeter catheter in aortic stenosis and aortic regurgitation. Am J Cardiol 62: 917–922
10. Sabbah HN, Stein PD (1988) Reduction of systolic coronary blood flow in experimental left ventricular outflow tract obstruction. Am Heart J 116: 806–811
11. Pyle RL, Lowensohn HS, Khouri EM, Gregg DE, Patterson DF (1973) Left circumflex coronary artery hemodynamics in conscious dogs with congenital subaortic stenosis. Cir Res 33: 34–38
12. Folts JD, Rowe GG (1974) Coronary and hemodynamic effects of temporary acute aortic insufficiency in intact anesthetized dogs. Circ Res 35: 238
13. Falsetti HL, Carroll RJ, Cramer JA, Lenth RA (1979) Total and regional myocardial blood flow in aortic regurgitation. Am Heart J 97: 485
14. Karp RB, Roe BB (1966) Effect of aortic insufficiency on phasic flow patterns in the coronary artery. Ann Surg 164: 959–966

15. Mizutani T (1973) A study of coronary circulation in experimental aortic insufficiency with special reference to phasic coronary flow pattern. Jpn Circ J 37: 123–133
16. Nakao S, Nagatomi T, Kiyonaga K, Kashima T, Tanaka H (1987) Influences of localized aortic valve damage on coronary artery blood flow in acute aortic regurgitation: an experimental study. Circulation 76: 201–207
17. Downey JM, Kirk ES (1974) Distribution of the coronary blood flow across the canine heart wall during systole. Circ Res 34: 251
18. Harvey WP, Segal JP, Hufnagel CA (1957) Unusual clinical features associated with severe aortic insufficiency. Ann Intern Med 47: 27–38
19. Judge TP, Kennedy JW, Bennett LJ, Wills RE, Murray JA (1971) Quantitative hemodynamic effects of heart rate in aortic regurgitation. Circulation 44: 355–367
20. Folts JD, Rowe GG, Kahn DR, Young WP (1979) Phasic changes in human right coronary blood flow before and after repair of aortic insufficiency. Am Heart J 97: 211

Coronary Circulation in Patients with Typical Angina Pectoris but Normal Coronary Angiograms

WOLFGANG WAAS, DIETER OPHERK, GERHARD SCHULER, and WOLFGANG KÜBLER[1]

Summary. To gain further insight into the mechanism of angina in patients with normal coronary angiograms (Syndrome X), a standard exercise test was performed and coronary blood flow and myocardial lactate metabolism were assessed at rest and during dipyridamole-induced vasodilation in 32 patients with this syndrome. In 25 out of these 32 patients, the response of pulmonary artery pressure and left ventricular ejection fraction to dynamic exercise was determined. Results were compared with those obtained in 30 patients without detectable heart disease. Patients with Syndrome X had a positive exercise test, minimal coronary resistance was twice as high as in controls, and significant lactate production occurred upon infusion of dipyridamole. At similar workload, pulmonary artery pressure was significantly higher in patients with typical angina than in controls, and left ventricular ejection fraction did not increase, in contrast to control patients, who showed an increase in left ventricular ejection fraction of more than 5% in response to exercise. These results indicate that symptoms in patients with Syndrome X are due to myocardial ischemia, which may precipitate left ventricular dysfunction during exercise.

Key words: Syndrome X—Coronary vascular reserve—Dipyridamole—Myocardial ischemia—Radionuclide ventriculography

Introduction

Syndrome X [1–3] or microvascular angina [4] characterize the clinical picture in patients with typical exercise-induced angina pectoris, but normal coronary angiograms. It is still a matter of debate whether symptoms in those patients may be ascribed to esophageal disease [5] and behavioral disorders [6] or to myocardial ischemia.

To differentiate between cardiac and extracardiac causes for anginal pain in the absence of even minimal coronary artery disease, the following studies were performed in addition to an upright exercise test and left ventricular and coronary angiography: (1) coronary blood flow was measured at rest and during dipyridamole-induced vasodilation, (2) myocardial lactate metabolism was assessed under the same conditions, (3) pulmonary artery pressure was obtained at rest and during maximal exercise, and (4) left ventricular ejection fraction was similarly obtained.

[1]Department of Cardiology, University of Giessen, Klinikstr. 36, 6300 Giessen, Federal Republic of Germany

Methods

Patients

Control group (group A). Thirty patients, 10 females and 20 males, were evaluated. Their average age was 46.7 ± 1.75 years, range 37–65 years. All patients suffered from atypical chest pain, which was long lasting and not sensitive to nitroglycerin. They had a normal ECG at rest and a normal exercise test. These patients were referred for catheterization and coronary angiography to definitely exclude cardiac disease as the cause of their symptoms.

Syndrome X (group B). This group consisted of 32 patients, 11 females and 21 males, with an average age of 41.1 ± 1.17 years, range 29–60 years. These patients had typical angina pectoris, mostly during strenuous activities and rarely at rest, but in every case, promptly relieved by nitroglycerin.

In all patients from groups A and B, an echocardiographic and Doppler sonographic study was performed with normal result. No patient had valvular heart disease or left ventricular hypertrophy. Moreover, no patient gave a history of arterial hypertension, diabetes mellitus, or collagen disease.

Cardiac Catheterization

Before invasive studies, all cardiac medications were withdrawn for 2 days except short-acting nitrates. During catheterization, patients were in a fasted, nonsedated state. Right and left ventricular pressures were recorded using a fluid-filled pressure-transducer system. Left ventricular angiograms were obtained in left and right anterior oblique projections and selective coronary angiography was performed in multiple, including angulated views.

Coronary blood flow at rest and after intravenous application of 0.5 mg/kg dipyridamole was determined by the argon tracer technique (for details, see [7]). This method correctly measures coronary blood flow even at high flow rates [8]. Coronary resistance was calculated as the quotient of the difference between mean aortic pressure and right atrial pressure and coronary blood flow. Coronary reserve was obtained by comparing coronary resistance before and after dipyridamole.

In 20 patients of the control group and in 19 patients with Syndrome X, arterial and coronary blood samples were drawn at rest and after dipyridamole for determination of lactate by an enzymatic method. Coronary blood flow and the arterio-coronary venous difference of lactate were used to calculate net cardiac lactate uptake or release.

Pulmonary artery pressure was determined at rest and during maximal supine bicycle exercise in 11 patients of the control group and in 25 patients with Syndrome X.

In addition, *left ventricular ejection fraction* was determined at rest and during maximal supine bicycle exercise by radionuclide ventriculography. Erythrocytes were labeled by technetium-99 (25 mCi) and scintigraphy was performed in the left anterior oblique view. Ejection fraction was calculated from the left ventricular time-activity curve.

All tests were performed within 3 weeks. Data were analyzed with a two-tailed *t*-test for unpaired data or with Wilcoxon's rank sum test. Differences were considered to be significant if the probability (*P*) value was less than 0.05. All values are presented as mean ± standard error of the mean.

Results

Electrocardiography

ECG and exercise tests in group A patients were normal. In group B, 3 patients showed left bundle branch block, and in 8 patients, minor changes of ST-segments or T-waves were seen at rest. During exercise, the 3 patients with left bundle branch block and 5 other patients experienced increasing typical anginal pain, severe enough to stop the exercise test. Five patients with a normal ECG at rest showed rate-dependent left bundle branch block at higher workload. In the other 19 patients horizontal or down-sloping ST-segment depressions (> 0.1 mm) were seen in precordial leads.

Angiographic Studies

In all patients coronary angiography did not reveal atherosclerotic lessions nor any minor irregularities. Left ventricular angiograms showed no abnormalities, ejection fraction was 65% or more. Left ventricular enddiastolic pressure was slightly higher in group B patients (9.2 ± 1.4 mmHg) than in group A (8.3 ± 1.3 mmHg), but this difference was not statistically significant.

Coronary Blood Flow

In control patients coronary blood flow at rest was 79.7 ml/100 g·min, not different from 83.3 ml/100 g·min, measured in group B patients (Fig. 1a). After dipyridamole, coronary blood flow increased in both groups. Maximal flow rate in control patients, however, was twice as high as in patients with typical angina and normal coronary angiograms (291.8 vs 147.7ml/min·100 g; $P < 0.01$) (Fig. 1a).

At rest, an identical coronary resistance (Fig. 1b) was found in both groups (1.08 mm Hg/ml/100 g·min). Following dipyridamole, coronary resistance decreased to 0.30 mmHg/ml/100 g·min in controls, but remained significantly elevated in patients of group B (0.58 mmHg/ml/100 g·min, $P < 0.01$) (Fig. 1b). Hemodynamic parameters like aortic pressure and heart rate were comparable in both groups at rest and after dipyridamole. In addition, myocardial oxygen consumption at rest was not different (10.1 vs 10.2 ml/100 g·min), but after dipyridamole it increased in group A to 14.0 ml/100 g·min and decreased slightly in group B (9.24 ml/100 g·min).

Coronary reserve (Fig. 2) ranged from 2 to 6 for group A patients with a mean of 4.1. In patients with typical angina but normal coronary angiograms, coronary reserve was 3.0 or less with a mean of 2.0 ($P < 0.01$). However, as shown in Fig.

2, there is a considerable overlap between both groups in the range of 2–3.

Lactate metabolism was assessed in 20 patients out of the control group and in 19 patients with Syndrome X. At rest, both groups have comparable rates for lactate uptake. After intravenous administration of dipyridamole, lactate uptake significantly increased in control patients, whereas in group B patients, significant lactate production occurred (Fig. 3).

Within 3 weeks after diagnostic coronary angiography, *pulmonary artery*

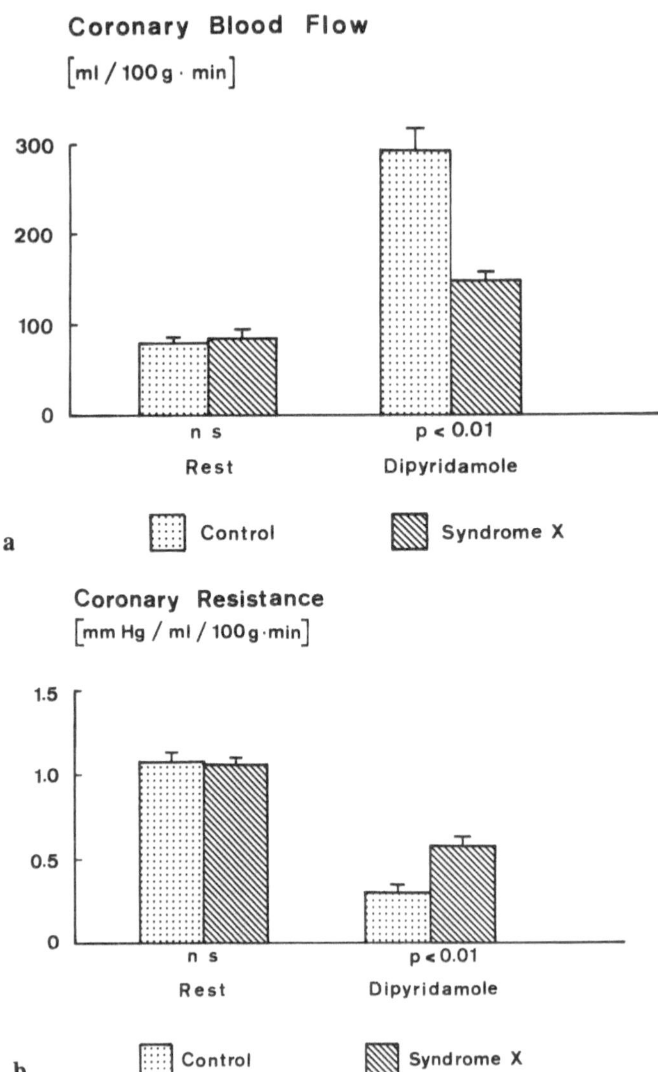

Fig. 1a, b. Response of **a** coronary blood flow and **b** coronary resistance to intravenous application of dipyridamole (0.5 mg/kg) in 30 control subjects and 32 patients with Syndrome X

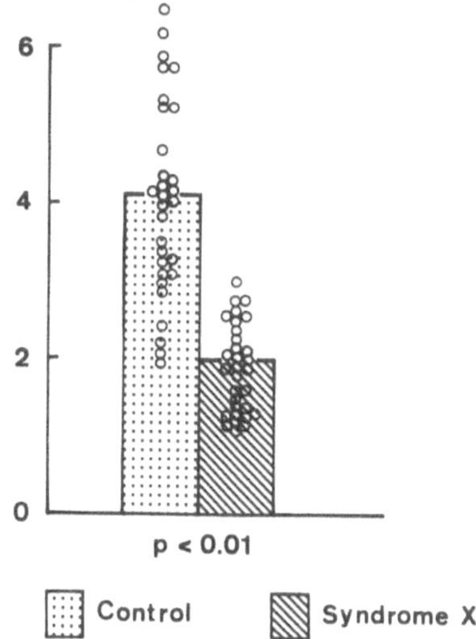

Fig. 2. Coronary vascular reserve in 30 control subjects and 32 patients with Syndrome X. Coronary reserve was calculated from coronary resistance at rest and minimal coronary resistance, assessed by infusion of dipyridamole

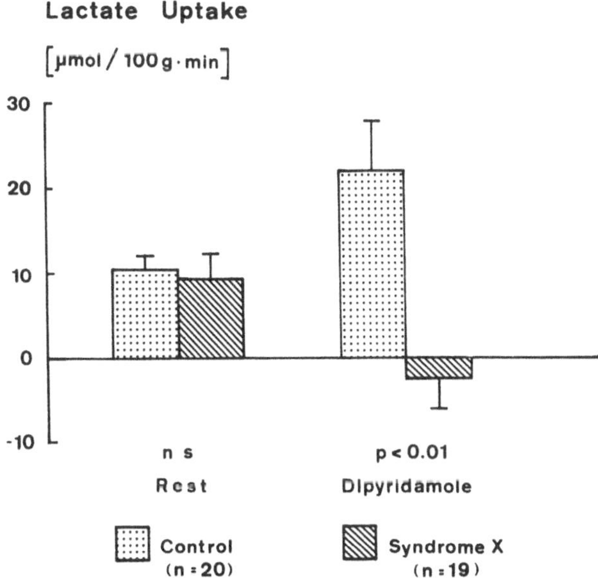

Fig. 3. Lactate uptake at rest and during dipyridamole-induced vasodilation in 20 patients of the control group and 19 patients with Syndrome X

pressure was determined with a balloon-tipped, flow-directed catheter at rest and during maximal, symptom-limited supine bicycle exercise in 11 control patients and in 25 patients of group B. Mean pulmonary artery pressure at rest (Fig. 4) was similar in both groups (16.5 mmHg, group A vs 15.4 mmHg, group B). Maximal workload was not different for both groups (93.5 ± 3.69 vs 93.0 ± 4.78 W). During exercise, mean pulmonary artery pressure increased slightly in group A from 16.5 to 21.1 mmHg. In this group, only 1 patient exhibited an increase in pulmonary artery pressure of more than 10 mmHg (Fig. 4).

In contrast, in 19 out of 25 patients from group B, mean pulmonary artery pressure increased more than 10 mmHg during exercise. In addition, the group average during exercise (29.8 mmHg) was significantly higher than for group A (21.1 mmHg). On the other hand, 6 patients with typical angina showed only a modest and normal rise in mean pulmonary artery pressure during exercise (Fig. 4).

Radionuclide left ventricular angiography at rest and during maximal exercise was performed in all patients in whom pulmonary artery pressure was determined, except in 2 patients from group A. All patients studied from group A showed a normal response to exercise, i.e., an increase in left ventricular ejection fraction of 5% or more (Fig. 5). In total, ejection fraction increased from 65.4% to 75.0% during exercise. The average maximal workload in those patients reached 94.7 ± 3.61 W and was not different from the maximal workload in group B patients (98.0 ± 4.40 W).

In group B, however, only 3 patients showed a normal response to exercise. In 7 patients, left ventricular ejection fraction was unchanged or increased less than 5%, and in the other 15 patients, ejection fraction actually decreased in response to exercise. As can be seen further in Fig. 5, mean left ventricular ejection fraction at rest was similar for both groups, but was significantly lower during exercise in group B than in group A.

Discussion

The results of this study clearly demonstrate pathophysiologic abnormalities in patients with Syndrome X, a clinical condition characterized by normal coronary angiograms, but typical, exercise-induced angina pectoris, responding to nitroglycerin. During exercise testing, these patients developed significant ST-segment depression, left bundle branch block, or increasing, disabling angina pectoris. Both medical history and the result of the exercise test suggest that myocardial ischemia may occur during strenuous physical activities.

Most commonly, myocardial ischemia is due to coronary artery disease. In these patients, however, coronary angiography, performed in multiple angulated views, failed to demonstrate even minor irregularities of epicardial coronary vessels. Another cause for myocardial ischemia may be coronary spasm. However, frank spasm in the absence of atherosclerotic lesions is a rare event [9] and does not play a major role in the patients under study, since angina pectoris was mostly related to physical activities. In addition, during the rare episodes of angina at rest, no ST-segment elevations were observed.

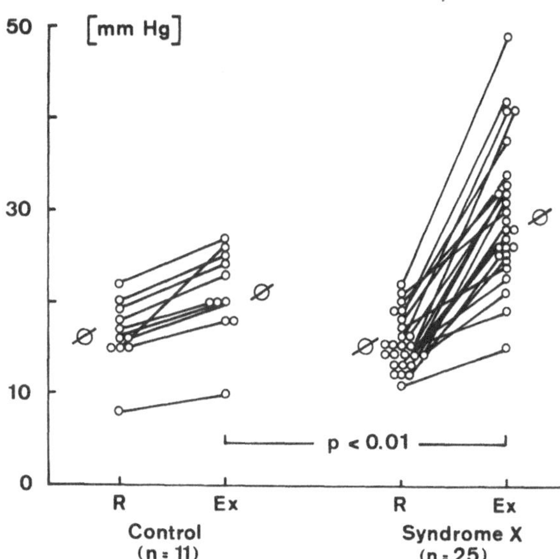

Fig. 4. Response of mean pulmonary artery (*PA*)-pressure to maximal, symptom-limited exercise in 11 control patients and 25 patients with Syndrome X. Maximal workload was not different for both groups (*R*, rest; *Ex*, exercise)

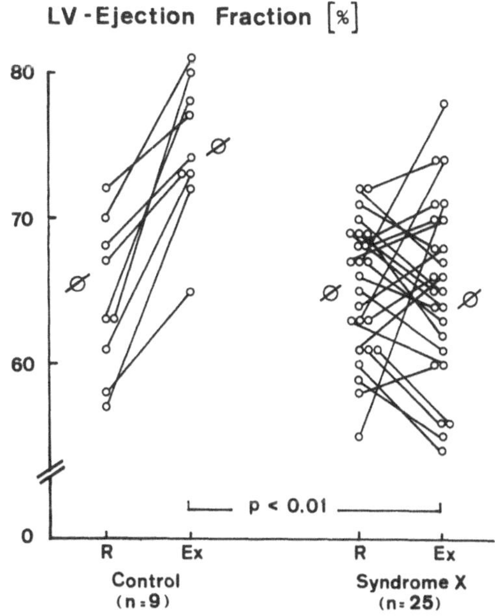

Fig. 5. Left ventricular (*LV*)-ejection fraction at rest (*R*) and during maximal, supine bicycle exercise (*Ex*) in 9 patients of the control group and 25 patients with Syndrome X

Other conditions which increase the extravascular component of coronary resistance and thereby may precipitate myocardial ischemia, e.g., left ventricular hypertrophy due to arterial hypertension, hypertrophic cardiomyopathy, or aortic valvular disease, (for review, see [10]), were excluded by history, clinical, echocardiographic, and Doppler sonographic evaluation prior to enrollment. Likewise, no patient had diabetes mellitus or collagen disease—well-known causes for microvascular abnormalities [11–13].

On the other hand, results of exercise tests may be false-positive. However, in our patients with angina and a positive exercise test, a markedly blunted rise in coronary blood flow and an increased coronary resistance following intravenous application of dipyridamole were observed. That reduced coronary vascular reserve indeed may precipitate ischemic episodes, is documented by the significant production of lactate following dipyridamole-induced vasodilation in those patients.

There is further, but indirect evidence, that myocardial ischemia actually develops during exercise in patients with Syndrome X. Despite a similar workload, a significantly higher increase in pulmonary artery pressure was observed in patients of group B, compared to controls, and 19 out of 25 patients with Syndrome X showed an inappropriate rise of pulmonary artery pressure during exercise. In accordance, left ventricular ejection fraction did not increase more than 5% or actually decreased in 22 out of 25 patients from group B. Hence, the results presented so far support the hypothesis that myocardial ischemia is the cause for anginal pain in patients with Syndrome X.

This view is further supported by an earlier study of Opherk et al. [7], who could demonstrate that in patients with Syndrome X myocardial lactate production may be precipitated by atrial pacing. More recent studies of Greenberg et al. [14] and Cannon et al. [15, 16] demonstrated an impaired coronary blood flow response to atrial pacing as the mechanism for pacing-induced ischemia. Because of a similar rate-pressure product during pacing in controls and in patients under study, impaired oxygen supply will be the determinant for ischemia in those patients.

The mechanism for impaired oxygen supply during dipyridamole-induced vasodilation, during exercise, and during pacing may be related to functional or morphological abnormalities of intramyocardial vessels. Yet, left-[7] and right-[17, 18] sided endomyocardial biopsies did not reveal any histologic abnormalities of arterioles and meta-arterioles. However, the absence of such abnormalities in biopsy specimens does not definitely exclude structural disease, since intermediate vessels (40–400 μm) are not accessible by this technique. On the other hand, biopsies showed alterations of mitochondria, e.g., swelling and small myelin figures, which may be due to repetitive phases of ischemia [19], but not the causative mechanism for ischemia.

The vasodilating action of dipyridamole strongly depends on transport-inhibition of adenosine [20, 21], a local hormone, which is thought to play a major role in local metabolic regulation of coronary blood flow [22]. Consequently, in patients with Syndrome X, the impaired coronary vascular reserve, assessed by dipyridamole, may be explained by abnormalities in adenosine metabolism or by a decreased response of smooth muscle cells to adenosine. A simple shift of the dose response curve for dipyridamole to the right can be excluded by the observation that such patients had proven ischemic episodes and a blunted rise in coronary blood flow also during atrial pacing.

Based on our results, an increased vascular tone or an increased sensitivity to vasoconstrictor stimuli, as suggested by Cannon et al. [4, 15, 16], cannot be excluded as a mechanism for ischemic episodes in the syndrome with chest pain and normal epicardial coronary arteries. However, an increased tone of prearteriolar vessels appears unlikely, since coronary resistance at rest was normal and in the same range for patients with Syndrome X and for controls. Consequently, a decreased sensitivity of smooth muscle cells to endogenous vasodilating substances, most probably adenosine, is the most likely pathophysiologic mechanism for ischemia in Syndrome X. The dysregulation of vascular tone may not be restricted to coronary vessels, since forearm hyperemic flow was significantly lower in patients with chest pain and normal coronary angiograms than in control patients [23].

However, ischemia and myocardial lactate production occur in Syndrome X in the presence of a limited, but clear, increase in global coronary blood flow following infusion of dipyridamole. Since myocardial oxygen consumption does not increase simultaneously, a homogenous dysregulation of arterioles would not lead to ischemia. Hence, vessels with normal and disturbed vasomotion must exist, either within different layers of the myocardium or within different vascular regions. Under such conditions, maximal vasodilation in the normal region may induce a "steal mechanism" for regions with an abnormal vasodilator response. This phenomenon was described in a study by Legrand et al. [24], favoring the concept of perfusion abnormalities within distinct vascular beds. The authors showed that patients with chest pain and angiographically normal coronary arteries have thallium-201 perfusion defects, an impaired regional wall motion during exercise, and an abnormal response of global left ventricular ejection fraction to exercise. In addition, regional coronary flow reserve, assessed by quantitative digital coronary angiography, was significantly lower in regions with abnormal radionuclide studies. These results are confirmed by further studies [15, 25–27] which showed regional wall motion abnormalities or thallium-201 defects during exercise in patients with normal or near normal coronary angiograms.

Despite an increasing number of publications dealing with Syndrome X, the literature adds little help to manage therapeutic and prognostic problems in such patients. At our institution, routine therapy of patients with Syndrome X starts with nitrates. By addition of calcium-channel blockers to medical therapy, further beneficial effects on symptoms and exercise tolerance can be obtained [28]. Yet, the possibility, that β-blockers may exert detrimental effects on symptoms [29] cannot be confirmed by our clinical experience.

Although patients with normal or near normal coronary arteries have an excellent prognosis [30], no data on the subgroup of patients with typical angina pectoris, but impaired vasodilator reserve in response to dipyridamole and atrial pacing have been published. In a recent study, Opherk et al. [31] report on the follow-up of patients with Syndrome X who were seen for at least 4 years. In contrast to the control group, which showed no change over the years, in patients with Syndrome X, symptoms worsened, left ventricular ejection fraction decreased, and mean pulmonary artery pressure increased in response to exercise during the follow-up. The most pronounced deterioriation of left ventricular function was seen in patients with left bundle branch block either at rest or during exercise. One of those patients developed overt heart failure within 3 years. Control angiography showed a resting left ventricular ejection fraction of 23%, compared to 66% at the time of the first evaluation.

In summary, patients with Syndrome X, characterized by typical exercise-induced angina pectoris, but normal coronary angiograms, show functional abnormalities which indicate that myocardial ischemia may be the cause for symptoms in those patients. Further evidence comes from the dipyridamole study, where significant lactate production was observed in conjunction with a decreased vasodilator reserve. Most probably, the impaired response of coronary resistance vessels to dipyridamole and exercise will be due to a decreased sensitivity of smooth muscle cells to vasodilating substances involved in local metabolic control of coronary blood flow, such as adenosine. This functional defect is not uniformly distributed within the myocardium but, rather, restricted to distinct vascular areas.

Medical history, standard exercise test, and coronary angiography are adequate to establish the diagnosis. In addition, either coronary blood flow or the response of pulmonary artery pressure or left ventricular ejection fraction to exercise should be assessed. None of these methods is superior to the other because of significant overlap between control patients and patients with Syndrome X. Patients with abnormal results need to be followed, since heart failure may develop, especially in the subgroup of patients with left bundle branch block.

Acknowledgments. This work was supported by a grant from the Deutsche Forschungsgemeinschaft (SFB 90-Cardiovascular System-University of Heidelberg) and a grant from the Fritz-Thyssen-Foundation, Cologne.

References

1. Likoff W, Segal BL, Kasparian H (1967) Paradox of normal selective coronary arteriograms in patients considered to have unmistakable coronary heart disease. N Engl J Med 276: 1063–1066
2. Kemp HG (1973) Left ventricular function in patients with anginal syndrome and normal coronary arteriograms. Am J Cardiol 32: 375–376
3. Kemp HG, Vokonas PS, Cohn PF, Gorlin R (1973) The anginal syndrome associated with normal coronary arteriograms: report of a six-year experience. Am J Med 54: 735–742
4. Cannon RO III, Epstein SE (1988) "Microvascular angina" as a cause of chest pain with angiographically normal coronary arteries. Am J Cardiol 61: 1338–1343
5. Vantrappen G, Janssens J, Ghillebert G (1987) The irritable oesophagus—A frequent cause of angina-like pain. Lancet 1: 1232–1234
6. Wieglosz AT, Fletcher RH, McCants CB, McKinnis RA, Haney TL, Williams RB (1984) Unimproved chest pain in patients with minimal or no coronary disease: a behavioral phenomenon. Am Heart J 108: 67–72

7. Opherk D, Zebe H, Weihe E, Mall G, Dürr C, Gravert B, Mehmel HC, Schwarz F, Kübler W (1981) Reduced coronary dilatory capacity and ultrastructural changes of the myocardium in patients with angina pectoris but normal coronary arteriograms. Circulation 63: 817–825
8. Opherk D, Finke R, Mittmann U, Müller JH, Wirth RH, Schmier J (1977) The influence of the size of acute ischemic myocardial lesions on coronary reserve and left ventricular function in the dog. Basic Res Cardiol 72: 402–410
9. Bertrand ME, LaBlanche JM, Tilmant PY, Thieuleux FA, DelFerge MR, Caril AG, Asseman P, Berzin B, Libersa C, Laurent JM (1982) Frequency of provoked coronary arterial spasm in 1089 consecutive patients undergoing coronary angiography. Circulation 65: 1299–1306
10. Bache RJ (1988) Effects of hypertrophy on the coronary circulation. Prog Cardiovasc Dis 31: 403–440
11. Zoneraich S (1988) Small-vessel disease, coronary artery vasodilator reserve, and diabetic cardiomyopathy. Chest 94: 5–7
12. Strauer BE, Brune J, Schenk H, Knoll D, Perings E (1976) Lupus cardiomyopathy: cardiac mechanics, hemodynamics and coronary blood flow in uncomplicated systemic lupus erythematosus. Am Heart J 92: 715–722
13. Nitenberg A, Foult JM, Kahan A, Perennec J, Devaux JY, Menkes CJ, Amor B (1986) Reduced coronary flow and resistance reserve in primary scleroderma myocardial disease. Am Heart J 112: 309–315
14. Greenberg MA, Grose RM, Neuburger N, Silverman R, Strain JE, Cohen MV (1987) Impaired coronary vasodilator resposiveness as a cause of lactate production during pacing-induced ischemia in patients with angina pectoris and normal coronary arteries. J Am Coll Cardiol 9: 743–751
15. Cannon RO III, Bonow RO, Bacharach SL, Green MV, Rosing DR, Leon MB, Watson RM, Epstein SE (1985) Left ventricular dysfunction in patients with angina pectoris, normal epicardial coronary arteries, and abnormal vasodilator reserve. Circulation 71: 218–226
16. Cannon RO III, Schenke WH, Leon MB, Rosing DR, Urquhart J, Epstein SE (1987) Limited coronary reserve after dipyridamole in patients with ergonovine-induced coronary vasoconstriction. Circulation 75: 163–174
17. Lösse B, Kuhn H, Rafflenbeul D, Kronert H, Hort W, Feinendegen LE, Loogen F (1980) Thallium-201 myocardial scintigraphy in patients with normal coronary arteries and normal left ventriculogram-comparison with hemodynamic, metabolic and morphologic findings. Z Kardiol 69: 523–530
18. Richardson PJ, Livesly B, Oram S, Olsen EGJ, Armstrong P (1974) Angina pectoris with normal coronary arteries. Transvenous myocardial biopsy in diagnosis. Lancet 2: 677–680
19. Kawamura K, Cowley MJ, Karp RB, Mantle JA, Logic JR, Rogers WJ, Russell RO Jr, Rackley CE, James TN (1978) Intramitochondrial inclusions in the myocardial cells of human hearts with coronary disease. J Mol Cell Cardiol 10: 797
20. Schrader J, Berne RM, Rubio R (1972) Uptake and metabolism of adenosine by human erythrocye ghosts. Am J Physiol 223: 159–166
21. Newby AC (1986) How does dipyridamole elevate extracellular adenosine concentration? Biochem J 37: 845–851
22. Berne RM, Rubio R (1979) Coronary circulation. In: Berne RM, Sperelakis N, Geiger SE (eds) The cardiovascular system. American Physiological Society, Bethesda, pp 873–952 (Handbook of physiology)
23. Sax FL, Cannon RO III, Epstein SE (1987) Forearm flow in patients with Syndrome X: evidence of a generalized disorder of vascular tone? N Engl J Med 317: 1366–1370
24. Legrand V, Hodgson JM, Bates ER, Aueron FM, Mancini GBJ, Smith JS, Gross MD, Vogel RA (1985) Abnormal coronary flow reserve and abnormal radionuclide exercise test results in patients with normal coronary angiograms. J Am Coll Cardiol 6: 1245–1253

25. Berger BC, Abramowitz R, Park CH, Desai AG, Madsen MT, Chung EK, Brest AN (1983) Abnormal thallium-201 scans in patients with chest pain and angiographically normal coronary arteries. Am J Cardiol 52: 365–370
26. Kaul S, Newell JB, Chesler DA, Pohost GM, Okada RD, Boucher CA (1986) Quantitative thallium findings in patients with normal coronary angiographic findings and in clinically normal subjects. Am J Cardiol 57: 509–512
27. Berger HJ, Sand MJ, Davies RA, Wackers FJT, Alexander J, Lachman AS, Williams BW, Zaret BL (1981) Exercise left ventricular perfomance in patients with chest pain, ischemic-appearing exercise electrocardiograms and angiographically normal coronary arteries. Ann Intern Med 94: 186–191
28. Cannon RO III, Watson RM, Rosing DR, Epstein SE (1985) Efficacy of calcium channel blocker therapy for angina pectoris resulting from small-vessel coronary artery disease and abnormal vasodilator reserve. Am J Cardiol 56: 242–246
29. Cannon RO III, Watson RM, Rosing DR, Epstein SE (1983) Detrimental effect of propranolol on anginal threshold of patients with angina and insignificant epicardial coronary artery disease (abstract). J Am Coll Cardiol 1: 595
30. Kemp HG, Kranmal RA, Vliestra RE, Frye RL, and participants in the coronary artery surgery study (1986) Seven year survival of patients with normal and near-normal coronary arteriograms: A CASS Registry Study. J Am Coll Cardiol 7: 479–483
31. Opherk D, Schuler G, Wetterauer K, Manthey J, Schwarz F, Kübler W (1989) Four-year follow-up study in patients with angina pectoris and normal coronary arteriograms ("Syndrome X"). Circulation 80: 1610–1616

Blood Velocity Waveforms in Coronary Bypass Grafts

TAKASHI FUJIWARA[1] and FUMIHIKO KAJIYA[2]

Summary. The characteristics of blood velocities were investigated at different sites in two types of coronary artery bypass graft, the life spans of which are known to be different, i.e., the internal mammary artery graft has the longer life span. The side-to-side anastomosis of the sequential saphenous bypass graft is longer than the end-to-side anastomosis. The blood velocities were measured by the 20 MHz, 80-channel ultrasound pulsed Doppler method during cardiac surgery. A parabolic velocity profile across the graft with a narrow spectrum was always observed in the mammary artery graft, while complex flow fields including the flow separation, the directional change in skewing of the velocity profile, and a wider spectrum were recognized in the vein graft. The patterns of blood flow seem to be a contributory factor in determining the fate of the graft. Therefore, these findings may provide insight into the underlying mechanisms.

Key words: Sequential saphenous vein graft—Internal mammary artery graft—High-frequency ultrasound pulsed Doppler velocimeter—Wide velocity spectrum—Skew of velocity profile

Introduction

The internal mammary artery and saphenous vein grafts are frequently used for coronary revascularization surgery. Despite the symptomatic alleviation of angina pectoris afforded by the procedure, there has been a problem with an appreciable incidence of graft occlusion after operation. Long-term follow up studies have shown that the internal mammary artery graft (IMAG) has a longer life span compared with the saphenous vein bypass graft (SVG) [1–5]. Several factors, including technical problems, determine the early fate of coronary bypass grafts [6]. However, progressive intimal hyperplasia and atherosclerosis are emerging as the major determinants of long-term viability [7, 8]. It is suggested that atherosclerosis is influenced not only by such global factors as abnormal lipoprotein levels but also by local blood-flow patterns in the graft and in the native coronary artery distal to grafting [9, 10]. Interactions between the flow velocity and the vessel wall of the graft and coronary artery may be related to the

Department of Thoracic Surgery[1], Department of Medical Engineering and Systems Cardiology[2], Kawasaki Medical School, 577, Matsushima, Kurashiki, 701-01 Japan

development of intimal hyperplasia and atherosclerosis. One variation of the SVG is a sequential saphenous vein graft (SSVG) in which a side-to-side anastomosis of the vein graft into a coronary artery is performed in addition to the end-to-side anastomosis. In cases of SSVG, the patency rates of side-to-side anastomoses (SSA) have been reported to be better than these of end-to-side anastomoses (EDA) [11, 12]. The patency difference between SSA and EDA may again indicate the importance of hemodynamic factors on the fate of the graft.

Recently we developed a 20-MHz 80-channel pulsed Doppler velocimeter to study velocity waveforms as well as the velocity profile across a vessel. By using this velocimeter, we evaluated characteristics of blood velocities in SVG, SSVG, and IMAG.

Measuring Techniques

In order to achieve high spatial resolution, a transducer with a 20-MHz carrier frequency was used as indicated by Hartley and Cole [13]. The signal-processing system of our velocimeter is shown in Fig. 1 [14–16]. The transducer consists of a $\pi \times 0.5^2$-mm^2 piezoelectrical crystal with a 20-MHz carrier frequency. The pulse repetition period is 20 μs, and the sampling pulse width is 0.25 μs. The depth resolution is about 0.2 mm. The system has 80 sampling gates, and Doppler signals from the multigated circuit are analyzed by a zero-cross method. In addition, a fast Fourier transform (FFT) is performed for an optional channel by the

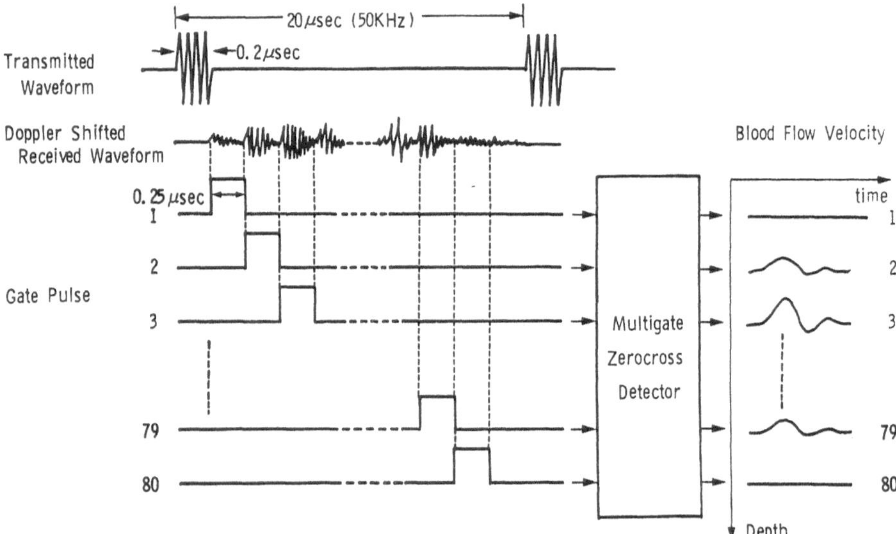

Fig. 1. Schematic drawing of the 80-channel 20-MHz pulsed Doppler velocimeter. Doppler signals from 80 channels are detected by a multigated zero-cross method and a Doppler signal from one optional channel is analayzed by a fast-Fourier transform method (FFT). From [15] with permission of the American Heart Association

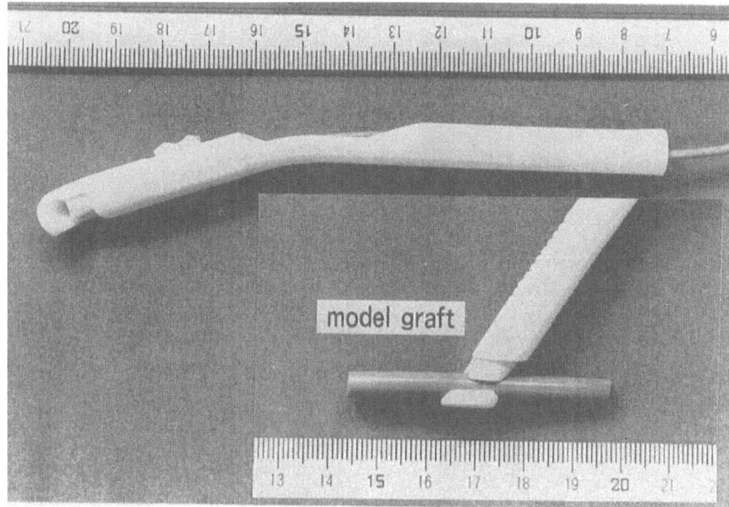

Fig. 2. Specially designed pencil-type probe holder that was used to place the transducer on the graft. See text for details. From [17] with permission of the American Heart Association

hardware to detect the spectrum broadening seen in disturbed velocity fields. These procedures are performed in real time.

A specially designed pencil-type probe holder with a slight curvature was used to measure blood-flow velocities in the graft (Fig. 2) [17]. This holder was made of Teflon, and a transducer was mounted in it at an angle of 60° to the central axis of the cuff. The size of the cuff is controlled by the sliding of a knob with the operator's index finger. The holder is slightly curved to obtain better accessibility to the graft. A plate-type probe holder was also used [15] when necessary. The holder was placed perpendicular to the myocardial surface. Blood-flow velocity measurements were performed when hemodynamic variables became stable after the completion of bypass grafting and weaning from the cardiopulmonary bypass. The maximum diameter of a graft during a cardiac cycle was roughly estimated by the disappearance of the Doppler signals.

Velocity Profile in a Saphenous Vein Graft

Figure 3 shows a typical recording of pulsatile flow velocities near the axial region of the vein grafts as well as velocity profiles across the grafts of the left and right coronary arteries [14]. The velocity waveforms near axial regions were obtained by the FFT method and the velocity profile corresponding to the peak velocity was displayed by the zero-cross method. The velocity waveforms of left and right grafts exhibited a diastolic-predominant pattern, which is a characteristic of coronary artery flow. In the right bypass graft, however, considerable systolic blood flow was present, especially in the proximal portion. The systolic forward-flow component, in general, was larger in the proximal portion of vein grafts. The blood velocity profiles across the graft showed a skewed pattern in

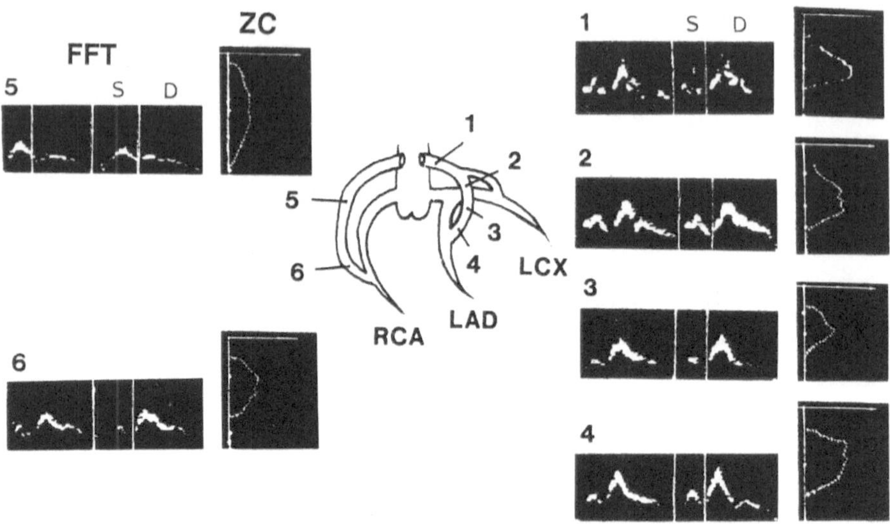

Fig. 3. A representative recording of blood flow velocity waveforms at different points in a vein bypass graft. Each measuring point is indicated in the drawing at the center of this figure. The number of each panel corresponds to that of the drawing. *S*, systole; *D*, diastole; *ZC*, zero-cross (method); *FFT*, fast-Fourier transform (method); *RCA*, right coronary artery; *LAD*, left anterior descending (artery); *LCX*, left circumflex (artery). From [15] with permission of the American Heart Association

the proximal location and distal end (*panels 1* and *4* in Fig. 3). The profile was, in general, nearly symmetrical in the mid-graft location, although the velocity configuration was not necessarily recognized as parabolic by a statistical judgement (AIC: an information theoretical criterion) because of the relatively flat configuration. AIC represents a measure of the fitness of a model to a given set of statistical data. Rittgers et al. [18] studied the hemodynamics and histopathology of autologous vein bypass grafts between the external iliac arteries in dogs. They observed skewed velocity profiles in the proximal location in all grafts, while velocity profiles in the middle region were more symmetric. Our results are consistent with their observations and also with those by Freed et al. [19]. Rittgers et al. [18] also reported that there was an inverse correlation between the apparent fluid shear rate and degree of intimal proliferation, indicating the occurence of the greatest proliferation in the region experiencing the lowest shearing forces.

Velocity Profiles in a Sequential Saphenous Vein Graft

In 6 patients, the SSVG was anastomosed to the major diagonal branch by side-to-side anastomosis and to the left anterior descending coronary artery (LAD) by end-to-side anastomosis [17]. The blood flow velocities in the SSVG were measured at two regions, i.e., just proximal to the graft-diagonal branch anastomosis (side-to-side) and the bridge portion between the graft-diagonal and graft-LAD anastomoses (Fig. 4).

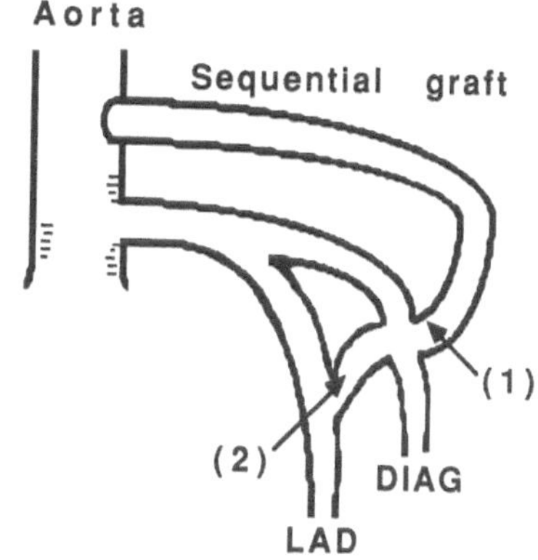

Fig. 4. Diagram of two portions of blood-flow velocity measurements in the sequential saphenous vein graft. (*1*) Portion just proximal to the graft-DIAG anastomosis (side-to-side anastomosis: SSA); (*2*) Bridge portion between graft-DIAG and graft-LAD anastomosis (end-to-side anastomosis: ESA). Unlike the diagram, the anastomosis angle between the graft and the coronary artery was actually small for both SSA and ESA. *DIAG*, major diagonal branch; *LAD*, left anterior descending artery. From [17] with permission of the American Heart Association

Table 1. Characteristics of velocity profile in sequential saphenous vein bypass graft

	Just proximal to SSA		Bridge portion[a]
Patient	Skewing toward SSA side wall	Reverse flow near the wall opposite to anastomosis	Skewing toward the wall opposite to anastomosis
1	+	+	+
2	+	−	−[b]
3	+	+	+
4	+	+[c]	−[b]
5	+	+	+
6	+[d]	+[c]	+

SSA, side-to-side anastomosis
[a] Bridge portion between side-to-side and end-to-side anastomosis.
[b] Symmetric pattern.
[c] Transient and small reverse flow.
[d] Double-peaked (M-shaped) pattern.

Figure 5 shows a typical example of the waveform just proximal to the side-to-side anastomosis. The velocity profile was skewed toward the anastomosis side wall in all patients (Table 1), especially during diastole, but the spectrum of the peak velocity was narrow in this region. The flow velocity near the wall opposite the anastomosis was reversed (Fig. 5, *dark shading*) during diastole in 5 of 6 patients, and was close to the zero-velocity line in one patient. The early systolic reverse flow seen in this patient was observed frequently. In the bridge portion, the shape of the profile showed a symmetric (2 of 6 patients) or a skewed pattern (4 patients) toward the wall opposite the anastomosis (Fig. 6). A reverse flow was noticeable in early systole. The spectrum of the peak velocity was narrow, as in the region just proximal to the side-to-side anastomosis. The peak diastolic velocity in the bridge portion was 25.4 ± 5.8 cm/s, significantly lower than that (46.6 ± 12.3 cm/s) just proximal to the side-to-side anastomosis ($P < 0.05$).

One of the prominent features of the velocity profile just proximal to the side-to-side anastomosis in the SSVG was the skewing of the profile toward the

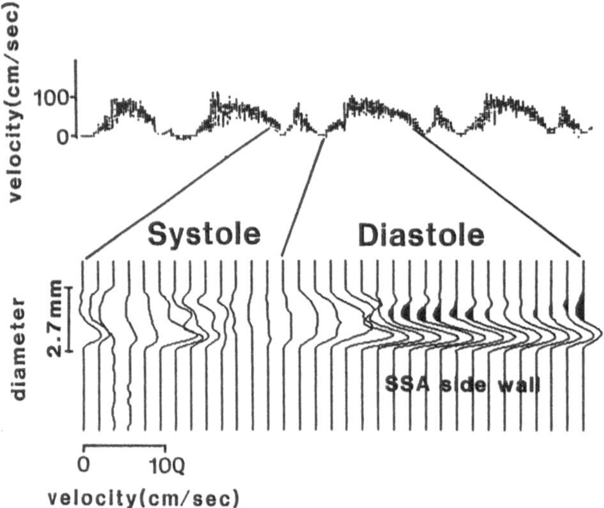

Fig. 5. Tracings of blood-flow velocities just proximal to a side-to-side anastomosis (*SSA*) in the sequential saphenous vein graft. *Top*: velocity waveform at the central axial region of the vessel by fast-Fourier transform (FFT). *Bottom*: velocity profiles in a three-dimensional display composed of coordinates of velocity (zero-cross), radial position, and cardiac cycle. Reverse flow velocities during diastole near the probe side wall (opposite to the side-to-side anastomosis) are indicated by the *dark shading*. From [17] with permission of the American Heart Association

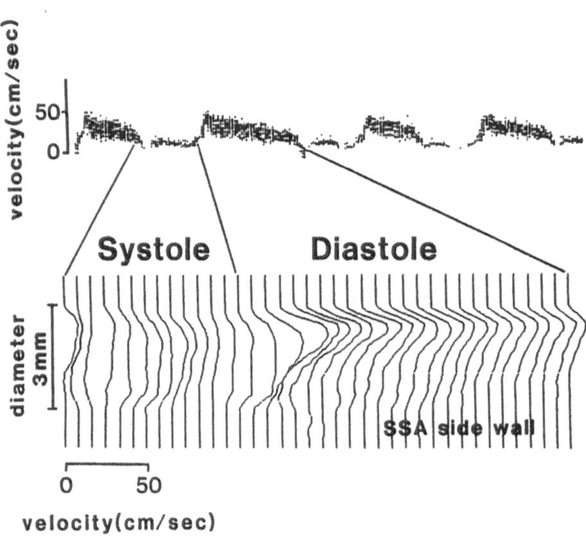

Fig. 6. Tracings of typical blood-flow velocity at the bridge portion in the sequential saphenous vein graft. *Top*: velocity waveform at the central axial region of the vessel by fast-Fourier transform (FFT). *Bottom*: velocity profiles in a three-dimensional display composed of coordinates of velocity (zero-cross), radial position, and cardiac cycle. *SSA*, side-to-side anastomosis. From [17] with permission of the American Heart Association

anastomosis side wall, suggesting high velocities into the diagonal branch. Despite the skewing, the configuration of the velocity profile was smooth, and the velocity spectrum was narrow. Watts et al. [20] studied the velocity patterns in distensible model tubes with different types of anastomoses by using a bloodflow visualization technique. They observed that the flows at a kiss anastomosis (side-to-side) were less agitated than those at an end-to-side anastomosis, supporting our present results. These flow configurations may contribute to the better patency rate of side-to-side anastomosis.

The reverse flow near the wall opposite the anastomosis suggests the existence of flow separation and recirculation in this region. The change in the shape of the velocity profile between side-to-side and end-to-side anastomoses was another characteristic of the velocity pattern in the SSVG; that is, the skewed pattern toward the anastomosis side wall at the region just proximal to the side-to-side anastomosis changed to a skewed pattern toward the wall opposite the anastomosis (4 of 6 patients) or a symmetric pattern (2 patients) in the bridge portion. This directional change in the velocity profile also suggests a complex flow field. Such alterations in the velocity patterns may contribute to the poor patency rate of the end-to-side anastomosis, since it is suggested that plaque formation correlates with low mean shear stress and oscillation in the direction of shear stress.

Velocity Profiles in the Internal Mammy Artery Graft

IMAG anastomoses were performed in 10 patients. The blood flow velocities were measured at two sites, i.e., several centimeters proximal and just proximal to the IMAG-LAD anastomosis [17].

Figure 7 shows a representative recording of the IMAG flow velocity in the region several centimeters proximal to the graft-LAD anastomosis. The velocity waveform had a normal coronary artery flow pattern, that is, diastolic predominant, and the velocity spectrum was narrow. The peak diastolic velocity in IMAG was 2.6 ± 2.0 cm/s. The velocity profiles across the graft were judged to be parabolic in 8 out of 10 patients by a statistical analysis (AIC). The axial velocity waveform and velocity profiles across the graft in the region just proximal to the graft-LAD anastomosis were almost the same as those in the region several centimeters proximal to it. An early systolic reverse flow was frequently observed at both measuring sites.

It has become increasingly clear that the long-term patency of the IMAG group versus that of the SVG group was apparent within 5 years and has increased with time [5]. This is probably due to the relative immunity of the internal mammary artery to atherosclerosis. The location of the vasa vasorum [21], perfectly formed internal elastic lamina [22], and intact vascular smooth muscle [23] may contribute to the higher patency of the IMAG. Among these properties, the intact vascular smooth muscle allows the IMAG to retain a flexible caliber and supply a blood flow dictated by myocardial demand [23]. As for the internal elastic lamina, it has been suggested that if it has deficiencies early, progressive intimal thickening will occur [22]. Recent histological study of the vasa vasorum of the internal mammary artery by Landymore et al. [21], however, demonstrated that the vasa vasorum were confined to the adventitia, sug-

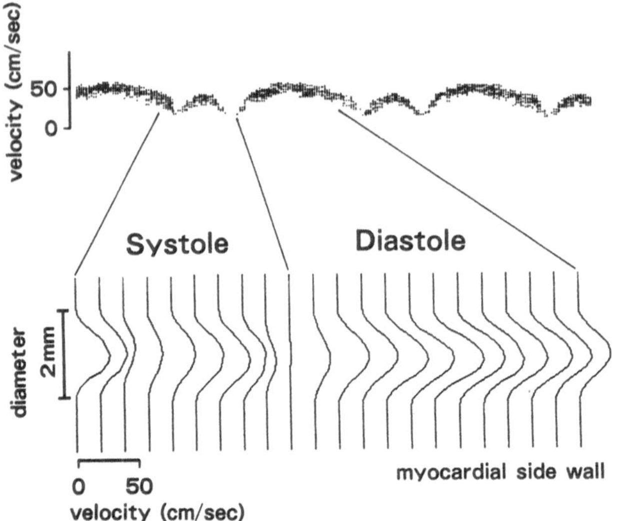

Fig. 7. Tracings of typical blood-flow velocity patterns in the internal mammary artery graft. *Top:* velocity waveform at the central axial region of the vessel by fast-Fourier transform (FFT). *Bottom*: velocity profiles in a three-dimensional display composed of coordinates of velocity (zero-cross), radial position, and cardiac cycle. Parabolic, nondisturbed velocity configuration is characteristic of internal mammary artery graft flow. From [17] with permission of the American Heart Association

gesting that its nutritional role is relatively unimportant. The mismatch of compliance between the graft and the native coronary artery beyond the graft insertion may cause the reduction of energy available for distal circulation. In this sense, the IMAG may be better than the SVG.

Compared with the velocity profiles in SVG including SSVG, those in the IMAG appeared to be more parabolic and regular, based on both the statistical (AIC) and subjective judgements. Although the peak-diastolic velocities between IMAG and SSVG did not differ, the caliber of the IMAG was significantly smaller than that of the SSVG, indicating relatively high shear in the IMAG. This nondisturbed, well-ordered velocity configuration and relatively high shear may help to explain why the patency of the IMAG is higher than that of the SVG.

Concluding Remarks

The blood-flow velocities in SSVG and IMAG were analyzed by an 80-channel 20-MHz pulsed Doppler velocimeter during aortocoronary bypass graft surgery. The velocity profile across the graft in IMAG was more parabolic with a narrower spectrum than that in the SSVG several centimeters proximal to the graft-diagonal branch anastomosis. Complex flow fields, including flow separation and directional changes in skewing of the velocity profile, were recognized along the SSVG just proximal to the side-to-side and the end-to-side anastomosis. These characteristics of the blood-flow velocity patterns seem to be a contributory factor in determining the fate of the graft.

Acknowledgments. We thank Dr. M. J. Lever, Imperial College of Science, Technology and Medicine, London for his critical comments. We are grateful to American Heart Association, Inc. for permission to reproduce in part our paper published in Circulation 78(5), 1988 [17].

References

1. Galbut DL, Traad EA, Dorman MJ, DeWitt PL, Larsen PB, Weinstein D, Ally JM, Gentsch TO (1985) Twelve-year experience with bilateral internal mammary artery grafts. Ann Thorac Surg 40: 264–270
2. Lytle BW, Loop FD, Cosgrove DM, Ratliff NB, Easley K, Taylor PC (1985) Long-term (5–12 years) serial studies of internal mammary artery and saphenous vein coronary bypass grafts. J Thorac Cardiovasc Surg 89: 248–258
3. Barner HB, Standeven JW, Reese J (1985) Twelve-year experience with internal mammary artery for coronary artery bypass. J Thorac Cardiovas Surg 90: 668–675
4. Sauvage LR, Wu H-D, Kowalsky TE, Davis CC, Smith JC, Rittenhouse EA, Hall DG, Mansfield PB, Mathisen SR, Usui Y, Goff SG (1986) Healing basis and surgical techniques for complete revascularization of the left ventricle using only internal mammary arteries. Ann Thorac Surg 42: 449–465
5. Loop FD, Lytle BW, Cosgrove DM, Stewart RW, Goormastic M, Williams GW, Golding LAR, Gill CC, Taylor PC, Sheldon WC, Proudfit WL (1986) Influence of the internal-mammary-artery graft on 10-year survival and other cardiac events. N Engl J Med 314: 1–6
6. Hutchins GM, Bulkley BH (1977) Mechanisms of occlusion of saphenous vein-coronary artery "jump" grafts. J Thorac Cardiovasc Surg 73: 660–667
7. FitzGibbon GM, Leach AJ, Keon WJ, Burton JR, Kafka HP (1986) Coronary bypass graft fate. Angiographic study of 1,179 vein grafts early, one year, and five years after operation. J Thorac Cardiovasc Surg 91: 773–778
8. Campeau L, Enjalbert M, Lesperance J, Vaislic C, Grondin CM, Bourassa MG (1983) Atherosclerosis and late closure of aortocoronary saphenous vein grafts: Sequential angiographic studies at 2 weeks, 1 year, 5 to 7 years, and 10 to 12 years after surgery. Circulation 68 (Suppl II): II-1
9. Schettler G, Nerem RM, Schmid-Schonbein H, Morl H, Diehm C (eds) (1983) Fluid dynamics as a localizing factor for atherosclerosis. Springer-Verlag, Berlin
10. Caro CG, Lever MJ, Parker KH, Fish PJ (1987) Effects of cigarette smoking on the pattern of arterial blood flow: Possible insight into mechanisms underlying the development of arteriosclerosis. Lancet 2: 11–13
11. Kieser TM, FitzGibbon GM, Keon WJ (1986) Sequential coronary bypass grafts: Long-term follow-up. J Thorac Cardiovasc Surg 91: 767–772
12. Bigelow JC, Bartley TD, Page US, Krause AH Jr (1976) Long-term follow-up of sequential aorto coronary venous grafts. Ann Thorac Surg 22: 507–514
13. Hartley CJ, Cole JS (1974) An ultrasound pulsed Doppler system for measuring blood flow in small vessels. J Appl Physiol 37: 626–629
14. Ogasawara Y, Hiramatsu O, Kagiyama M, Tsujioka K, Tomonaga G, Kajiya F, Yanashima T, Kimura Y (1984) Evaluation of blood velocity profile by high frequency ultrasound pulsed Doppler velocimeter by a multigated zerocross method together with a Fourier transform method. IEEE Comput Cardiol 447–450
15. Kajiya F, Ogasawara Y, Tsujioka K, Nakai M, Goto M, Wada Y, Tadaoka S, Matsuoka S, Mito K, Fujiwara T (1986) Evaluation of human coronary blood flow with an 80 channel pulsed Doppler velocimeter and zero-cross and Fourier transform methods during cardiac surgery. Sequential saphenous vein grafts and internal mammary artery grafts. Circulation 74(Suppl III): III-53–60
16. Kajiya F, Tsujioka K, Ogasawara Y, Wada Y, Matsuoka S, Kanazawa S, Hiramatsu O, Tadaoka S, Goto M, Fujiwara T (1987) Analysis of flow characteristics in post stenotic regions of the human coronary during bypass graft surgery. Circulation 76: 1092–1100

17. Fujiwara T, Kajiya F, Kanazawa S, Matsuoka S, Wada Y, Hiramatsu O, Kagiyama M, Ogasawara Y, Tsujioka K, Katsumura T (1988) Comparison of blood-flow velocity waveforms in different coronary artery bypass grafts. Circulation 78: 1210–1217
18. Rittgers SE, Karayannacos PE, Guy JF, Nerem RM, Shaw GM, Hostetler JR, Vasko JS (1978) Velocity distribution and intimal proliferation to autologous vein graft in dogs. Circ Res 42: 792–801
19. Freed DB, Hartley CJ, Noon GP, Short D (1984) Intraoperative velocity profiles in human coronary saphenous vein grafts (abstract). Circulation 70 (Suppl II): II-384
20. Watts KC, Marble AE, Sarwal SN, Kinley CE, Watton J, Mason MA (1985) Simulation of coronary artery revascularization. J Biomech 19: 491–499
21. Landymore RW, Chapman DM (1987) Anatomical studies to support the expanded use of the internal mammary artery graft for myocardial revascularization. Ann Thorac Surg 44: 4–6
22. Sims FH (1987) The internal mammary artery as a bypass graft? Ann Thorac Surg 44: 2–3
23. Singh RN, Beg RA, Kay EB (1986) Physiological adaptability: The secret of success of the internal mammary artery grafts. Ann Thorac Surg 41: 247–250

Index of Key Words